Single Biomolecule Detection and Analysis

This collection discusses various micro/nanodevice design and fabrication for single-biomolecules detection. It will be an ideal reference text for graduate students and professionals in diverse subject areas including materials science, biomedical engineering, chemical engineering, mechanical engineering, and nanoscience.

This book-

- Discusses techniques of single-biomolecule detection, their advantages, limitations, and applications.
- Covers comprehensively several electrochemical detection techniques.
- Provides single-molecule separation, sensing, imaging, sequencing, and analysis in detail.
- Examines different types of cantilever-based biomolecule sensing, and its limitations.

Single Biomolecule Detection and Analysis covers single-biomolecule detection and characterization using micro/nanotechnologies and micro/nanofluidic devices, electrical and magnetic detection technologies, microscopy and spectroscopy techniques, single biomolecule optical, and nanopore devices. This text covers key important biosensors-based detection, stochastic optical reconstruction microscopy-based detection, electrochemical detection, metabolic engineering of animal cells, single-molecule intracellular delivery and tracking, terahertz spectroscopy-based detection, total internal reflection fluorescence (TIFR) detection, and fluorescence correlation spectroscopy (FCS) detection. This text will be useful for graduate students and professionals in diverse subject areas including materials science, biomedical engineering, chemical engineering, mechanical engineering, and nanoscience. Discussing chemical process, physical process, separation, sensing, imaging, sequencing, and analysis of single-molecule detection, this text will be useful for graduate students and professionals in diverse subject areas including materials science, biomedical engineering, chemical engineering, mechanical engineering, and nanoscience. It covers microscopy and spectroscopy techniques for single-biomolecule detection, analysis, and biomedical engineering applications.

Single Biomolecule Detection and Analysis

Concepts, Applications, and Future Prospects

Edited by
Tuhin Subhra Santra
Fan-Gang Tseng

CRC Press
Taylor & Francis Group
Boca Raton London New York

CRC Press is an imprint of the
Taylor & Francis Group, an **informa** business

Front cover image: Christoph Burgstedt/Shutterstock

First edition published 2024
by CRC Press
2385 NW Executive Center Dr, Suite 320, Boca Raton, FL 33431

and by CRC Press
4 Park Square, Milton Park, Abingdon, Oxon, OX14 4RN

CRC Press is an imprint of Taylor & Francis Group, LLC

ISBN: 978-0-367-75430-3 (hbk)
ISBN: 978-1-032-52967-7 (pbk)
ISBN: 978-1-003-40947-2 (ebk)

DOI: 10.1201/9781003409472

Typeset in Times
by codeMantra

Contents

Acknowledgements

We are thankful to all our colleagues who contributed to the chapters and provided their expertise that greatly assisted in arranging this book.

We are also grateful to Dr. Ashwini Shinde for arranging the chapters and all related documents for the publication of this book.

This work was supported by DBT/Wellcome Trust India Alliance Fellowship grant number IA/E/16/1/503062 awarded to Dr. Tuhin Subhra Santra.

Preface

Biological and microscopic developments are inextricably intertwined. In the seventeenth century, Leeuwenhoek applied a microscope for biological analysis, which started the modern biology era and later triggered many technical developments in optics related to microscopy. Traditionally, experimental research in life science has been performed on an average population level, such as cell culture. The general cell population though genetically identical but is intrinsically heterogeneous. Thus, different cells possess various physical, chemical, and biological properties. This population is employed to investigate the physical or chemical state of various cellular parameters. Since this technique averages out the observations of minor and abnormal cells in the population, the 'noise' is smoothed down. Therefore, much valuable information could be hidden in this 'noise.' As a result, there is a high risk of losing valuable data relevant to biological heterogeneity. Population heterogeneity may result from changes in the environment that may have an impact on biological materials' soft matter and genetic variances that may have an impact on gene expression, which may lead to changes in the activity of different cellular components. Therefore, single-cell experiments should be performed under precisely controlled environment.

At the molecular level, a simple unicellular organism is also representing heterogeneity. An average analysis of the molecular properties may result in information loss regarding any molecular heterogeneity. This will eventually lead to misinterpretation of the physiological significance of molecular subpopulation. Thus, we should focus on molecules as the smallest functional unit in biological systems, leading to single-molecule biophysics research. Such biological investigation will impact various fields such as immunology, systems, and synthetic biology. Modern techniques have enabled cellular probing of signal transduction networks, facilitating the precise understanding of biological processes such as gene expression, cell differentiation, and immune responses.

All this has led up to the development of high-resolution fluorescence microscopy. In 1988, the first precise single-molecule fluorescence microscopy (SMFM) was demonstrated, showing molecular memory-based activity in flavin adenine dinucleotide (FAD) behavior. Since then, SMFM has uncovered many fundamental molecular scale biological processes, which previously were limited due to the use of population methods in studies such as DNA replication and modeling, protein folding, biopolymer mechanism, oxidative phosphorylation, cell division mechanism, viral infection process, endocytosis and exocytosis, cell wall synthesis, and structural dynamics of DNA.

Along with microscopy, other analytical tools such as electrochemical methods, spectroscopy, mass spectrometry, and micro/nanofluidics are trying to resolve and detect smaller quantities of materials faster and in a shorter time. These analytical techniques are now divided into two sub-categories – single-cell and single-molecule-oriented techniques. When compared to the measurements from thousands of cells, single-cell and single-molecule measurements provide far more qualitative information. Single-cell and single-molecule information have the capability to reveal the heterogeneity and continuous processes within biological systems. One of the primary requirements for single-molecule analysis is detecting the most minor signals over relatively large noise levels. Higher levels of photon signals can now be detected with a smaller sample volume, thanks to the integration of various cutting-edge techniques such as sensitive camera detectors and improved photostability engineering. However, single entity measurement in complex systems has its challenges both from technical and data interpretation points of view. We steer the readers to the recent technological advancement that will discuss different approaches for single-biomolecule detection and analysis.

This book comprises eleven chapters and broadly covers several aspects of single biomolecule detection and analysis using different technologies. Below is a short description of each chapter. Hartanto et al. have discussed the recent advances of microfluidic platforms in applications such as DNA extraction, detections using DNA separation, and detection using affinity probes. Further, giving envision microfluidic-based DNA detection will provide a revolutionary solution readily available for point-of-care detection of DNA. Lee et al. have emphasized highly sensitive single-molecule detection using solid-state nanopores. The authors have explained the ultrathin solid-state nanopore sensing, from the fabrication processes to fundamental transport phenomena and sophisticated electrokinetic phenomena during molecule translocation events. Wang et al. have explained the basic principles of the fluorescent laser scanning confocal microscope (FLSCM), microscopic imaging and spectral scanning-based FLSCM, stimulated emission depletion, reversible saturable optical fluorescence transitions, and ground state depletion, label-free laser scanning confocal microscope, and its types, and fluorescence resonance energy transfer. Finally, the applications of FLSCM on individual DNA-protein detection and measurement of single-molecule fluorescence resonance energy transfer in live cells have also been depicted. James F. Leary has discussed the basics of flow cytometry detection and cell subpopulation separations by cell sorting at the molecular level. This chapter also discusses the biological and clinical applications involving DNA detection and multicolor detection of cell subpopulations, including immunophenotyping of leukemias and lymphomas and monitoring of gene expression patterns at the single-cell level using phospho-specific antibodies. Kuo and Lin aimed to discuss the current techniques for single-molecule separation, such as aqueous two-phase systems, capillary electrophoresis, chromatography, protein crystallization, hydrodynamics, and/or electrohydrodynamics-based technology, and microfluidics/nanofluidics-based devices. Yen-Liang Liu has discussed various single-molecule tracking/single-particle tracking (SMT/SPT) microscopy techniques and elaborated on their potential for gaining insights into biology. E. O. Puchkov has briefly described fluorescence and atomic force microscopy techniques for determining the location and movement of biomolecules and biomolecular complexes. This will provide unique data on molecular microorganisms' structural and functional organization at the molecular level. Kaur et al. have described the principal and methodology of the pull-down assay for analyzing protein–protein or protein–ligand interactions. The author has also discussed the other known methods such as enzyme-linked immune-sorbent immunoassay (ELISA), digital ELISA, single molecule pull-down assay, detection of alpha-synuclein (α-SYN) protein, and oligomer detection. Illath et al. demonstrated different atomic force microscopy (AFM) approaches and their working principles, requirements for single-molecule imaging, force kinetics, and their illustrative examples for understanding of structural characterization, mechanism, and interactions of biomolecules. Meenakshi et al. have summarized the fundamental concepts and mechanisms behind surface-enhanced Raman spectroscopy (SERS) and their significance in detecting a single molecule at ultra-low concentrations for early disease diagnosis. Bhupathi et al. have discussed single-molecule detection techniques aided by imaging modalities, fluorescent probes, and labeling methods. This chapter has emphasized the quantitative aspects of the imaging modality, image formation, processing, and different analytical estimation techniques for visualization.

We hope this book will be fascinating for academic and industrial researchers, especially those working on various aspects of single biomolecule detection and analysis. Moreover, this book will be helpful for undergraduate and graduate students in their advanced-level courses about single-cell and single-biomolecules detection and analysis.

Editors
Tuhin Subhra Santra
Fan-Gang Tseng

Editors

Tuhin Subhra Santra is an Associate professor in the Department of Engineering Design at Indian Institute of Technology Madras, India. Dr. Santra is a tenure track "Honorary Visiting Professor" at National Tsing Hua University" Taiwan from 2018 to 2023, and he was a "Visiting Professor" at University of Cambridge, UK in 2019. Dr. Santra received his PhD in Bio-NEMS from National Tsing Hua University (NTHU), Taiwan in 2013 and he was a post-doctoral researcher at the University of California, Los Angeles (UCLA), USA, from 2015 to 2016. His areas of main research interests include Bio-NEMS, MEMS, single-cell technologies, single-biomolecule detection, bio-micro/nano fabrication, biomedical micro/nano devices, nanomedicine, etc.

Dr. Santra served as a Guest Editor for *Frontiers of Bioengineering and Biotechnology* 2021, *Cells, MDPI* 2021, 2020; *International Journal of Molecular Science* (IJMS) 2018, 2017, 2015; *Sensors Journal* 2016; *Journal of Molecules* 2016; and *Journal of Micromachines* 2020, 2019, 2013, among others. Dr. Santra has received many honors and awards such as "Wellcome Trust/DBT India Alliance Fellowship" in 2018; Honorary Research Fellow from National Tsing Hua University, Taiwan in 2018; IEEE-NEMS best conference paper award in 2014; etc. Dr. Santra has published more than 45 international journals, has 25 US, Taiwan, and Indian pending/approved patents, 8 books, 20 book chapters, 20 proceedings, in his research field.

Fan-Gang Tseng Fan-Gang (Kevin) Tseng received his PhD in Microelectro Mechanical System (MEMS) from University of California, Los Angeles, USA (UCLA). He became an Assistant Professor in Department of Engineering and System Science, National Tsing Hua University (NTHU), Taiwan from August 1999. Currently, he is a Professor in the Department of Engineering and System Science (ESS). He is a Deputy Director of Biomedical Technology Research Center, NTHU, form 2009. He is also an affiliated Professor in Institute of Nanoengineering and Microsystem (NEMS), NTHU, and Research Fellow in Academia Sinica, Taiwan. Dr. Tseng focuses his research area in MEMS, Bio-MEMS, Nano-biotechnology, Nanomedicine, Fuel Cell, MEMS packaging and Integration. Dr. Tseng is in Editorial board of Applied Sciences from 2010, open Micromachine Journal 2009, etc. Dr. Tseng received many awards such as outstanding in research award from National Science Council, Taiwan 2010–2013, iCAN Taiwan competition award, Ton-Yuan innovative competition award, the best project award, NTHU outstanding in research award, 2005–2007, National innovation award, First class research award from National Research Council 2005–2008, outstanding teaching award 2003, new faculty research award and many more. Dr. Tseng has published 8 books/book chapters, more than 40 US and Taiwan patents. Also he has published more than 200 international Journal papers and 350 International Conference paper/proceedings related to this field.

List of Contributors

Amer Alizadeh
School of Engineering, University of
    Birmingham
Edgbaston, Birmingham,
    United Kingdom

Gowri Annasamy
Indian Institute of Information
    Technology, Design and
    Manufacturing
Kancheepuram, Chennai, India

Bhupathi Arun
Department of Orthodontics
    Vishnu Dental College
Vishnupur, Bhimavaram, Andhra
    Pradesh, India.

Ting-Hsuan Chen
Department of Biomedical Engineering
City University of Hong Kong
Kowloon, Hong Kong, SAR

Yih Bing Chu
Department of Electrical and Electronic
    Engineering
FETBE, UCSI University
Kuala Lumpur, Malaysia

Chenyu Cui
Department of Biomedical Engineering
City University of Hong Kong
Kowloon, Hong Kong, SAR

Hirofumi Daiguji
Department of Mechanical Engineering
The University of Tokyo
Hongo, Bunkyo-ku, Tokyo, Japan

Hogi Hartanto
Department of Biomedical Engineering
City University of Hong Kong
Kowloon, Hong Kong, SAR

M. Hema Brindha
Department of Biomedical Engineering
SRM Institute of Science and
    Technology
Chennai, India

Chengjun Huang
E-Health Center, Institute of
    Microelectronics of the Chinese
    Academy of Sciences
Beijing, China

Wei-Lun Hsu
Department of Mechanical Engineering
The University of Tokyo
Hongo, Bunkyo-ku, Tokyo, Japan

Kavitha Illath
Department of Engineering Design
Indian Institute of Technology
Madras, Chennai, India

Tianyi Jiang
Department of Biomedical Engineering
City University of Hong Kong
Kowloon, Hong Kong, SAR

P. Kaavya
Department of Biomedical Engineering
SRM Institute of Science and
    Technology
SRM, Chennai, India

Amandeep Kaur
Central Scientific Instruments
    Organisation (CSIR-CSIO)
Chandigarh, India

Ganapathy Krishnamurthi
Department of Engineering Design
Indian Institute of Technology
Madras, Chennai, India

N. Ashwin Kumar
SRM Institute of Science and
  Technology
Chennai, India

Ching-Te Kuo
Department of Mechanical and Electro-
  Mechanical Engineering
National Sun Yat-sen University
Kaohsiung, Taiwan

James F. Leary
Departments of Basic Medical Sciences
  and Biomedical Engineering
Purdue University
West Lafayette, Indiana

Chun-Yen Lee
Department of Mechanical Engineering
The University of Tokyo,
Hongo, Bunkyo-ku, Tokyo, Japan

Yu-Chia Lin
Department of Mechanical and Electro-
  Mechanical Engineering
National Sun Yat-sen University
Kaohsiung, Taiwan

Yen-Liang Liu
Master Program of Biomedical
  Engineering, Research Center for
  Cancer Biology
China Medical University
Taichung City, Taiwan

Xinchao Lu
E-Health Center
Institute of Microelectronics of the
  Chinese Academy of Sciences
Beijing, China

M. Muthu Meenakshi
Vel Tech Rangarajan Dr. Sagunthala
  R&D Institute of Science and
  Technology, Avadi
Chennai, India

Moeto Nagai
Department of Mechanical Engineering
Toyohashi University of Technology
Toyohashi, Japan

Satish Pandey
Department of Biotechnology
Mizoram University
Aizawl, India

E. O. Puchkov
All-Russian Collection of
  Microorganisms, G.K. Skryabin
  Institute of Biochemistry and
  Physiology of Microorganisms of
  the Russian Academy of Sciences,
  Pushchino Center for Biological
  Research of the Russian Academy of
  Sciences
Pushchino, Russia

Tuhin Subhra Santra
Department of Engineering Design
Indian Institute of Technology, Madras
Chennai, India

Ashwini Shinde
Department of Engineering Design
Indian Institute of Technology, Madras
Chennai, India.

Suman Singh
Central Scientific Instruments
  Organisation (CSIR-CSIO)
Chandigarh, India

Xue Wang
E-Health Center, Institute of
  Microelectronics of the Chinese
  Academy of Sciences
Beijing, China

1 Microfluidics-Based DNA Detection

Hogi Hartanto, Tianyi Jiang, Chenyu Cui, and Ting-Hsuan Chen
City University of Hong Kong

CONTENTS

DOI: 10.1201/9781003409472-1

1.1   INTRODUCTION

Microfluidics is the science of fluids within microchannels. As an emerging multidisciplinary technology, it aims at miniaturizing the processes of biochemical reactions, from the initial sampling to result outputs, onto a centimeter-scale microchip [1,2]. In recent decades, microfluidics has progressed significantly. Through the fundamental transformation from macro to microscale, the miniaturized and automated settings of microchannels with dimensions compared to that of biomolecules supplies a unique circumstance for cellular and genetic assays, and reduction of reagent/sample consumption and manual operation. From then on, point-of-care (POC) equipment exploitation became research hotspots, and the concept of micro total analysis system (μTAS) appears to represent the ideal modality for microfluidics [3,4].

Deoxyribonucleic acid (DNA) is the building block of living organisms. Located in the nucleus of each cell, DNA functions as a central dogma encoding genetic information. During gene activation, DNA is transcribed into ribonucleic acid (RNA), which will later be used as a template during translation for the protein synthesis that regulates the development, processes, growth, and reproduction of all forms of life. As the essential biomacromolecules, DNA can be found in every cell in all known life. As such, the detection of DNA allows us to understand the macromolecular machinery for life important for development, diseases, and treatment. In this chapter, we first discuss the basic information of microfluidics and DNA, followed by the microfluidic techniques used for DNA extraction. Next, we discuss the state-of-art techniques and progress of microfluidics detection using separation and affinity binding, including preparation of DNA probes, signal generation, and signal amplification.

1.2   DESIGN OF MICROFLUIDICS DEVICE

A microfluidic chip is a device with patterns of microchannels engraved upon it [5]. It generally consists of channels, chambers, pillars, junctions and holes to accomplish liquid-handling tasks, including mixing, transportation, storage, metering, etc. The microchannel inlet/outlet is interfaced through the devices for connecting multiple micro modules and exterior flow control (Figure 1.1). For the set goal of specific assays, the fabrication materials are selected with appropriate properties, and the fabrication procedures must be efficient and dependable [6].

Exploiting the miniaturized systems is more than simply mimicking macro assays within a microchip but also to provide significant performance enhancements. Microfluidic devices are generally designed after comprehensive considerations. The techniques employed and the systemic calculation and simulation to predict on-chip functionality are important. Several aspects also must be considered through: (i) whether there is a low-cost mass manufacturing method suitable for the working principle; (ii) what process flow to ensure the successful assembly of the final device, and (iii) how to find the balance point among performance, ease of fabrication, and cost [5,7].

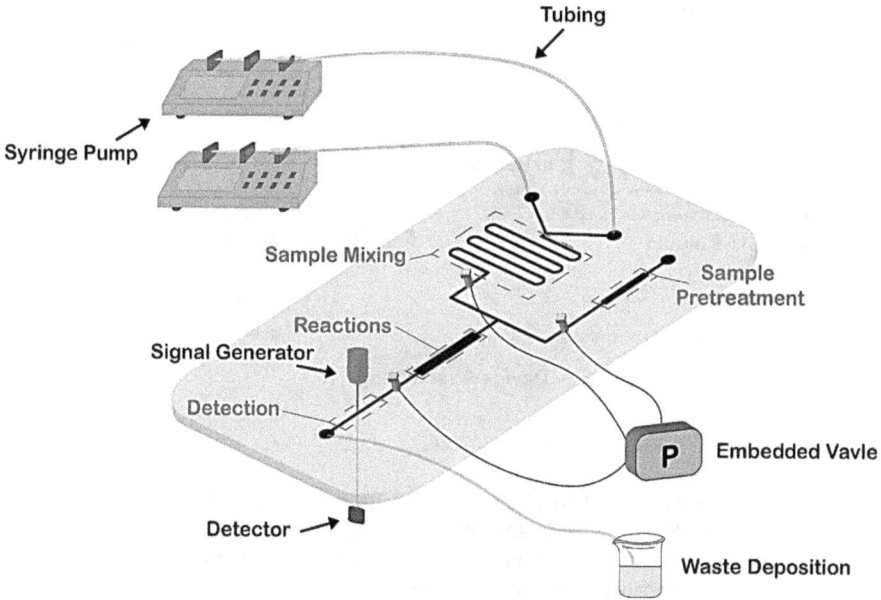

FIGURE 1.1 Structures of typical microfluidics detection that incorporate sample pretreatment, mixing, reaction, and signal generation.

The design of a microfluidics device starts with portraying microchannels by design software (AutoCAD, CorelDRAW, SOLIDWORKS, etc.). There are detailed guidelines for the layouts, and fabrication frequently imposes additional structural constraints. Later, the pattern is transferred to a photomask for photolithography [8]. Alternatively, the manufacturing can be also processed by micro-milling or laser-cutting. The basic information about fabrications is introduced in the following section.

1.3 MATERIALS FOR MICROFLUIDICS

Silicon and glass are the original choices when the concept of microfluidics emerged [9,10]. However, they suffer from high cost and are challenging to handle due to their hardness and brittleness. In addition, the complicated fabrication protocols in wet/dry etching and vapour deposition also make them less popular [11]. In contrast, other materials such as polymers, composite materials, and papers are widely used for microfluidics [11]. Polymer-based material is the current mainstream option for their inexpensiveness and accessible [12]. Elastomers are one type of amorphous polymers generated by light crosslink chain reactions. Because of the weak intermolecular forces and low Young's modulus, recuperable elastic deformations happen under external pressure. Polydimethylsiloxane (PDMS) is the most widespread polymer for microchip prototyping, fabricated by soft lithography to replicate the moulds achieved from micromachining or photolithography

1 : SU-8 Coating

SU-8 Photoresist

Silicon Wafer

2 : Photolithography

UV Exposure

Mask

3 : SU-8 Mould

3 : PDMS Stamp

PDMS Prepolymer

4 : PDMS Peel Off

5 : Plasma Bonding

O₂ Plasma

6 : Device Assembling

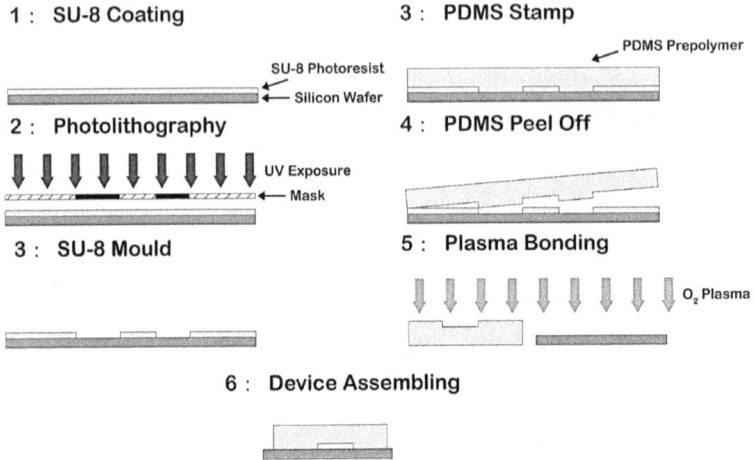

FIGURE 1.2   Process flow chart of standard soft lithography in PDMS fabrication based on SU-8 mould. SU-8 photoresist is first fabricated by a standard lithography process. Next, uncured PDMS precursor is poured on to the SU-8 master. After curing, the PDMS replica can be peeled off and bonded with glass for making the microfluidic device.

(Figure 1.2) [13,14]. Its porous nature and gas-permeable character provide excellent compatibility with cellular assays. On the other hand, thermoplastic is a rigid polymer generated by a high level of crosslinking [15]. The status at room temperature retains robustness and airtight block, making it suitable for solvent bonding, micromachining and hot embossing. The melting matrix is remouldable only when its glass transition temperature is reached. Due to the excellent support for micromechanical processing, polyethene (PE), polycarbonate (PC), poly-methyl methacrylate (PMMA), etc., are practical solutions for the commercial mass production of microdevices [16]. In recent years, paper is receiving increasing attention because it is flexible, economic, and disposable [17,18]. Moreover, its biocompatibility and modifiability using surface chemistry also make it popular for biomedical applications. Paper microfluidics mostly relies on passive capillary mechanisms for driving flow and colourimetric reactions in rapid biological detections [19]. Other common materials such as hydrogels [20] and ceramic [21] also possess unique advantages and applications.

Nowadays, microfluidics technology has been incorporated into DNA analysis, from extraction, separation, signal generation and amplification [22,23]. It simplifies the operation procedures between each step, reduces the contamination risk, and offers the possibility of developing the multi-step μTAS.

1.4   DNA STRUCTURE

The name deoxyribonucleic acid comes from the component and structure (Figure 1.3). DNA is composed of monomers of sugar molecules with a phosphate group as a backbone and a nucleic acid base sticking out of it [24–27]. During

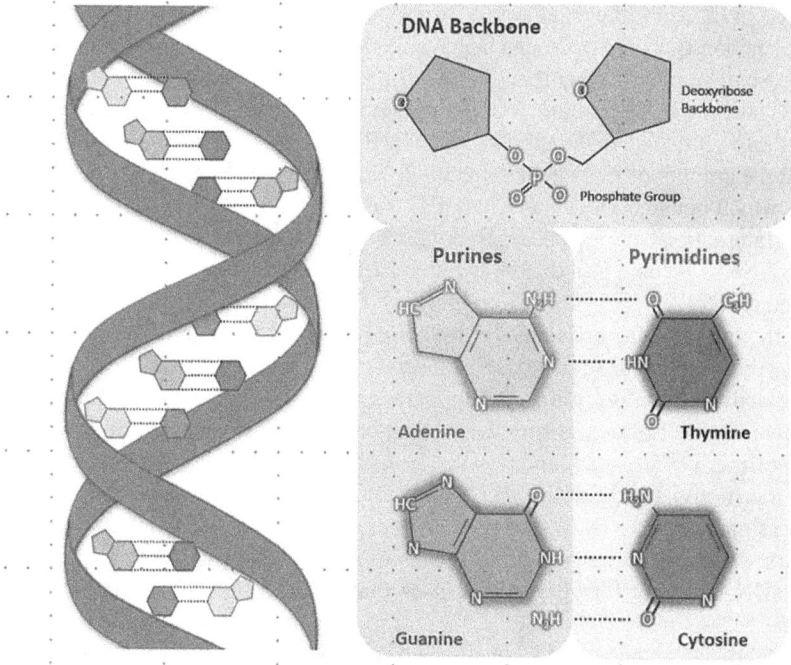

FIGURE 1.3 DNA double helix structure, backbone structure and structures of purine nucleobases adenine, guanine and pyrimidine nucleobases thymine, and cytosine.

polymerization, the nucleotide monomers are connected through covalent bonds between phosphate groups, forming a long chain of linear or filamentous macromolecules of polynucleotides. The nucleobase can be differentiated into four types: adenine (A), thymine (T), cytosine (C), and guanine (G). Based on the structure of their heterocyclic compound, the nucleotides can also be further classified into two groups. Adenosine (A) and Guanine (G) are considered purines because they have a combined five and six-heterocyclic structure, whereas thymine (T) and cytosine (C) are classified as pyrimidines due to their six-membered heterocyclic structure. Uracil (U), the replacement of thymine in RNA, is also considered as a pyrimidine [25]. The only difference between uracil and thymine is the absence of the methyl group in one of the heterocyclic rings. Importantly, using the nucleobase as the building blocks, the sequence of their combination encodes all of the information inside all organisms.

As their name implies, single-strand DNA (ssDNA) is the one-strand of polynucleotides, whereas the double-strand DNA (dsDNA) is two complementary chains of ssDNA linked together. Notably, dsDNA is formed through the pairing of nucleobases of ssDNA using Watson–Crick rule, i.e., adenine (A) only bind to thymine (T), and guanine (G) only bind to cytosine (C), forming the A-T and C-G base pairs through hydrogen bonding [26]. As purines contain two nitrogen rings, it's unavailable for purine–purine bondage in the double helix structure of three nitrogen rings spacing in total. By the same token, two pyrimidines are too close to constitute the hydrogen bonds. Three hydrogen bonds hold the bondage between guanine and cytosine base, which is stronger than the two hydrogen bonds between thymine and adenine. Thus, the DNA containing a higher percentage of G-C pairings is more stable than those with A-T abundant regions [27]. Based on such pairing in double-strand DNA, the amount of A is equal to that of T, while the amount of C is the same as that of G, which became known as Chargaff's rule. Eventually, using a delicate balance of the interactions among hydrogen bonds, bases, and surrounding water molecules, the dsDNA is formed as a double-helix construction where two strands of DNA coil around each other.

Although ssDNA and dsDNA are composed of the same four nitrogenous bases, the pairing of nucleobases causes several discrepancies. Because of this stable and stiff configuration, the appearance of dsDNA is akin to being linear or filamentous. At the same time, the ssDNA behaves highly flexible with a stellate-like outline. Different secondary structures are accessible for ssDNA under multiple thermodynamic conditions. In addition, dsDNA is more resistant to foreign aggression, whereas ssDNA is more vulnerable to chemicals and environmental factors. As a result, the genetic materials in the vast majority of living organisms are dsDNA, except for very few viruses that possess ssDNA as their hereditary substance.

Notably, the sequence of DNA nucleobase is used to encode genetic information that can influence the phenotypic characteristics of an organism. Chromosomes carry genes in the form of dsDNA. When being activated, a segment of the dsDNA structure is unwinded, revealing a sequence of nucleotides of one DNA strand as a template for the synthesis of single-strand mRNA molecule with complementary sequences. After being released to the cytoplasm, the mRNA becomes a template with translating codons in groups of three nucleotides to initiate the synthesis of corresponding amino acids in the peptide chain. After initiation, extension and termination, the genetic code is eventually translated to protein synthesis. Therefore, the DNA sequence is the main property to be used for detection.

1.5 DNA EXTRACTION

As DNA primarily locates in the cell nucleus, it is essential to extract or harvest pure DNA from original samples for genetic analysis. Typically, the entire extraction is of three basic steps. First, cell boundary and nuclear membrane are disrupted through cell lysis, releasing nucleotides and intracellular components. Second, the DNA fragments are isolated among cell debris and are extracted by proper screening protocols. Finally, the DNA sample is eluted and purified [28]. At present, many commercial kits for bench-top DNA extraction are well-established [29]. However, depending on the origin of raw materials, e.g., clinical samples (human tissues, blood/body fluids), environmental samples (plant tissues, water, and soil), food, bacteria/fungi and forensic samples (dried blood stamps or fingerprints) etc., the extraction efficiency and yield vary [30,31]. Thus, the choice of extraction techniques should be considered based on their specialties and restrictions.

1.5.1 CELL LYSIS

Cell lysis is the approach to dissociate the cell membrane such that the intracellular substance can be released into the solution [32]. For DNA extraction, the destruction of the cytoplasmic membrane, cell wall, and the nuclear envelope must be accomplished. Of note, plant cells and most prokaryotic cells, including bacteria, are surrounded by the cell wall, which should be specially considered. Categorized by operating principles, the cell lysis can be achieved using physical or chemical approaches. Nowadays, many microdevices for cell disruption have been demonstrated. There is review literature that detailed and summarized the function and performance of the micro platforms designed explicitly for cell lysis [32–34].

For DNA extraction, chemical lysis using buffer and protease enzymes is the primary approach. Detergents such as sodium dodecyl sulfate (SDS), cetyltrimethylammonium bromide (CTAB) and Tris solubilize cell membranes and walls [35,36]. In contrast, lysozyme, glycanase, and protease break the amino acids' bonds and accelerate protein digestion. In general, the lysis components are pumped into the microchamber behind raw samples injection. After sufficient reactions, subsequent techniques screen the desired sequences. This approach highly relies on systematic fluidic control, and additional modules like valves and sample mixers. In addition, the removal of residue chemicals is also critical.

On the other hand, physical methods are also widely used. Electrochemical cell lysis is a reagent-free method [37]. The electrolyzed hydroxide ions at the cathode vicinity cause the destruction of the fatty acid-glycerol ester bonds of the phospholipid, inducing permeabilization of the cell membrane. Typically, the lysis is realized within 10 mins driven by a low voltage roughly equal to 5 volts, which is moderate as compared to other physical approaches such as electroporation method with much higher voltage [38]. Some microdevices also employed the cation/negatively charged exchangeable polymer separating the anode and cathode chambers to maintain the high pH requirements and prevent the interferences from the anode electrolytic products like hypochlorous acid towards downstream DNA analysis [37,39]. In addition, thermal microfluidics disrupted cells directly by a rising temperature generated by attached resistive heaters. This process is highly consistent with the subsequent PCR process with heating [40,41]. Unfortunately, an extended period of heating may damage DNA configurations as well as other proteins and enzymes. Thus, the usage scenarios are limited.

1.5.2 DNA PURIFICATION

1.5.2.1 Liquid-Phase Purification

The liquid-phase purification utilizes the solution chemistry to select DNA molecules from the sample matrix [42]. The most common method is phenol chloroform isoamyl alcohol (PCI) due to better purity and efficiency (Figure 1.4). Liquid–liquid method isolates the biological molecules relying on their solubility [43]. After mixing the two immiscible phases, the polar DNA molecules dissolve into the polar chloroform water phase in a pH-dependent manner. At the same time, the non-polar phenol digests the protein and lipids. Due to the higher density of phenol over water,

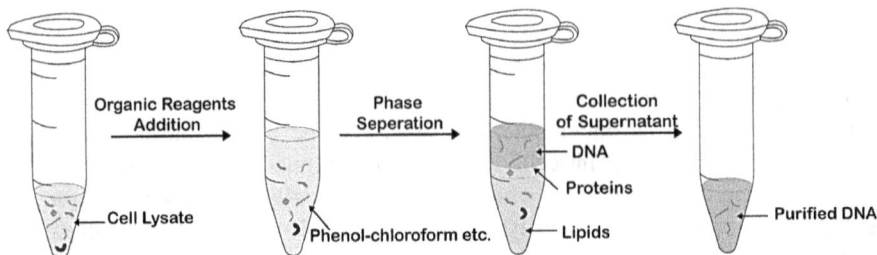

FIGURE 1.4 Protocol of PCI liquid-phase purification. DNA extraction is achieved using the higher dissolvability of polar DNA molecules in the polar chloroform water phase, while non-polar phenol digests the protein and lipids. Thus, DNA is isolated in the upper aqueous supernatant after centrifugation and concentrated by ethanol precipitation.

the DNA in the upper aqueous supernatant is collected after centrifugation and concentrated by ethanol precipitation.

It is indeed challenging to implement the above extraction method onto a microfluidic platform. Zhang et al. fabricated PDMS on-chip wells to store the sample matrix in an aqueous phase. Using an upper continuous-flow organic phase to carry the impurities away, DNA within the wells is purified. In their research, an 85%–120% nucleic acid recovery from both Gram-positive and negative bacteria is received, and the pH and flow rate of the organic phase are the critical factors closely related to DNA recovery [44].

In contrast, microdroplets have been used as the increased interfacial area contributes to mass transportation between the two liquid phases. Moreover, inner conservative flow recirculation also accelerates the movement of interior molecules across the liquid interface. Morles et al. achieved a 92% plasmid DNA recovery rate depending on their droplet-based microsystems. However, an evident drawback is the difficulty of two-phase separation at the microchip outlet, undermining its applications in practical DNA analysis [45]. Since the use of organic solvents may introduce toxicity and complicate fluid manipulation at the microscale, only very few microdevices are available for liquid-phase DNA extraction [46–48].

1.5.2.2 Solid-Phase Purification

Solid-phase purification uses the affinity differences between the analyte and the rest of the interferents when a solid adsorbent contacts the liquid matrix [49]. The silica-based technique is the most common one among all. Since the 1980s, even before the appearance of the microfluidic system, people have started to purify the negatively charged DNA by positively charged glass beads in the presence of high-concentration chaotropic agents such as guanidinium and sodium iodide [50]. Later, the glass beads were replaced by silica columns in most commercial kits [51]. Conventionally, the DNA is attached to the silica substance in the high ionic buffer, and the potential contaminants from sample matrixes are wiped off simultaneously by centrifugation. Later the DNA is dissolved into a buffer with low ionic strength and concentrated to a small volume for further usage. In contrast to the liquid-based strategies, collecting the aqueous phase and precipitation is unnecessary, demonstrating a unique superiority.

Some early microsystems realized the above experimental process by directly packing the silica beads within the chamber on-chip [52,53]. Micro weirs confine the silica beads to adapt the chamber geometry while using a syringe force toe flow matrix through them. However, the extraction rate could be inconsistent because the instability of beads leads to an unexpected flow pattern and increased back pressure through the microfluidic channel. Clogging frequently happens as well [54] (Figure 1.5a).

The silica-based monoliths (silica sol-gel) are viable alternatives to anchor the silica beads within the microdevices [55,56] (Figure 1.5b). A continuous silica network is fabricated either through gelation chemistry [57] or through a polycondensation process controlled by UV irradiation [58], resulting in a porous monolith for DNA extraction. The pores in the monolith allow the flow to pass through with low pressure and an improved surface-to-volume ratio. Tetraethoxyorthosilicate (TEOS) monomer was initially employed for monolith preparation, and the gel was mechanically unstable over time and found to be dried and cracked [55]. Later, tetramethyl orthosilicate (TMOS) was demonstrated to be an optimal material for monolith fabrication or just as a layer of surface coatings for increased DNA binding capacity [57 58]. Sol-gel monoliths suffer from shrinkage when the gel is aged, causing voids between the matrix and channel walls. This shrinking phenomenon leads to a significant variation in extraction efficiency.

Fabricating silica microstructures within the microchannels is a straightforward solution to promote extraction, and it is generally realized by patterning silicon or glass substrates by reactive ion etching [59,60] (Figure 1.5c). It is to increase the surface-to-volume ratio, which determines how well the DNA interacts with the silica materials. Since the microstructure and the main chip body are integrated, a fast flow rate and high throughput can be achieved compared to packed silica beads and sol-gel-based protocols. Silica microstructures microdevice has been widely integrated into multiple-steps µTAS proceeding DNA amplification and detection [61]. However, the shape of the microstructure also introduces flow disturbance, influencing the mixing efficiency within the channel, and creating complications for DNA extraction.

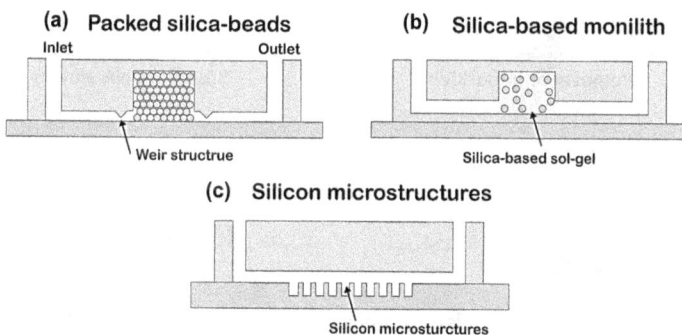

FIGURE 1.5 Protocol of silica-based purification based on (a) silica beads, (b) silica-based monoliths, and (c) silica microstructures.

Similar to silica membranes, aluminum oxide membranes (AOMs) bind DNA at high salt conditions. The advantages of AOMs over silica membranes are that AOMs don't impede the PCR reactions. Also, the elution reagent bovine serum albumin and Taq polymerase could be added into PCR mixtures. This attractive character increases the compatibility of AOMs towards downstream reactions [40].

On the other hand, the usage of paramagnetic beads in DNA extraction microassays has been well-established over the past decade [62]. Low cost, easy fabrication, stable flow rate and large-scale sample volume accommodation of this technique are advantageous. Through surface modifications, paramagnetic beads are freely suspended within the solution absorbing non-specific or targeted specific DNA. Thus, the interaction between the DNA and absorbent is enhanced. In addition, there is no risk of channel clogging by the impurities since no weir structures and tiny matrices pores are present here. Due to the excellent compatibility with other downstream processes, some point-of-care µTAS platforms have been developed based on this.

Silica-coated paramagnetic beads employed the interaction between silica and DNA (Figure 1.6). Typically, the discrete beads are grabbed together by the external magnetic field after capturing DNA. Then, a flowing washing solution rinsed all the impurities away. Chung et al. fixed beads from the beginning and forced the sample solution to flow back and forth to improve DNA capturing [28]. Their research shows that at low nucleotide concentrations, the extraction efficiency was 100–1000-fold better than that of free-moving beads, which is more appropriate for rare DNA samples. Other alternatives include the application of stationary solutions by processing buffers are separately sealed in Tygon tubing with an air gap in between. A permanent ring magnet outside the tube transfers the magnetic particles from one solution to another. This portable setting doesn't rely on an external flow controller and is suitable for the resource-limited area [63].

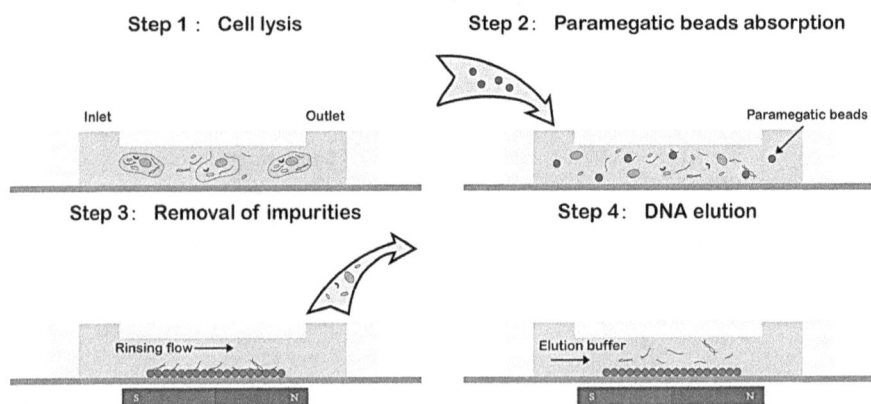

FIGURE 1.6 Paramagnetic-beads-based purification. Cells are first lysed using a conventional lysis buffer (Step 1). After loading the paramagnetic beads to capture the target DNA (Step 2), the beads are collected together by the external magnetic field (Step 3), which allows the removal of the impurities through a flowing washing solution.

The surface coating on paramagnetic beads can be diverse and versatile. In addition to silica coating, numerous commercial kits for charge switchable magnetic beads are available and ready for micro-assay usage [64,65]. Depending on the solution pH value, a low pH value contributes to a positive charge environment for DNA adsorption, while a high pH benefits DNA elution. Chitosan-coated particles are one type of switchable magnetic beads, which attract DNA at a pH value of 5 and release it at a pH value of 9. More importantly, the PCR inhibitors such as chaotropic salts and 2-propanol can be avoided, making it more suitable for the two-stage µTAS proceeding sample extraction and PCR [66].

If a specific DNA sequence is targeted, one reasonable approach is to modify the beads' surface with its complementary sequence [67]. Such specific DNA hybridization allows a low detection limit (LOD) without further purification and contamination. Here, the concentration and density of DNA probes need to be carefully optimized. Too sparse or too dense of the probes decrease the extraction efficiency by the non-specific hybridization or electrostatic repulsion. Various µTAS have been published focusing on sensitive detection for definite sequences in a wide range of sample types [68]. Instead of using microbeads, the oligonucleotide probe can also be immobilized directly onto the polymer matrix surface. Omitting beads provide one-stage microsystems with high integrity. However, the synthesis of polymer is labor-intensive and introduces solvents that may inhibit DNA hybridizations [69].

1.5.2.3 Validation of DNA Extract

The purity of the extracted DNA is mostly validated by ultraviolet absorbance. The optical absorbance for nucleic acid peaks at a wavelength of 260 nm. In contrast, proteins, the most common contaminants, absorb light at 280 nm. Thus, the ratio between the absorbance at the wavelength of 260 nm and 280 nm is used to assess the purity of nucleic acids. Commonly, the ratio of the absorbance at 260 and 280 nm (A260/280) of 1.8 is considered pure DNA. Also, other methods like Qubit assay, agarose gels and real-time quantitative PCR (qPCR) are routinely used [70,71].

1.5.2.4 Summary

Reliable yet straightforward methods of obtaining high-quality DNA are prerequisites before successfully implementing them in any analytical process. Microfluidics DNA extraction platforms provide unique perspectives and upgrade performance from multiple aspects. Since many techniques are available, the parameter demands include crude sample types, reaction speed, DNA purity, fabrication strategy, etc. are considered comprehensively. The combination of upstream and downstream processes is also worthy of attention. Nowadays, only limited µTAS can handle the entire flow from crude material to visible detection results, which will be the future of DNA extraction microdevices.

1.6 MICROFLUIDIC DNA DETECTION USING DNA SEPARATION

After extraction, DNA separation appears to be an ideal approach that classifies DNAs based on their physical properties, which is essential in pathogen identification, forensic analysis, genome sequencing, and genetic assays of disease diagnostics

[72,73]. DNA has a length ranging from nanometer to microscales with diverse spatial structures. In addition, DNA molecules are negatively charged due to their sugar-phosphate double helix backbone. This structural nature enables the separation of DNA molecules mainly by their inherent electrophoretic mobility, although affinity chromatography is seldom employed [74]. The history of electrophoretic mechanisms for molecular separation dates back to the 1930s, and it developed rapidly in the last two decades, closely relating to the rise of the microfluidics concept [72].

Gel electrophoresis (GE) is a golden standard bench-top protocol evaluated vis fluorescence staining after 0.5-2 h processing in porous polyacrylamide or agarose slab gels. Because of the similarity of charge-to-mass ratio among different DNA molecules, the length of DNA, or the number of nucleobases, becomes the main factor affecting electrophoretic mobility. The capillary electrophoresis (CE) replaces the sieving matrix of the slab gel with polymer/gel liquid filled in a capillary tube (Figure 1.7). It's more advanced at the microscale over GE because of the higher resolution, rapid analysis (<10 minutes) and convenient laser-induced fluorescence detection aroused by the high surface-to-volume ratio of capillary tubes.

Li et al. demonstrated an agarose-based microfluidic platform incorporating on-chip cell lysis and electrophoresis reaction to investigate DNA damage. The main body of the chip is moulded by curing agarose gel on a SU-8 pattern by photolithography. After cell loading, the microchannels are sealed with ultra-low melting point agarose to build a cap. The reagents are directly added through the agarose cap. Their device can manipulate 10,000 individual cells simultaneously, and this throughout is 100 times higher than that of traditional comet assay [75].

On the other hand, electroosmotic flow (EOF) is a byproduct during electrophoresis and is undesired as it impedes the electrophoretic motions. To suppress it, polyester–toner (PeT) was used as it exhibits a low EOF velocity compared to that

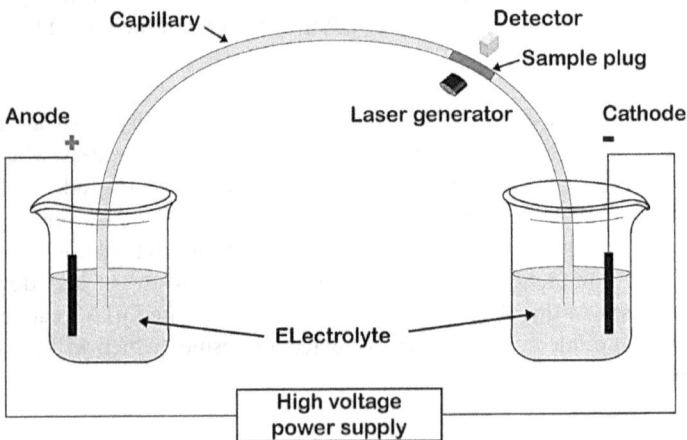

FIGURE 1.7 Protocol of CE separation. A capillary tube filled with polymer/gel liquid is used to replace the slab gel. When a high voltage is applied between the anode and the cathode, electrophoresis inside the capillary enables the separation of DNA into plugs detectable by laser-induced fluorescence.

of glass, PDMS and other popular fabrication compounds. As such, the PeT-based microfluidics platforms provide better efficiency in DNA separation. Carrilho's group utilized polymer gel-filled PeT chips separating the DNA fragments between 100 and 1000 bp in 4 min [76].

In a CE system, high resolution is achieved using high voltage supplies (generally 500–1500 V). The same condition is required for the microfluidic capillary electrophoresis (MCE) system and prevents it from being portable [77]. The separation matrix composition controls the injection difficulty, separation resolution and device lifetime. Typically, a high-molecular-weight matrix suffers from high loading pressure and time consumption. In contrast, the low-molecular-weight matrix can't afford a sufficient resolution due to the limited interactions between DNA fragments and the matrix. Multiple viscosity-turntable materials have been developed in some early works for this issue. The media is injected at low viscosity conditions and recovered to high viscosity within the tube by adjusting the temperature or shear force. Later, the crosslinked polyacrylamide gels are replaced by non-cross-linked polyacrylamide (LPA) and polydimethylacrylamide (PDMA), facilitating the versatility of MCE devices [78].

The pre-concentration of DNA samples is another crucial issue to a highly accurate and reproducible MCE assay, especially for diluted samples. Isotachophoretic (ITP) is a powerful technique regulating samples between a high ionic strength leading buffer and a low ionic strength terminating buffer in a discontinuous buffer system [79] (Figure 1.8). Through the unique self-sharpening character of ITP, the DNA molecules are stacked as consecutive, non-overlapped zones in the order of size. ITP requires no solid surface, surface modification, or mobile component in

1. Sample loaded

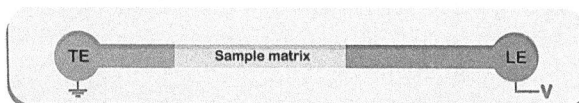

2. Seperation

3. Isokinetic motion

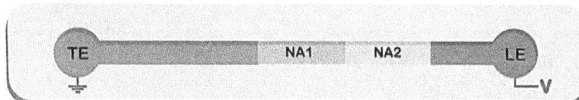

FIGURE 1.8 Strategies of ITP in DNA separation. The oligonucleotide sample matrix with intermediate ionic mobility is introduced between a leading electrolyte (LE) zone with high ionic mobility and a terminating electrolyte (TE) zone with low ionic mobility. Under an external electric field, all sequential zones migrate at equal velocity. The nucleic acid (NA) molecules are classified by electrophoretic mobility to maintain velocity equilibrium, resulting in a clear boundary between different NAs.

microfluidic systems. Thus, plenty of micro platforms were designed using ITP for performing DNA extraction, separation and purification [80] or the ITP hyphenated GE [81].

Dielectrophoretic (DEP) is a technique to induce the migration of polarizable objects through the external inhomogeneous electric field. As a label-free and non-invasive strategy, DEP assays have been well established in biomolecule separation, especially for cell sorting. Like other micromoles, DNA polarizability closely relies on the frequency range, the AC voltage, and the medium's conductivity. Separation happens at the corners or sharp edges of the microchannel. Hsing group is the pioneer that demonstrated the separation of 48.5 kbp and 2.868 kbp linear DNA using microelectrode-embedded intermittent-flow devices [82]. Later, the size gap between the DNA targets is gradually reduced [83]. Through continuous flow with high throughput, Viefhues et al. successfully separated the 6.0 kbp and 5.0 kbp DNA fragments equivalent to a size difference of 16.7% through a strong DEP force generated by a 670 nm nanoslit [84]. DEP likewise achieves topological isolations determined only by the conformation (linear or spiral) for the same size fragments [85]. Thus, after a systemic study of the dielectrophoretic migrating behaviours, DEP is capable of separating the target DNAs.

Other than the active methods, the use of micropillars also has great potential in DNA separation. The Brownian ratchet assay and periodic micropillars were initially employed in the large DNA fragments (> 40 kbp) separation, depending on their diffusion coefficients [86]. Later, Deterministic Lateral Displacement (DLD) and its derivatives called electric fields DLD (eDLD) found DNA separation applications [87] (Figure 1.9). DLD is a size-based molecule sorting technique, which depends on the steric interactions between the particles and micropillar arrays to regulate the particle's trajectories through different streamlines. Austin's group exploit this phenomenon to separate 61 and 158 kbp DNA into two bands in an asymmetric microarray driven by electrophoresis [88]. A recent work by Wunsch et al realized the separation between 100–10000 bp DNA with a resolution of 200 bp in a velocity-dependent nanoDLD platform [89].

1.7   MICROFLUIDIC DNA DETECTION USING AFFINITY PROBES

Compared to DNA detection using separation, which relies on the physical properties and is non-specific, making use of the natural hybridization between two ssDNA is by far the most effective method to capture a DNA target with a specific sequence. Such a method lies in the synthesis of DNA probes, signal generation based on hybridization, and signal amplification using a DNA circuit.

1.7.1   SYNTHESIS OF DNA PROBES

The first attempt for oligonucleotide synthesis can be dated back to the 60s when Scheuerbrandt and Khorana successfully fabricated a 77-nucleotide alanine tRNA utilizing phosphoramidite chemistry.

Small   Large
Molecules Molecules

Micropillar
Array

Flow Direction

Migration Angle

Steamlines

Zigzag
Mode

Bump
Mode

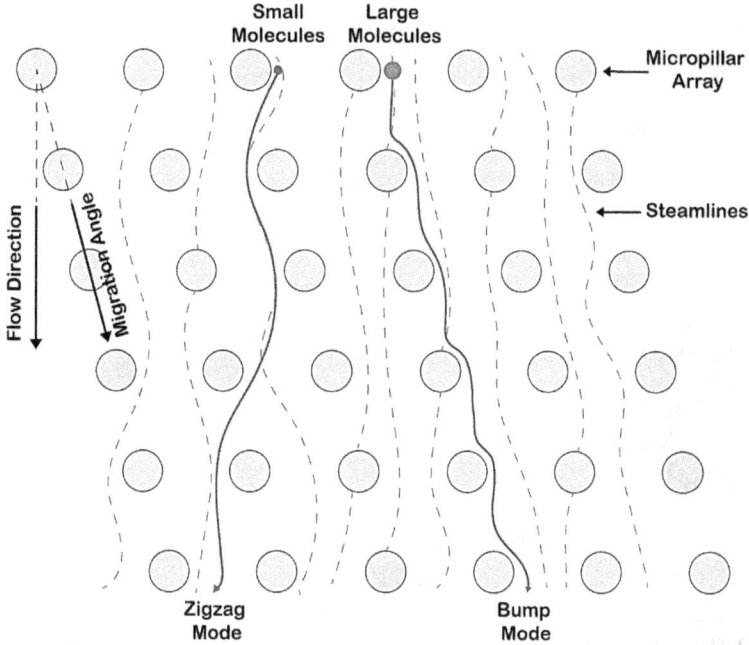

FIGURE 1.9 Separation of DNA using DLD separation. The molecules whose dimension is smaller than the critical diameter moves in a zigzag mode within the initial streamline, while large molecules are bumped to the micropillar arrays and traverse multiple streamlines. Thus, DNA with length differences can be well separated.

1.7.1.1 Phosphoramidite Method

The phosphoramidite method is a common chemical reaction that is used for the synthesis of oligonucleotides. The name phosphoramidite comes from the use of nucleoside phosphoramidite modified with dimethoxytrityl (DMT), which serves as basic building blocks for oligo synthesis to stabilize and protect the nucleoside [90,91]. The reaction consists of a four-step cycle reaction to add a nucleotide base for every cycle (Figure 1.10).

- Step 1: De-protection
    The first step of the cycle is to remove the DMT group from the phosphoramidite. To do this, a solution of trichloroacetic acid is introduced to the system which detritylating DMT-blocked oligonucleotides to activate the 5'-terminal hydroxyl group of the phosphoramidite.
- Step 2: Coupling
    The coupling step involves the introduction of another nucleoside base in the form of a DMT-protected phosphoramidite. It will be coupled to the 5'-hydroxyl group from the previous nucleoside phosphoramidtite forming a phosphitetriester.

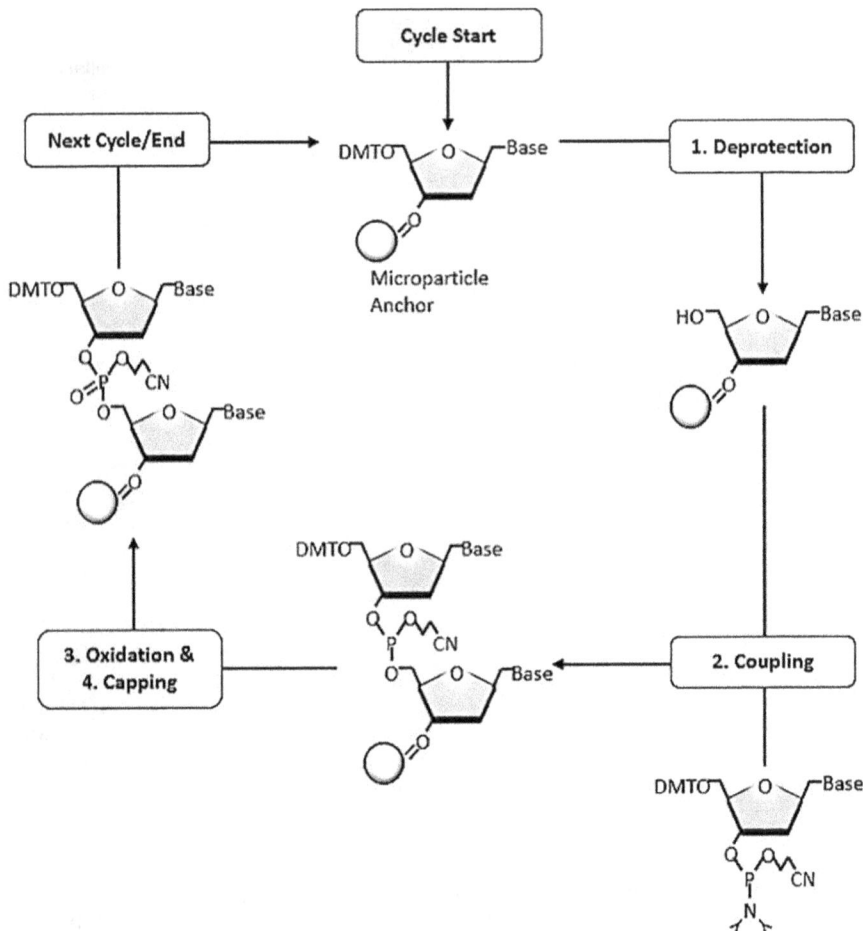

FIGURE 1.10 Nucleoside phosphoramidite oligonucleotide synthesis cycle consists of (1) deprotection step, (2) coupling, (3) oxidation, and (4) capping.

- Step 3: Capping
    Capping step is used to block the remaining 5'-hydroxyl group which is not coupled to phosphoramidite to inhibit the formation of the non-specific oligonucleotide. The capping/blocking is done by acylation by acetic anhydride and 1-methylimidazole.
- Step 4: Oxidation
    The last and final step of the cycle is to convert the phosphite triester linkage between the two nucleoside phosphoramidite into a phosphate linkage. By introducing an iodine solution, the phosphate triester is oxidized into tetracoordinated phosphate triester, which is a more stable precursor for the naturally occurring phosphate diester bond.

1.7.1.2 Column-Based Oligonucleotide Synthesis

Numerous platforms have been used for this chemical method. The most traditional platform is by using a column-based oligonucleotide synthesis. Different DMT-stabilized nucleoside phosphoramidite bases are contained in separate vials (Figure 1.11). During the synthesis, these different bases are introduced to a column which is the main container unit for the reaction using the phorphoramidite method [91]. The column is filled with CPG beads as a substrate to support the oligonucleotide elongation. As a result, one type of oligonucleotide will be synthesized for each column. Recently, this method has now been able to fabricate 100 nucleotide bases oligonucleotide with a price of $0.05–$0.15 per nucleotide. Commercially available synthesizers can achieve the creation of 96–768 oligonucleotides with a concentration from 10 nM–2 μM simultaneously [92]. While the method is heavily studied and improved, the low throughput and high cost still limit it from bulk manufacturing [93].

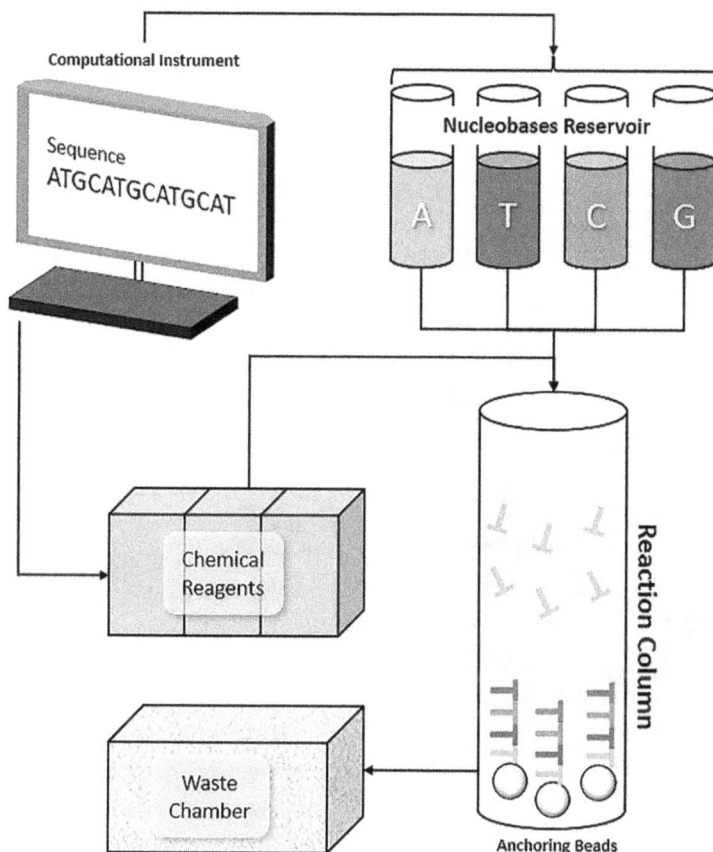

FIGURE 1.11 Column-based oligonucleotide synthesis method where a computer-controlled nucleobase reservoir is introduced to a column with anchoring beads for DNA probe synthesis.

1.7.1.3 Microarray Based Oligonucleotide Synthesis

Microarray is another approach to chemical-based synthesis. In this method, the DMT stabilizing molecule used in the phosphoramidite chemical method is replaced by another stabilizing and light-sensitive molecule called the photolabile protective group [93]. The method requires a light source which can target a specific location on the microarray (Figure 1.12). When the light exposes nucleotide bases, the photo-labile protecting group is removed, allowing coupling, capping, and oxidation steps. After one cycle, the light will expose to another set of nucleotide bases for the next cycle of elongation. As such, each oligonucleotide can be synthesized with a specific sequence. The early methods utilize a photomask that is aligned to each microarray. Nowadays, ink jet printing technologies have been reliably used for specific light targeting which can remove all the unnecessary bothersome procedures. In comparison to the column-based oligonucleotide, the microarray-based synthesizing technology could synthesize multiplex oligonucleotide sequences at the same time where one microarray will produce 1 specific oligo sequence with a lower price per nucleotide of $0.00001 to $0.0001. The disadvantage of this method is, however, that the overall concentration of the synthesized oligonucleotides is lower, making it less compatible with a large-scale production.

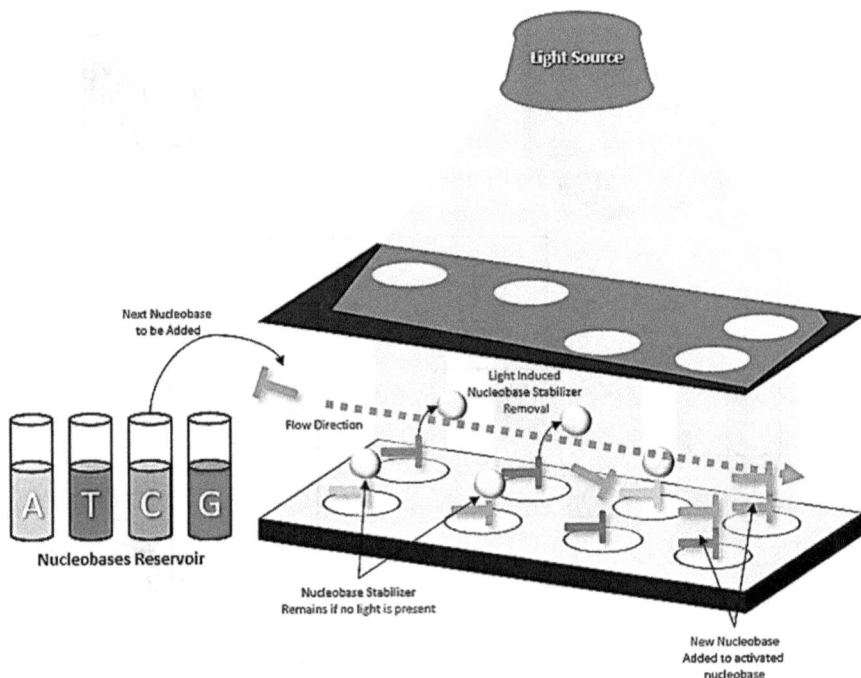

FIGURE 1.12 Microarray oligonucleotide synthesis method. A microarray is fitted with a photomask where specific spots are allowed with light to trigger the de-protection of the photolabile nucleobase stabilizer before the new nucleobase is introduced from the reservoir. Thus, multiplexed oligonucleotide sequences can be synthesized at the same time.

1.7.2 Signal Generation

After all the processes to isolate the target DNA and synthesis of DNA probes, it is essential to quantitatively detect the target DNA. For specific and sensitive detection of target DNAs, numerous methods can be applied using microfluidic platforms. In this section, detection methods to measure DNA will be discussed and will be categorized mainly into the type of signal reporting.

1.7.2.1 Optical

Optical detection is a very popular method to be utilized for highly sensitive and selective microflow detection of nucleic acid targets due to the simplicity, robustness, and versatility of the approach. Optical signals such as colorimetric and fluorescence can be measured either by the naked eye or by simple equipment such as microscope, handheld spectrometer, and now even smartphones.

1.7.2.1.1 Colorimetric

Colorimetric detection revolves around the detectable changes in color. This method is commonly used in many detections. For example, the litmus paper reports the acidic/basic condition by showing the color of the paper after being dipped in a solution. For the pregnancy test, colorimetric detection also allows the appearance of a red band which indicates the presence of human chorionic gonadotropin (HCG), the pregnancy hormone in urine.

Colorimetric detection assays can be done through numerous approaches. Either by using metal indicators, dyes, and gold nanoparticles (AuNPs) which can indicate the presence of nucleic acid target through a change in color which can be seen by the naked eye. Metal indicators work by pairing a metal ion indicator with another molecule which facilitates the color change of the material (Figure 1.13a). A study done by Oh et al. utilized Magnesium ion (Mg^{2+}) indicator that reacts with a pyrophosphate ion ($P2O7^{4-}$) created during the loop-mediated isothermal amplification (LAMP), resulting in a magnesium pyrophosphate ($Mg_2P_2O_7$) molecule which will change the color of the originally purple LAMP solution into a light blue color. Utilizing their centrifuge-facilitated microfluidic device to do the LAMP reaction, they are able to detect foodborne bacteria pathogens such as *E. coli, Salmonella typhimurium, Vibrio parahaemolyticus*, and *Listeria monocytogenes* with a limit of detection of 10 bacterium cell level in 65 minutes. [94]

Another principle is by using gold nanoparticles (AuNPs). AuNPs possess the property that allows them to change color depending on their size or their aggregation (Figure 1.13b). A study by Alhasan et al. utilized AuNPs modified with ssDNA probes which will hybridize to the target DNA named as Scanometric Array-based Multiplexed Detection of miRNA Species (Scano-miR). Using this methodology, they are able to profile prostate cancer markers miR-270 with a LOD of 1 fM and an accuracy of 98.8% compared to the standard fluorescence-based miRNA assay [95].

Another approach by Zhao et al. demonstrates a detection using a pair of magnetic microparticle (MMPs) and non-magnetic microparticle such as polystyrene microparticles (PMPs). Both MMPs and PMPs are modified with ssDNA probes that will bind

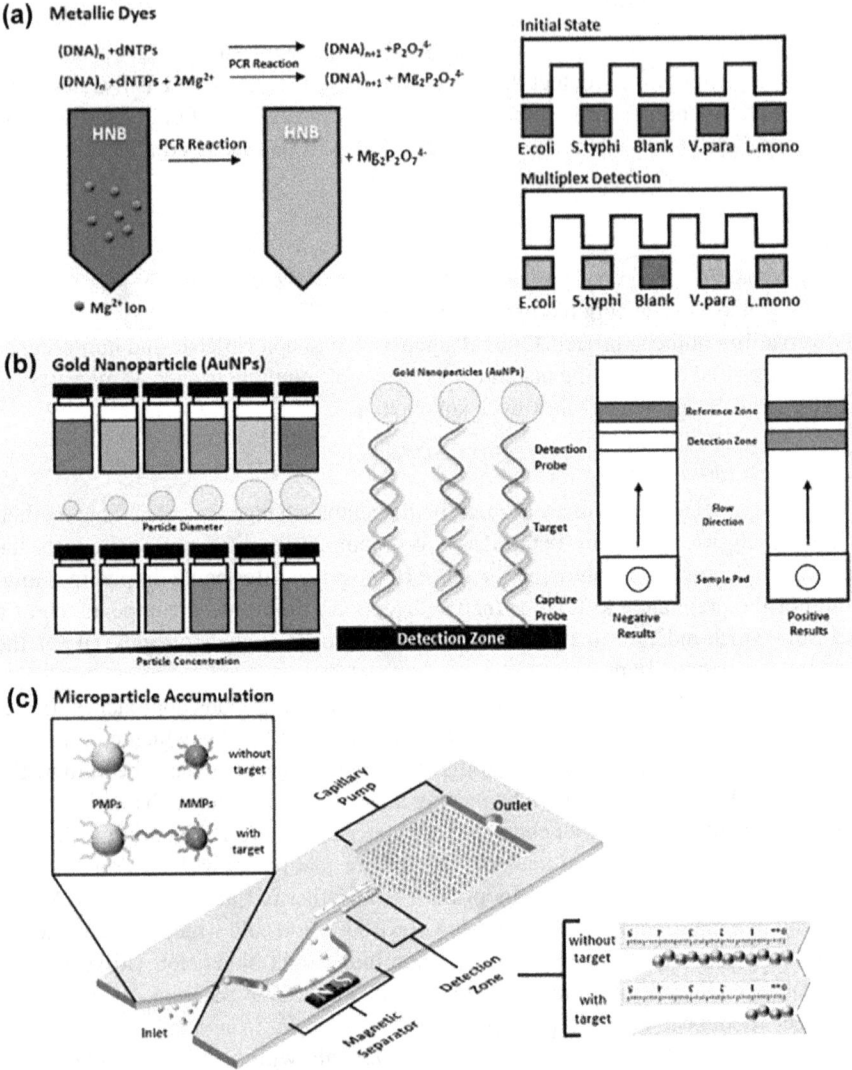

FIGURE 1.13 Optical microfluidic detection with (a) metallic dyes of magnesium which will react with the pyrophosphate generated by the reaction to produce magnesium pyrophosphate that changes the color of the solution, (b) gold nanoparticles which changes color and intensity with different particle size and concentration respectively, and (c) microparticle accumulation where concentration of DNA targets can be measured with the different lengths of free PMP accumulation which are not bonded to their MMP counterpart with the absence of a target in the magnetic separation zone.

to each end of the target DNA sequence (Figure 13c). As such, when the target DNA is present, the target DNA will join the MMP and PMP together making the whole complex become magnetic. In contrast, when target DNA is absent, the microparticles will not connect. Such altered connection can be visualized by a specific microfluidic device. After magnetic separation, the non-magnetic nanoparticle would be trapped and form a visible particle accumulation at a particle trapping area, which allows the quantification of the target concentration by the length of the trapped PMPs. Their microfluidic design is tested with the MB155 sequence which is the DNA counterpart of the breast cancer marker miR-155. Through this method, they achieved a limit of detection of 12.2 nM or 2.5 µg/L [96].

1.7.2.1.2 *Fluorescence*

Fluorescent detection is a very popular approach for DNA detection. Fluorescence detection revolves around fluorophores which fluoresces when being excited to a higher energy state by an electromagnetic wave (Figure 1.14). Upon excitation, the fluorophore will emit the excess energy in the form of fluorescence light with a longer wavelength. To detect DNA using fluorescence, a fluorophore is modified onto a nucleotide probe which complements itself towards the target DNA. The conjugation between the fluorophore to the nucleotide is facilitated through a chemical reaction typically by carboxyl-amine interaction. The fluorophore will serve as a label for the target DNA when it is inserted into a measuring apparatus such as a fluorescence spectrometer or a fluorescence microscope. Early studies utilize an integrated fluorescence system in a PDMS-based microfluidic chip by using optical fibers in 2001 by Chabinyc et al. [97].

Utilizing fluorescence comes with the advantage for being compatible with in vitro processes, meaning that it will not interfere with biological processes happening inside the body or artificially created in a laboratory environment, making it suitable for real-time observation in a biological environment. A study done by Li et al. utilizes fluorescence detection to do a real-time measurement of Hepatitis B virus (HBV) DNA under a PCR amplification setup in a microfluidic chip. This setup configuration gave them the ability to detect their HBV gene target with a limit of detection of 1×10^{12} copies/µL with a correlation coefficient of 0.9966. The total time of their reaction until the target gene could be visible is 30 minutes. Their detection method is very sensitive to a point that they can detect faint fluorescence signal from a volume as low as 100 pL [98].

The most common method to achieve fluorescence DNA detection is by using fluorescence in situ hybridization (FISH) where the DNA target is introduced into the microflow device and isolated inside the microfluidic chip. The DNA probes modified with fluorophore are then introduced into the device making the probe hybridize with the target sequence and fluorescence. A study by Nguyen et al. published in 2017 uses FISH to detect DNA strand markers [99]. Even when the FISH method for DNA detection results in a very sensitive detection method, one limitation of the method is that it does not allow for multi-fluorescence detection [100]. For multiplexed DNA detections, Peng et al fabricated a microfluidic device which has a substrate of silver nanoparticles which are modified with capture probes for Enterbact3a and Eco3a DNA sequences. Silver nanoparticles are used

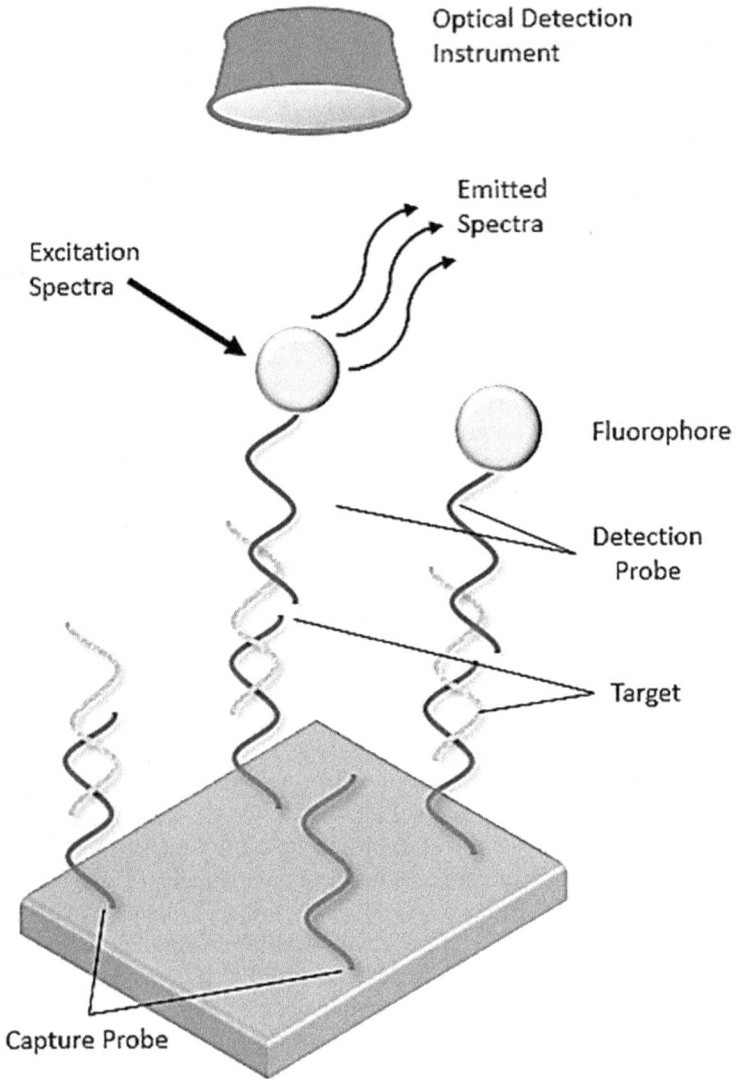

FIGURE 1.14 Microfluidic fluorescence detection with a capture DNA probe that binds with the DNA target and conjugated again to a detection probe modified with a fluorophore. When the fluorophore is exposed to an excitation spectrum, fluorescence signal will be emitted and detected.

in the setup due to the fluorescence quenching properties of fluorophores. Using a hairpin structure on the capture probe, where the fluorophore will be near the silver nanoparticles substrate and quenched when the DNA target is absent, the presence of target DNA will on the other hand bring the fluorophore away to render the fluorescence signal. Using this setup, they are able to obtain a sensitivity of 500 pM [100].

1.7.2.1.3 Chemiluminescence

The working principle of chemiluminescence is similar to fluorescence in the way that a molecule, commonly luminol, is put into a higher excited energy state by a chemical reaction. The reaction happens when luminol is mixed in an environment where hydrogen peroxide is present. Such a reaction can be catalyzed by the horse-radish peroxidase (HRP) enzyme, which will remove the amino group from one of the benzene rings and put the molecule in an excited (higher energy) electronic state. This excessive energy in their atomic structure will be released to bring the molecule back to a stable state in the form of a detectable electromagnetic wave (Figure 1.15).

The chemiluminescent approach has been implemented in numerous assays and microfluidic integrations. The detection scheme is exactly the same as fluorescence-based detection method except for the changes to the HRP enzyme being conjugated to the capture probe to replace the fluorophore, and the luminol introduction in a hydrogen peroxide environment instead of an excitation light [101]. In addition, another approach uses a quenching mechanism through another material to amplify their detection sensitivity. Recently, Graphene oxide (GO) has been a popular method for a chemiluminescence detection amplification scheme. Since GO quenches the luminol activity, when the capture probe is immobilized by the π-π bonding between the nucleic acid bases and the π rich GO, the chemiluminescent activity could be greatly reduced. However, the formation of the dsDNA when the target is present will break the π-π bonding and remove the capture probe from the GO interface. Since the quencher is not in an effective range to weaken the chemiluminescent signal, a signal increase could be detected [101].

1.7.2.1.4 Surface Plasmon Resonance

Optical quantification of oligonucleotides can also be approached without the use of labels as mentioned in the previous part. Surface Plasmon Resonance (SPR) has been regarded as the most well-known method. SPR happens when the surface plasmon polarization, which is a non-radiative electromagnetic wave, propagates in a direction towards an interface between the negative and positive permittivity of the material (Figure 1.16a). Since the wave is located at the interface between the material and the outside medium, the oscillations are very sensitive to small changes at the interface such as the molecules being adsorbed to the surface [102].

Even though SPR methodologies can provide a very sensitive detection, the problem remains in the requirement of additional equipment such as the optical prism and grating coupler for the excitation of propagating plasmons. Thus, localized surface plasmon resonance (LSPR) was created. LSPR provides not only improved miniaturization and cost-effectiveness but also the advantages of reducing background noise based on the localized field phenomenon (Figure 1.16b). A study done by Soares et al. utilizes LSPR in combination with gold nanotriangles (AuNTs). ssDNA probes are attached to the AuNTs which allows them to capture DNA targets through the probe. With this methodology, they were able to achieve a sensitivity of 468 nm per RIU and detectability of 20 nucleotide targets. It also shows superiority compared to using standard AuNPs, as AuNTs give a higher LSPR shift of 22 nm compared to 3 nm shift using AuNPs [103].

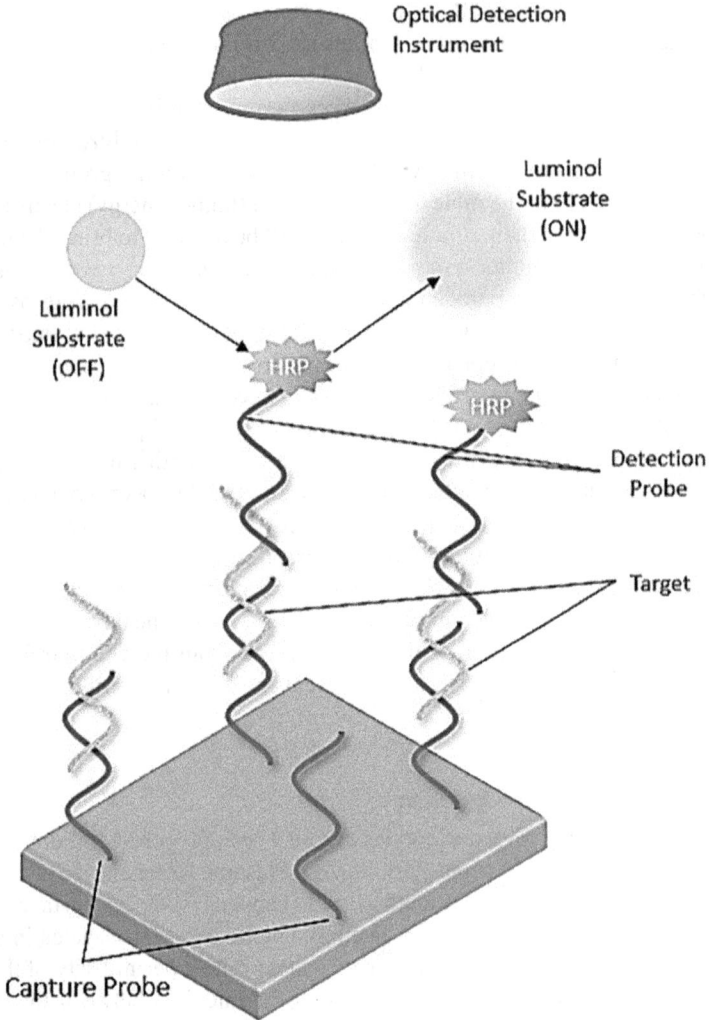

FIGURE 1.15 Microfluidic chemiluminescence detection utilizing a DNA capture probe to catch the DNA target and equip it with another detection DNA probe modified with HRP enzyme that will react with luminol to produce luminescence.

Another variation of surface modification plasmonic is Surface-enhanced Raman spectroscopy (SERS). Similar to the SPR method, SERS also measures slight surface changes. However, it is measured by the enhanced Raman scattering (Figure 1.16c). A study done by Tian et al. in 2017 utilizes aluminum nanocrystals which can differentiate the Raman intensity shifts between 4 different bases of nucleic acids. Thus, not only can we detect different DNA sequences by their length, but also the composition of their nucleic acid's bases [104]. Furthermore, He et al. used silicon nanowires as the base structure for SERS and their device

(a) Surface Plasmon Resonance (SPR)

Capture Probes • Target • Surface Plasmon Wave

Metallic Substrate

Prism • Target Shifted Surface Plasmon Resonance Angle

Initial Surface Plasmon Resonance Angle

(b) Localized Surface Plasmon Resonance (LSPR)

Electron cloud

Laser Induced Electric Field

Metallic Nanoparticle (Au, Ag, etc.)

(c) Surface Enhanced Raman Spectroscopy (SERS)

Excitation Light

SERS-Light

Rayleigh Scattered Light

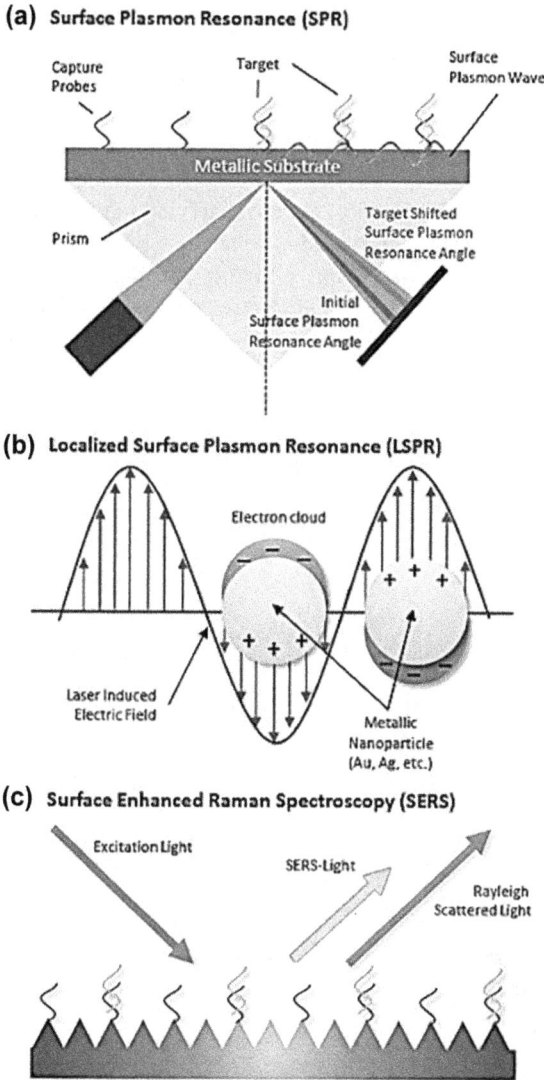

FIGURE 1.16 Signal generation method by (a) surface plasmon resonance, (b) localized surface plasmon resonance, and (c) surface-enhanced Raman spectroscopy.

can achieve an ultra-sensitive DNA detection of approximately 1fM. Nowadays, the target of SERS-based microfluidic devices can facilitate multiplexed detection of DNA targets [105].

1.7.2.2 Electrochemistry

The requirement of optical instruments such as a spectrometer or a microscope has made it complicated for measuring optical signals. In contrast, there are attempts using electrochemistry for the detection of DNA in the microfluidic platform. Electrochemical biosensors rely on electric circuitry to quantify the amount of target DNA, which can be easily integrated into the microfluidic platform and hence, making it a popular detection alternative.

Electrochemical detection is a redox method that usually involves three different electrodes: the working electrode, the counter electrode, and the reference electrodes. The working electrode (WE) is the place for carrying out the main electrochemical reaction driven by an electrical potential relative to the reference electrode, where capture probes are immobilized onto, allowing accompaniment of a probe with a redox molecule to bind and trigger the electrochemical reaction. Once the reaction is activated, the electric current reflecting the amount of target DNA is only through the counter electrode (CE), so the electrical potential can be maintained stable (Figure 1.17) [106].

A study done by Yeung et al. demonstrated multiplexed detection of genetic information on food pathogens from two different bacteria, *E. coli* and *B. subtilis*. In their

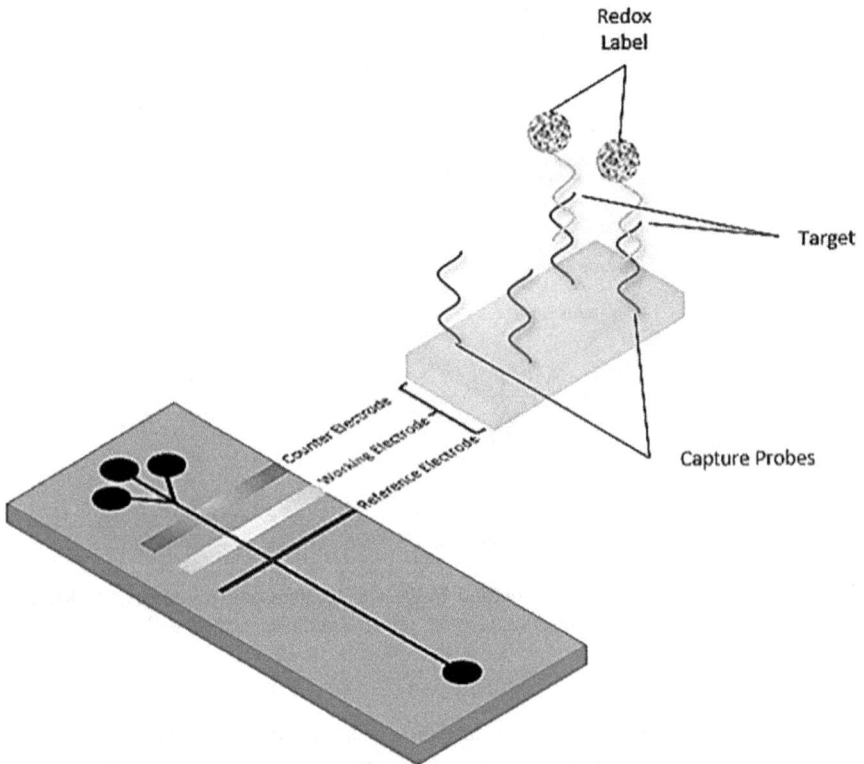

FIGURE 1.17   Basic microfluidic electrochemical detection setup with three electrodes. The working electrode contains DNA probes to capture target DNAs, which allows the attachment of a redox label to produce the electrochemical signals through the counter and reference electrode.

four-electrode design, two are fitted with capture probes for *E. coli* and the other two are with *B. subtilis* capture probes. Specific probe immobilization could be achieved by utilizing an electrochemical reaction of pyrrole and pyrrole-oligonucleotide. Specific capture probe immobilization could be achieved by applying a voltammetric scan to the electrodes while the other electrodes are either disconnected or grounded. Their device could achieve multiplexed detection with the LOD of 1fM [107]. On the other hand, a study by Liu et al from 2004 fabricated a microfluidic device which combines an in-situ PCR amplification with electrochemical detection by a ferrocene labeled signaling. Using their device, they can measure 106 *E. coli* K12 cells from 1 μL of rabbit blood and DNA detection of the human HFE-C gene from 1.4 μL of human blood with a LOD around 2 fM [108].

To further enhance the signal-to-noise ratio, Ferguson et al. utilizes the methylene blue redox label and attached it to a hairpin capture probe. When the target DNA is absent, the hairpin will place the redox label in proximity to the electrode, which increases the current in an ACV (Alternating Current Voltammetry) measurement. On the other hand, if the target DNA is present, the hairpin will be displaced due to hybridization, which moves the redox label away from the electrode such that the current in the measurement decreases. Using this method, they are able to detect DNA from *Salmonella enterica* serovar Typhimurium LT2 with a limit of detection below 10aM [109].

Together, electrochemical detection gives a very superior sensitivity compared to optical detection-based methods. However, it also requires a separate reader or potentiometers that can sensitively pick up these delicate changes.

1.7.2.3 Other Signaling Platforms

Other types of signals could also be used as a tool for DNA detection although not as common. Examples are magnetic-based and mass-based detection of DNA targets. The magnetic-based detection principle uses the magneto-resistive method. By conjugating the target DNAs with magnetic nanoparticles, their binding to the immobilized probes on the chip will cause a detectable change in resistance. Using this method with a combination of nuclear magnetic resonance (NMR) and magnetic tunneling junction (MTJ), multiple devices have been able to achieve DNA detection of food pathogens such as *M. tuberculosis* with high sensitivity within the nM range.

Mass-based signal transduction is also used for DNA detection. Methods such as quartz crystal microbalance (QCM) and surface acoustic waves (SAW) utilize the change in mass when the target binds to the capture probes atop the substrate, which in turn will shift the oscillation frequency of the substrate to measure and induce this oscillation change. The uniqueness of this method is that no label or tag is conjugated to the DNA capture probe. An example of this sensor was fabricated by Hong et al. which utilized the QCM principle to detect viral hemorrhagic septicemia (VHS) that achieve an LOD of 1.6 nM with high stability where the DNA probes survived 32 repeated regeneration without substantial signal reduction [110]. In addition, Zhang et al. presented another mass-based method which amplifies the mass signal generation by using enzyme-based

DNA elongation and AgNPs addition to enhance the mass change with an LOD of 0.8pM [111].

1.7.3   SIGNAL AMPLIFICATION

Low abundance of target DNA is a common problem in most biochemical analyses and biomedicine. Thus, it is of great significance to develop simple, rapid, efficient and cheap amplification technology. At present, signal amplification mechanisms can be divided into enzyme-mediated methods and enzyme-free methods.

1.7.3.1   Enzyme-Mediated Methods with Thermocycles

Polymerase chain reaction (PCR) and ligase chain reaction (LCR) are classic methods for detecting biological molecules such as DNA with good sensitivity and accuracy. Here, the mechanisms and applications of these two enzyme-mediated amplification methods are introduced.

1.7.3.1.1   Polymerase Chain Reaction (PCR)

In 1983, American scientist Kary Mullis raised an assumption about polymerase chain reaction technology. Later, reports and patents on PCR were published officially in 1986 [112]. Principally, PCR can amplify a single DNA molecule into millions of copies of DNA molecules through only a few thermal cycles, achieving exponential amplification. PCR thermal cycling usually includes three different steps, including denaturation, annealing and extension (Figure 1.18). In the denaturation step (about 95°C, high-temperature zone), the double-stranded

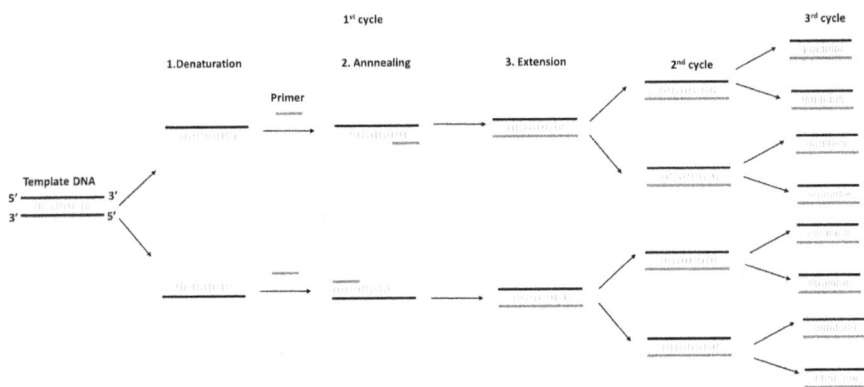

FIGURE 1.18   Denaturation, annealing and extension in PCR procedure. In the denaturation step (about 95°C, high-temperature zone), the double-stranded DNA is separated into two single-stranded DNAs. Next, the temperature is lowered to perform an annealing step (about 56°C, low-temperature zone) which promotes primers (short DNA sequences) to hybridize with the melted single-stranded DNA. Finally, under the assistance of polymerase, the free nucleotides in the solution are synthesized into a second complementary strand of DNA for the next cycle.

DNA is separated into two single-stranded DNAs. Next, the temperature is lowered to perform an annealing step (about 56°C, low-temperature zone) which promotes primers (short DNA sequences) to hybridize with the melted single-stranded DNA. Finally, with the assistance of polymerase, the free nucleotides in the solution are synthesized into a second complementary strand of DNA. Of note, because the polymerase is more active at 72°C, this temperature needs to be maintained during extension. By repeating three temperature steps, a large amount of DNA can be replicated, and the concentration of target double-stranded DNA can be multiplied exponentially.

First generation of PCR technology combines, the amplification product with gel electrophoresis or capillary electrophoresis to provide qualitative detection and analysis [37]. This PCR technology has an irreplaceable position in the preliminary preparations of studies such as cloning genes and genome sequencing. However, the first-generation PCR cannot achieve quantitative analysis. To solve it, the second-generation PCR was developed in 1992. Called the real-time quantitative PCR, it adds fluorescent substances to the PCR system. By utilizing the accumulation of fluorescent signals to monitor the entire PCR process, a time-varying fluorescence intensity curve reflecting the amount of DNA is used to compare the relative expression of specific genes among different samples. As a result, this method is widely used in pathogen detection [113], clinical diagnosis [114], food safety testing [115], and environmental microbial testing [116].

With the development of MEMS, scientists have made great progress in microfluidic PCR. Microfluidic PCR chip was mainly divided into three types: chamber-PCR, continuous PCR and thermal-convection-driven PCR [117]. Proposed by Northrup in 1993, chamber PCR miniaturizes traditional PCR by keeping the sample in the chamber with temperature changes. The whole system was composed of chamber chips and temperature control device. However, chamber PCR is difficult to combine with other functional modules and the reaction rate is constrained by the heat capacity of heater and sensor [118].

Continuous PCR was developed in 1998 by Manz. The sample was moved according to the snakelike channel of three temperature zones which keeps isothermal during the reaction process and decreases reaction time. Compared with chamber PCR, there are no complex valve settings and easier to integrate with other functional zones [119]. Li developed an all-in-one microfluidic chip based on continuous PCR and electrophoresis biochip for periodontal pathogens detection where the amplification time is 2′31″ and detection time is 3′43″ [120].

The thermal-convection-driven PCR is based on the principle of Rayleigh-Bénard convection. Such devices require no external force but only buoyancy to drive the PCR reagent through different areas. These kinds of chips are simple to manufacture, inexpensive and have higher temperature conduct velocity [117]. Chung developed a portable PCR chip based on natural convection to drive PCR reagents along the closed loop channel. The chip showed good performance in reaction time and uses fewer samples which can be applied to many fields [121].

1.7.3.1.2 Ligase Chain Reaction (LCR)

LCR is an exponential amplification technique similar to PCR [122]. This technique requires the thermally stable ligase for the reaction so it is also called ligase amplification reaction [122]. The reaction mechanism of LCR is shown in Figure 1.19. Two pairs of complementary probes and target sequences are denatured and melted under high-temperature conditions. Then, in the presence of thermophilic DNA ligase, the target sequence is used as a ligation template, where the phosphorylated part of the 5' of one probe is connected with the hydroxyl part of the 3' of the other probes to form a phosphodiester bond. After the ligation product is denatured, it can be used as a template for the next step of ligation [122]. Repeatedly, many double-stranded ligation products are formed. The ligation products can be directly analyzed by gel electrophoresis. Compared with PCR, the biggest advantage of LCR is that it can accurately distinguish single bases, and the ligation products can be analyzed by electrophoresis. Besides, the amplification efficiency is almost equivalent to that of PCR. Therefore, LCR has been widely used in the detection of nucleic acid [123].

LCR can be applied to a microfluidic chip for DNA detection. A device integrating FRET (Fluorescence resonance energy transfer) and cyclic olefin copolymer chip

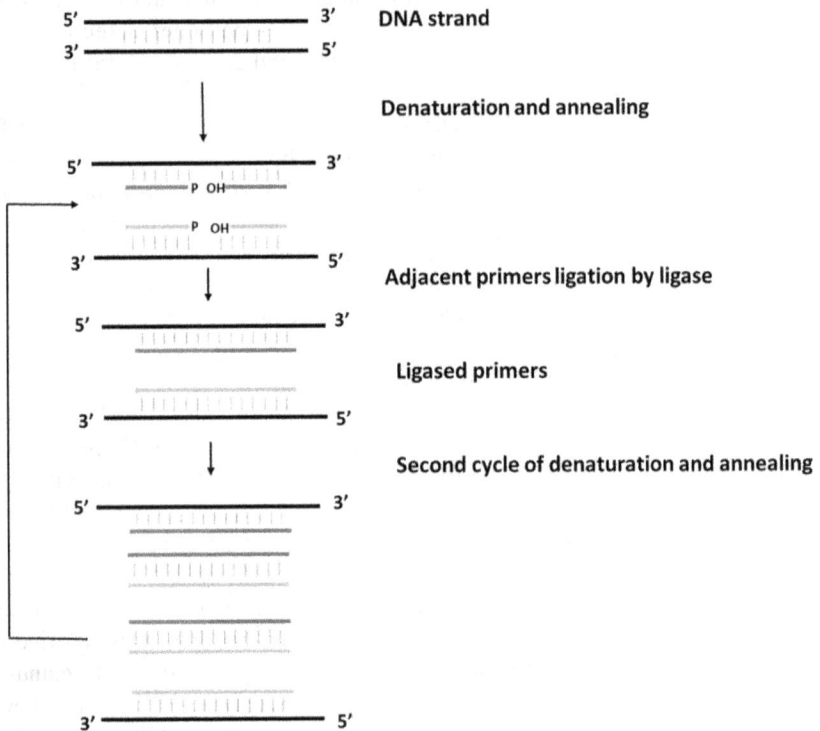

FIGURE 1.19 The LCR principle. The target DNA strands are denatured and hybridized to two pairs of complementary probes. Then, thermophilic DNA ligase ligates the phosphorylated part of the 5' of one probe to the hydroxyl part of the 3' of the other probes. After ligation and denaturation, it can be used as a template for the next step of ligation [122].

was established by Peng [124]. Two primers designed with 10 nt complementary sequence are modified with Cy5 or Cy5.5, respectively. Only when the target presents, the primers can hybridize with target and connected by DNA ligase to form a DNA single strand under 75°C. Because it can self-assemble into a stable stem-loop structure, FRET signal is produced, providing signals for *E. coli* O157:H7 analysis with a LOD of 0.2 cfu/μL.

1.7.3.2 Enzyme-Mediated Methods with Isothermal Process

Classic enzyme-mediated amplification methods usually require relatively complicated primer design and accurate temperature control. Besides, a dedicated and expensive thermal cycler is often required to achieve precise temperature control, which limits the application of these amplification methods. Here we introduce other enzyme-assisted isothermal nucleic acid signal amplification technologies that have gradually developed.

1.7.3.2.1 Rolling Circle Amplification (RCA)

RCA reaction is a common isothermal method for DNA or RNA amplification with good specificity and sensitivity [125]. The principle of RCA is shown in Figure 1.20. In RCA reaction, circular single-stranded DNA are usually used as an amplification template. After hybridizing with the forward primer, the forward primer can be elongated by DNA polymerase and displace the elongated product continuously, resulting in a long, linear DNA single-strands with repetitive sequence [126]. There are two main amplification strategies for RCA. One is to directly use a shorter target as a primer to amplify the circular template, and the other is to use a ligation reaction to loop a linear probe that is complementary to the target.

RCA can be integrated with the microfluidic platform to detect multiple forms of DNA. A device created by Hernandez-Neuta et al. combined RCA in a microfluidic platform which can detect DNA targets with a concentration as low as 0.1 fM. The device function is mainly facilitated by the utilization of magnetic beads which will capture and immobilize DNA targets followed by the introduction of the padlock probe and RCA ingredients. Through this method, they achieved a 93% of capture efficiency of target DNA and test a large volume of sample with a flow rate of 5 μL/min [127]. Another integration method is by utilizing flow control to sequentially introduce the RCA ingredients into the reaction chamber for the amplification of targets. Cao et al. realized the detection of microRNA directly from a biological

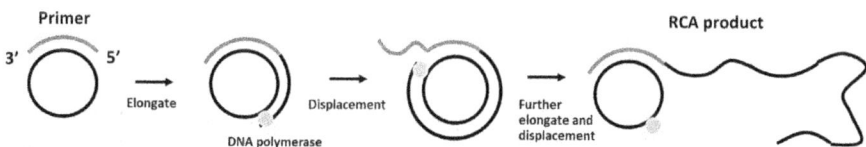

FIGURE 1.20 Rolling circle amplification (RCA). Circular single-stranded DNA is used as an amplification template. After hybridizing, the forward primer can be elongated by DNA polymerase and displace the elongated product continuously, resulting in a long, linear DNA single-strands with repetitive sequences.

sample with a detection limit of 5–8 zmol in a platform that they called MERCA (Microfluidic Exponential Rolling Circle Amplification) [128].

1.7.3.2.2 Strand Displacement Amplification (SDA)

The concept of SDA was first proposed by Walker et al. in 1992 [129]. As shown in Figure 1.21, the SDA reaction is a continuous cutting-polymerization/strand displacement process that occurs under the assistance of DNA polymerase and nickase [130]. Specifically, a single-stranded target DNA fragment is combined with a primer containing a HincII recognition site and then replicated by DNA polymerase using dCTP,

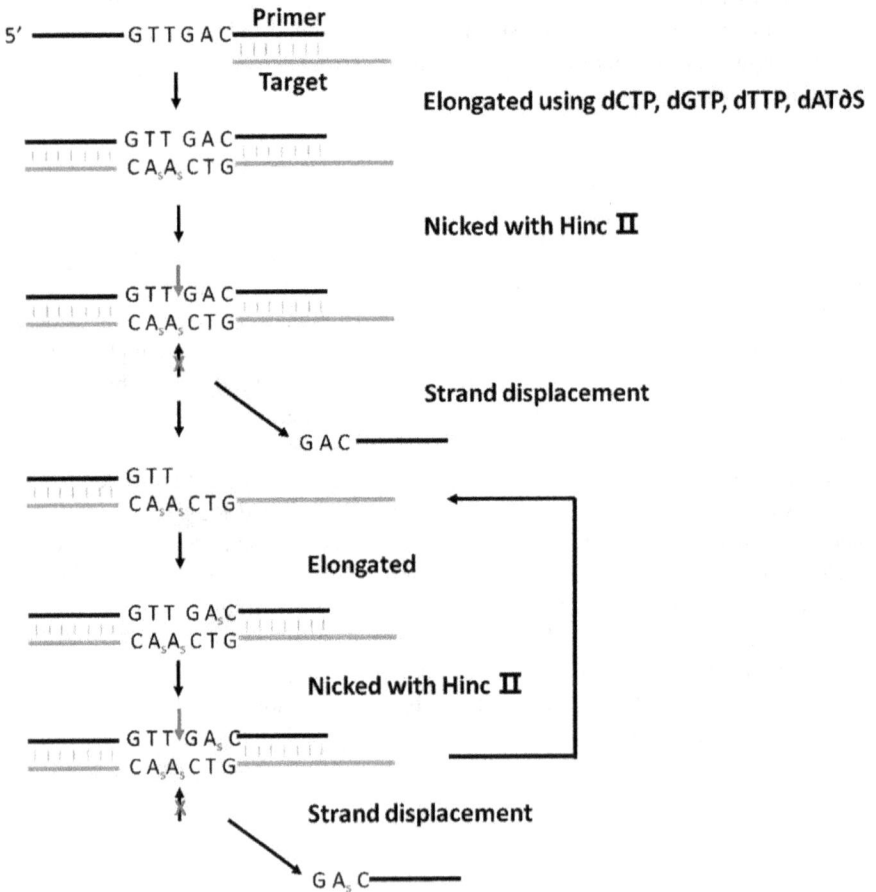

FIGURE 1.21 Scheme of SDA. A single-stranded target DNA fragment is combined with a primer containing a HincII recognition site and then replicated by DNA polymerase using dCTP, dGTP, dTTP, dATPαS as the building blocks. Since dATP is replaced by dATPαS, the newly synthesized nucleotide cannot be nicked by HincII. On the other hand, as HincII continues to cleave the counterparts without modified adenine, the DNA polymerase can work on the cleaved site, generating a new extension strand and displacing the extended strand at the same time.

dGTP, dTTP, dATPαS to produce the complete double-stranded DNA. Since dATP is replaced by dATPαS, the newly synthesized nucleotide having phosphorothioate-modified adenine, it cannot be nicked by HincII. On the other hand, as HincII continues to cleave the counterparts without modified adenine, the DNA polymerase can work on the cleaved site, generating a new extension strand and displacing the extended strand at the same time. As such, large amounts of the complete double-stranded DNA with nickase recognition sequence are produced and form the cycle to achieve the amplification [129].

Isothermal circular strand displacement polymerization (ICSDP) is an improved SDA method by utilizing a DNA probe with a hairpin loop with fluorescence and a quencher attached to both ends of it. Using this method in tandem with droplet microfluidics, Giuffrida et al. integrated the amplification scheme inside the micro-flow device to detect a modified roundup-ready soybean and a miR-210 with a concentration as low as 16.5nM [131].

1.7.3.2.3 Helicase-Dependent Amplification (HDA)

In 2004, to perform isothermal amplification of ultra-long gene sequences containing thousands of bases, Kong's research group proposed a new isothermal amplification method that mimics DNA replication in organisms—HDA reaction [132]. In organisms, the self-replication of DNA is achieved by DNA polymerase assisted by DNA helicase that helps to separate DNA duplexes. Similarly, the HDA reaction uses DNA polymerase and helicase to achieve DNA amplification. As shown in Figure 1.22, HDA utilizes the unwind activity of helicase under the existence of single-stranded

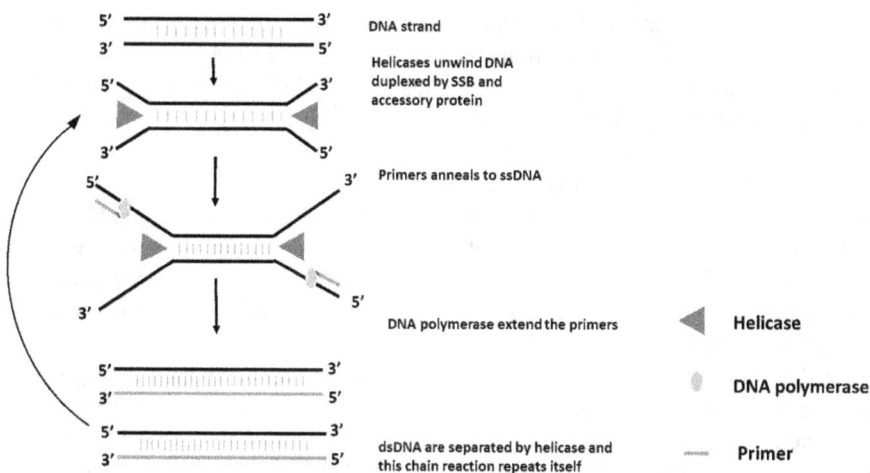

FIGURE 1.22 Helicase-dependent isothermal DNA amplification. Helicase first unwinds double-strand DNA with the assistance from single-stranded binding protein and accessory protein to open the double-stranded DNA. The forward primer and reverse primer hybridize with the single-stranded DNA and perform extension under DNA polymerase, producing the new double-stranded DNA. The new double-stranded DNA is opened by helicase and combined with single-stranded binding protein again, entering a new cycle of amplification.

binding protein and accessory protein to open the double-stranded DNA. The forward primer and reverse primer hybridize with the single-stranded DNA and perform extension under DNA polymerase, producing the new double-stranded DNA. The new double-stranded DNA is opened and combined with single-stranded binding protein again, entering a new cycle of amplification. HDA can achieve more than one million times amplification effects through the unwind and chain extension reaction under the same temperature [132]. HDA reaction has the obvious advantages of simple principle and time-saving operation.

Due to the isothermal nature of HDA, it is a compatible addition to a microfluidic system for DNA detection. The method for integrating HDA inside a microfluidic chip was first done by Ramalingan et al. in 2009. A microfluidic device was equipped with open reactions preloaded with primers for the amplification followed by the sequential introduction of the HDA mixture. With this setup, they are able to detect the BNI-1 fragments of SARS cDNA with a sequence length of 70–120 bp [133]. On the other hand, Mahalanabis et al. use a similar approach with the addition of a microfluidic-enabled solid phase extractor (μSPE) upstream of the reaction chamber which enables them to execute extraction and amplification in the microflow device. With their setup, as few as 10 colony-forming units of the *E. Coli* genetic material were detected [134].

1.7.3.2.4 Exponential Amplification Reaction (EXPAR)

EXPAR is an isothermal nucleic acid amplification method for exponentially synthesizing short DNA primers first proposed by the Galas research group in 2003 [135]. As shown in Figure 1.23, a typical EXPAR template consists of two repeated trigger chain (target chain) with complementary sequence to the target and a nicking enzyme recognition site in the middle. After the target strand hybridizes with the EXPAR template, it is extended along the template by a DNA polymerase with strand displacement activity, which generates the double-strand DNA containing a complete nicking enzyme recognition sequence. After that, the nicking enzyme recognizes and cuts the extended strand, creating a single-stranded gap. Thus, DNA polymerase can continue to extend the target chain and replace the trigger chain synthesized in the previous step. The trigger chain has the same sequence as the target chain, so it can bind to other free EXPAR templates in the system and initiate a new round of extension-cutting reaction [136]. In this way, two trigger chains will be generated from a single target chain in each amplification cycle of EXPAR, showing an exponential amplification trend within a few minutes.

In the microfluidic platform, the EXPAR principle can be applied through various means. One approach was through the use of paper microfluidics. Deng et al. employed quantum dots and the EXPAR amplification mixture inside a wax-printed foldable paper platform which serves as the chamber for both detection and amplification. Target bacterial DNA is firstly lysed and then introduced to paper fluidics, purified by filtration. Through the unfolding of the detection layer, they can optically detect multiplexed DNA targets with the detection limit of 3×10^6 copies [137]. Another integrated EXPAR microflow approach can be done through a droplet microfluidic setup where the target and the EXPAR mixture is introduced together through two separate inlets into the oil phase separator to create droplets of EXPAR

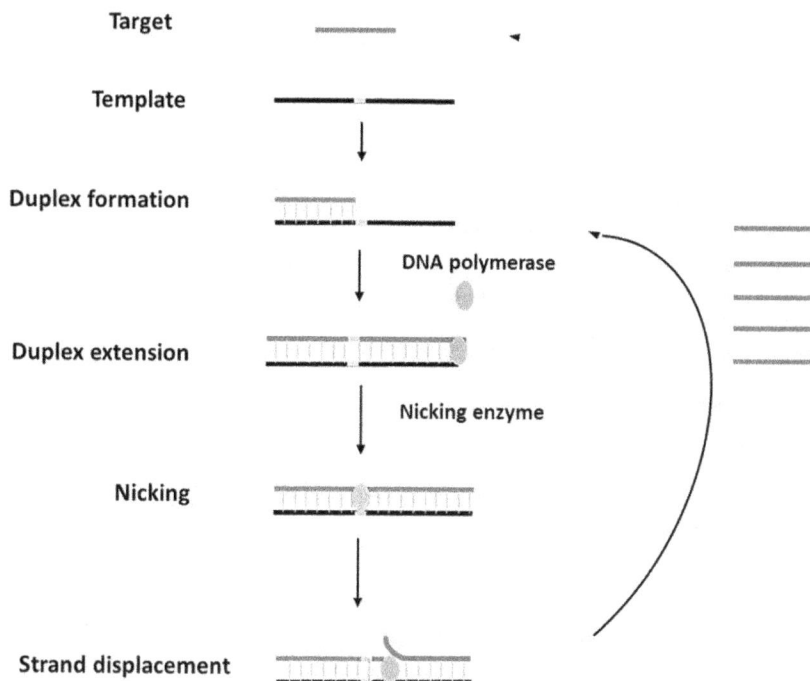

FIGURE 1.23 EXPAR principle. After the target strand hybridizes with the EXPAR template, it is extended along the template by a DNA polymerase, which generates the double-strand DNA containing a complete nicking enzyme recognition sequence. Nicking enzyme recognizes and cuts the extended strand, creating a single-stranded gap. Thus, DNA polymerase with strand displacement activity can continue to extend the target chain and displace the trigger chain synthesized in the previous step. The trigger chain has the same sequence as the target chain which can bind to other free EXPAR templates in the system and initiate a new round of extension-cutting reaction.

reactions. Through this setup, they achieved a limit of detection of 50 copies per mL of miRNA spiked in 100% raw plasma [138].

1.7.3.2.5 Recombinase Polymerase Amplification (RPA)

RPA method was published in 2006 by Piepenburg [139]. The principle is shown in Figure 1.24. The RPA uses recombinase protein, single-stranded binding protein and DNA polymerase with strand-displacement activity to avoid DNA denaturation that functions to repair and maintain DNA in living cells [140]. It combines with primer and forms a complex in the amplification process. When recognizing the target sequence, double-stranded DNA is opened in the presence of recombinase, and then primers insert. Subsequently, single-stranded binding protein combined with single-stranded DNA is used to avoid the melted DNA strands to hybridize together again. As such, primers are extended by DNA polymerase and enter the next round to achieve the amplification [141]. Implementation of the RPA method in microfluidics

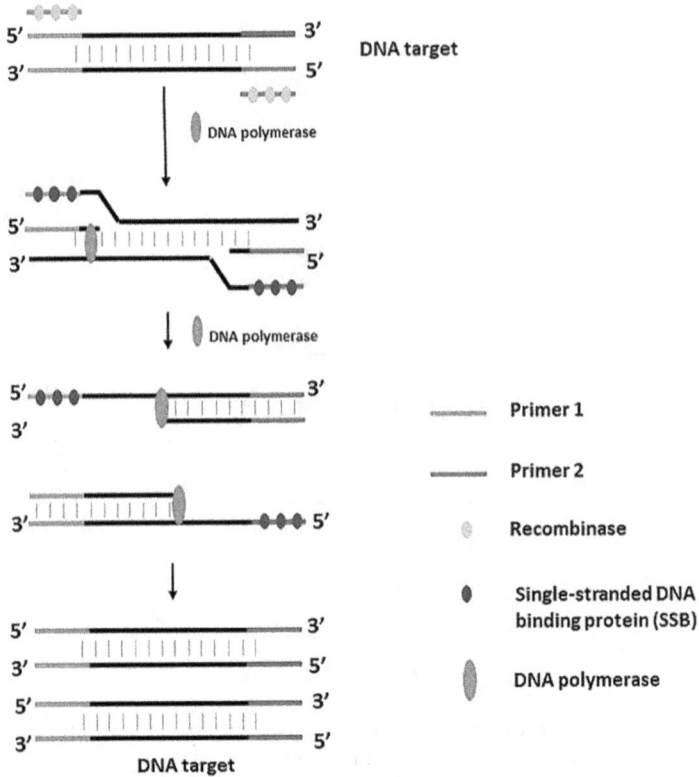

FIGURE 1.24 RPA principle. A double-stranded DNA is recognized and opened by a recombinase, allowing primers to insert and hybridize. Subsequently, single-stranded binding protein combined with single-stranded DNA is used to avoid the melted DNA strands to hybridize together again. As such, primers are extended by DNA polymerase and enter the next round to achieve the amplification.

can be done through multiple methods. A recent study by Liu et al. utilized the RPA method with a lateral flow microfluidic device for COVID-19 diagnosis through the detection of the SARS-CoV-2 armored RNA. Utilizing an optical dipstick signal generation, they achieved an LOD of 1 copy per μL sensitivity of 97% and a 100% specificity [142]. Other platforms come in the form of an active device which requires external stimulation such as using centrifugal disk-based microfluidic device to initiate the introduction and mixture of RPA mixture or even passive microfluidic devices to automatically regulate the flow profile for sequential introduction of RPA materials onto the target sample to achieve detection [143].

1.7.4 Loop-Mediated Isothermal Amplification

LAMP was first proposed by T. Botomi in 2000 [144]. The procedure is shown in Figure 1.25. There are four primers in the whole system, namely, forward inner

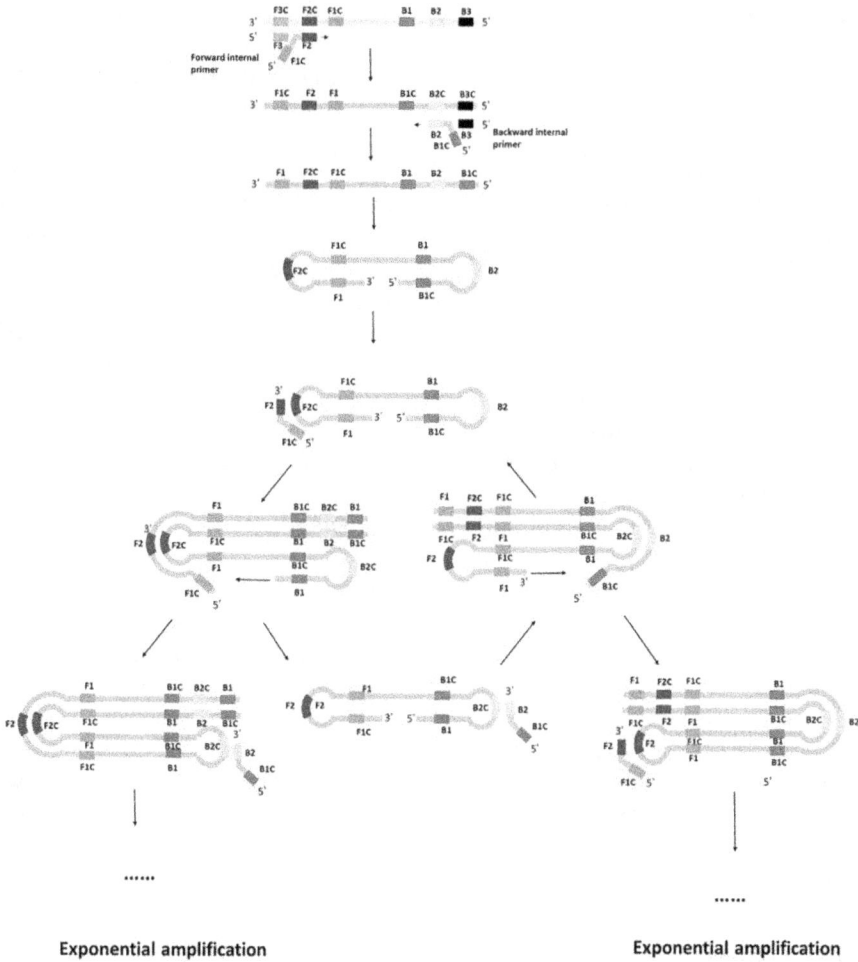

FIGURE 1.25 Loop-mediated isothermal amplification. F2 of FIP and B3 of BIP can respectively bind to F2C and B3C and be extended by Bst DNA polymerase. Importantly, the binding and extension of F3 and B3 would displace the product, releasing a newly synthesized strand with a dumbbell-shaped structure because of F1-F1C and B1-B1C. Next, FIP and BIP bind to the dumbbell-shaped structure as the template for extension and displacement. Importantly, 3' end pf F1 and B1 sequence are regarded as the starting point and used as a template to perform DNA extension. As a result of strand displacement, two new stem-loop structures of DNA strand including one dumbbell-shaped structure are formed. After iterations, the final product of the amplification is a mixture of DNA with different numbers of stem-loop structures with different lengths.

primer (FIP), backward inner primer (BIP), forward outer primer (F3) and backward outer primer (B3) [144]. Firstly, F2 of FIP hybridizes with F2C of template and extends the strand under the function of Bst DNA polymerase with strand displacement property. At the same time, F3 binds to template F3C, extending the strand and displaces the complementary single strand connected with FIP by Bst DNA polymerase. Using this chain as a template, the downstream primers BIP and B3 successively initiate the synthesis similar to FIP and F3. The newly synthesized strand can form a dumbbell-shaped structure because of complementary pair of F1-F1C and B1-B1C. Using the dumbbell-shaped structure as the template, FIP binds with loop F2C to start extension and strand displacement. At the same time, 3' end B1 sequence is regarded as the starting point and uses itself as a template to perform DNA extension. As a result of strand displacement, two new stem-loop structures of DNA strand including one dumbbell-shaped structure are formed. Similarly, the B2 sequence on the BIP primer also hybridizes to the dumbbell-shaped structure to initiate the same production of two new stem-loop structures of DNA strands. After iterations, the final product of the amplification is a mixture of DNA with different numbers of stem-loop structures with different lengths [144].

LAMP-on-chip has been a very popular tool in point-of-care applications. For example, the first digital LAMP chip was developed in 2012 [145]. The chip was fabricated on PDMS with a large number of wells. Sample loading depended on air pressure and the whole system was placed on a water reservoir for keeping an isothermal temperature, 65°C, for the performing LAMP reaction. In addition, LAMP is also applied to paper-based chips. Zhang proposed a microcapillary-based LAMP system for single nucleotide polymorphisms (SNPs) type screening which integrates sample pretreatment, extraction, amplification and detection [146]. Due to the low-cost and simple procedure of LAMP-on-chip, it has been applied in different fields such as food safety and clinical diagnostics [147].

1.7.4.1 Enzyme Free Methods

Enzyme-free amplification of nucleic acid signals is often achieved through nucleic acid circuits. Nucleic acid circuit is a type of DNA or RNA rationally designed self-assembly system. It uses easy-to-program DNA or RNA hybridization reactions to convert the activity and concentration of an analyte to an amplified signal to achieve a sensitive and quantitative analysis. The enzyme-free amplification method is widely used because of its low-cost and simple experiment procedure.

1.7.4.1.1 Hybridization Chain Reaction (HCR)

The concept of HCR was first introduced by Dirks in 2004 [148]. The whole system of HCR consists of two DNA hairpins and an initiator. DNA hairpins are composed of stem, loop and toehold region. The principle of HCR is shown in Figure 1.26. Two DNA hairpins exist stably in the solution. When a single strands initiator is added, it opens one hairpin from the toehold region, exposing a single strand which can open another hairpin and expose a single-strand region which has the same sequence with

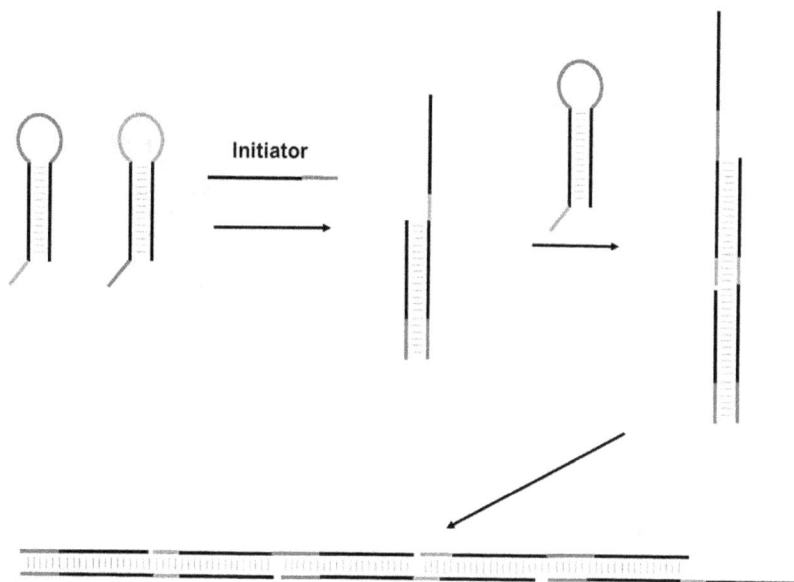

FIGURE 1.26 HCR scheme. Two DNA hairpins exist stably in the solution. When a single strands initiator is added, it opens one hairpin from the toehold region, exposing a single strand which can open another hairpin and expose a single-strand region which has the same sequence with initiator, resulting in the process moving in cycles to obtain a long double strand DNA copolymer and achieve signal amplification.

initiator, resulting in the process moving in cycles to obtain a long double strand DNA copolymer and achieve signal amplification.

By combining HCR with nanoparticles [149], fluorescence dye [150], electrochemical [151], etc, the product of HCR can be visualized and transferred to fluorescence, electrochemistry signal and other forms of signal. Besides, HCR is applied to many substances detection such as nucleic acid [151], protein [152], metal ions [153] and even bioimaging [154]. With the development of microfluidic, HCR is also combined with it to achieve DNA amplification without large-scale equipment and complex experiment procedures. For example, Guo et. developed a simple and continuous-flow microfluidics combining HCR to realize single cell miRNA detection. The single cell and two types of hairpins used for amplification were encapsulated in the water-in-oil droplets. After lysis, the released target miRNA will trigger the HCR to achieve cascade fluorescence signal output [151].

1.7.4.1.2 Catalyzed hairpin assembly (CHA)

In 2008, the Pierce research group first proposed the concept of CHA [155]. As shown in Figure 27, both H1 and H2 are in the stable hairpin structure because of their blocked stem part. When C1 comes, it can hybridize with H1 from the toehold region and open the H1, resulting in an exposure of a single strand sequence which can open the H2 through hybridization to the toehold regions on the H2. Next, the C1

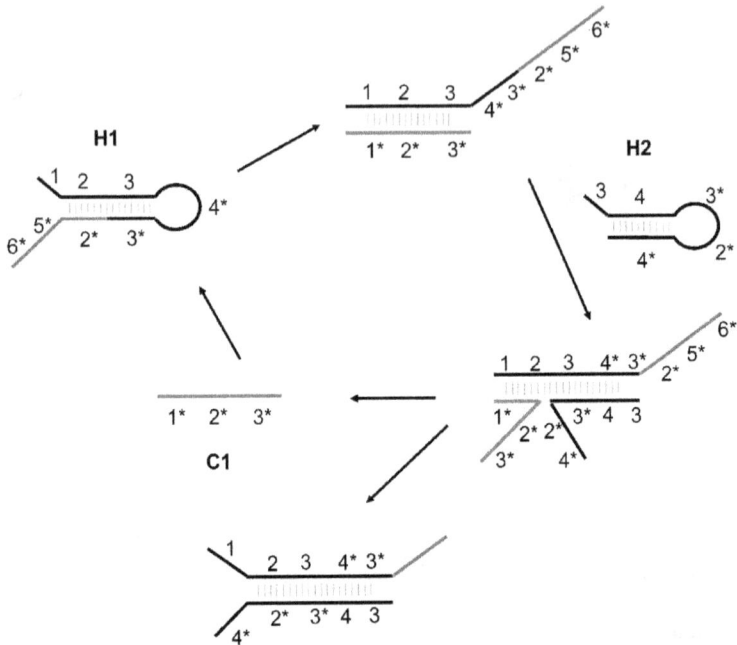

FIGURE 1.27 Schematic of CHA. C1 can hybridize with H1 from the toehold region and open the H1, resulting in an exposure of a single strand sequence which can open the H2 through hybridization to the toehold regions on the H2. Next, the C1 is displaced by H2 as the H1-H2 complex is more stable. Such cyclic reaction allows C1 to be reused as a catalyst and enter the next cycle of hairpin assembly, leading to massive production of the H1-H2 complex.

is displaced by H2 as the H1-H2 complex is more stable. Such cyclic reaction allows C1 to be reused as a catalyst and enter the next cycle of hairpin assembly, leading to massive production of the H1-H2 complex. CHA reactions enable efficient catalytic reaction and lower background signal compared with HCR. At present, through the combination of fluorescence, electrochemistry, and colorimetric measurement methods, CHA has been applied to the quantitative detection of nucleic acids [156], metal ions [157], proteins [158] and other analysts. Qian proposed a microfluidic chip based on the water permeability and capillary effect of agarose for exosome concentration and in situ exosomal miRNA detection. Besides, CHA was used for miRNA amplification which reaches LOD of ~10^3 exosome and early cancer detection [159].

1.8 CONCLUSIONS

DNA is not only the most important molecule in the human body for regulating all the biological processes but also the biomarker related to body anomalies and diseases. Microfluidics is a popular method to detect various molecules with advantages such as low sample requirement, miniaturized devices, and cost-effectiveness, which makes it an ideal detection method for DNA and their derivatives. To detect DNA in a microfluidic platform, we have summarized the essential steps to ensure the successful detection of DNA. First, DNA needs to be extracted from raw materials such as cells or blood samples followed by purification. Detection of DNAs can be based on DNA separation using their physical property or based on affinity probes that are able to capture specific target molecule. Short DNA probes are synthesized by a chemical reaction, followed by signal generation using optical, electrochemical, and other methods. The advantages and disadvantages as well as research examples to further illustrate the development and capabilities of each method were discussed. Moreover, target samples are amplified by various methods to maximize detection probability. Together, the integration between microfluidic and DNA detection has enabled much improved analytical performance with a simplified and miniaturized platform for enhanced portability.

Nowadays, smartphones have now become a mandatory possession for each individual due to how technology has developed into something accessible to everyone. From a future perspective, we believe that integration between microfluidics and smartphones would become an ideal platform for future POC diagnostics given its portability, availability, and multitude of features. For example, smartphones to take images with a camera which can be fitted with either a microscopic lens or become a computing power that can process raw signals. Also, it can be used as a thermal regulator which aids the DNA amplification process. Together, the integration of microfluidic with smartphones could be the direction in tandem with various technologies for achieving sensitive and specific detections of DNA targets in the future.

REFERENCES

1. Whitesides, G.M., The origins and the future of microfluidics. *Nature*, 2006. **442**(7101): 368–373.
2. Nguyen, N.-T., S.T. Wereley, and S.A.M. Shaegh, *Fundamentals and Applications of Microfluidics*. 2019: Artech house: Norwood, MA.
3. Sia, S.K. and L.J. Kricka, Microfluidics and point-of-care testing. *Lab on a Chip*, 2008. **8**(12): 1982–1983.
4. van den Berg, A. and T.S. Lammerink, Micro total analysis systems: microfluidic aspects, integration concept and applications. In: Manz, A., Becker, H. (eds.) *Microsystem Technology in Chemistry and Life Science*, 1998: pp. 21–49. Springer, Heidelberg.
5. Tabeling, P., *Introduction to Microfluidics*. 2005: Oxford University Press, Oxford.
6. Lake, M., et al., Microfluidic device design, fabrication, and testing protocols. *Protocol Exchange*, 2015. **10**. doi: 10.1038/protex.2015.069.
7. Erickson, D. and D. Li, Integrated microfluidic devices. *Analytica Chimica Acta*, 2004. **507**(1): 11–26.
8. Martinez-Duarte, R. and M. Madou, SU-8 photolithography and its impact on microfluidics. In: *Microfluidics and Nanofluidics Handbook*, 2011: 231–268. doi: 10.1201/b11188-11.
9. Gravesen, P., J. Branebjerg, and O.S. Jensen, Microfluidics-a review. *Journal of Micromechanics and Microengineering*, 1993. **3**(4): 168.
10. Iliescu, C., Microfluidics in glass: Technologies and applications. *Informacije Midem-Ljubljana*, 2006. **36**(4): 204.
11. Ren, K., J. Zhou, and H. Wu, Materials for microfluidic chip fabrication. *Accounts of Chemical Research*, 2013. **46**(11): 2396–2406.

12. Becker, H. and L.E. Locascio, Polymer microfluidic devices. *Talanta*, 2002. **56**(2): 267–287.
13. McDonald, J.C. and G.M. Whitesides, Poly (dimethylsiloxane) as a material for fabricating microfluidic devices. *Accounts of Chemical Research*, 2002. **35**(7): 491–499.
14. Kim, P., et al., Soft lithography for microfluidics: A review. *BioChip Journal*, 2008. **2**(1): 1–11.
15. Tsao, C.-W. and D.L. DeVoe, Bonding of thermoplastic polymer microfluidics. *Microfluidics and Nanofluidics*, 2009. **6**(1): 1–16.
16. Haiducu, M., et al., Deep-UV patterning of commercial grade PMMA for low-cost, large-scale microfluidics. *Journal of Micromechanics and Microengineering*, 2008. **18**(11): 115029.
17. Fobel, R., et al., Paper microfluidics goes digital. *Advanced Materials*, 2014. **26**(18): 2838–2843.
18. Hamedi, M.M., et al., Integrating electronics and microfluidics on paper. *Advanced Materials*, 2016. **28**(25): 5054–5063.
19. Li, X., D.R. Ballerini, and W. Shen, A perspective on paper-based microfluidics: Current status and future trends. *Biomicrofluidics*, 2012. **6**(1): 011301.
20. Cosson, S. and M. Lutolf, Hydrogel microfluidics for the patterning of pluripotent stem cells. *Scientific Reports*, 2014. **4**(1): 1–6.
21. Golonka, L.J., et al., LTCC microfluidic system. *International Journal of Applied Ceramic Technology*, 2006. **3**(2): 150–156.
22. Barry, R. and D. Ivanov, Microfluidics in biotechnology. *Journal of Nanobiotechnology*, 2004. **2**(1): 1–5.
23. Blair, E.O. and D.K. Corrigan, A review of microfabricated electrochemical biosensors for DNA detection. *Biosensors and Bioelectronics*, 2019. **134**: 57–67.
24. King, D., *DNA*, in *EME DNA*, EME Corporation, M. Educational, and C. Equipment, Editors. 2000, EME: Stuart, FL.
25. Neidle, S., *Principles of Nucleic Acid Structure*. 1st ed. 2008, Amsterdam: Boston: Elsevier Academic Press.
26. Nikolova, E.N., et al., A historical account of Hoogsteen base-pairs in duplex DNA. *Biopolymers*, 2013. **99**(12): 955–968.
27. Yakovchuk, P., E. Protozanova, and M.D. Frank-Kamenetskii, Base-stacking and base-pairing contributions into thermal stability of the DNA double helix. *Nucleic Acids Research*, 2006. **34**(2): 564–574.
28. Chung, Y.-C., et al., Microfluidic chip for high efficiency DNA extraction. *Lab on a Chip*, 2004. **4**(2): 141–147.
29. Lim, M.Y., et al., Comparison of DNA extraction methods for human gut microbial community profiling. *Systematic and Applied Microbiology*, 2018. **41**(2): 151–157.
30. Herrera, A. and C.S. Cockell, Exploring microbial diversity in volcanic environments: a review of methods in DNA extraction. *Journal of Microbiological Methods*, 2007. **70**(1): 1–12.
31. Tsai, Y.-L. and B.H. Olson, Rapid method for direct extraction of DNA from soil and sediments. *Applied and Environmental Microbiology*, 1991. **57**(4): 1070–1074.
32. Shehadul Islam, M., A. Aryasomayajula, and P.R. Selvaganapathy, A review on macroscale and microscale cell lysis methods. *Micromachines*, 2017. **8**(3): 83.
33. Nan, L., Z. Jiang, and X. Wei, Emerging microfluidic devices for cell lysis: A review. *Lab on a Chip*, 2014. **14**(6): 1060–1073.
34. Grigorov, E., et al., Review of microfluidic methods for cellular lysis. *Micromachines*, 2021. **12**(5): 498.
35. Mahalanabis, M., et al., Cell lysis and DNA extraction of gram-positive and gram-negative bacteria from whole blood in a disposable microfluidic chip. *Lab on a Chip*, 2009. **9**(19): 2811–2817.

36. Ayoib, A., U. Hashim, and S.C. Gopinath, Automated, high-throughput DNA extraction protocol for disposable label free, microfluidics integrating DNA biosensor for oil palm pathogen, Ganoderma boninense. *Process Biochemistry*, 2020. **92**: 447–456.

37. Wang, S., et al., Electrochemical cell lysis of gram-positive and gram-negative bacteria: DNA extraction from environmental water samples. *Electrochimica Acta*, 2020. **338**: 135864.

38. Geng, T., et al., Genomic DNA extraction from cells by electroporation on an integrated microfluidic platform. *Analytical Chemistry*, 2012. **84**(21): 9632–9639.

39. Lee, H.J., et al., Electrochemical cell lysis device for DNA extraction. *Lab on a Chip*, 2010. **10**(5): 626–633.

40. Oblath, E.A., et al., A microfluidic chip integrating DNA extraction and real-time PCR for the detection of bacteria in saliva. *Lab on a Chip*, 2013. **13**(7): 1325–1332.

41. Lee, C.-Y., et al., Integrated microfluidic systems for cell lysis, mixing/pumping and DNA amplification. *Journal of Micromechanics and Microengineering*, 2005. **15**(6): 1215.

42. Di Pinto, A., et al., A comparison of DNA extraction methods for food analysis. *Food Control*, 2007. **18**(1): 76–80.

43. Sambrook, J. and D.W. Russell, Purification of nucleic acids by extraction with phenol: Chloroform. *Cold Spring Harbor Protocols*, 2006. **2006**(1). doi: 10.1101/pdb.prot4455.

44. Zhang, R., et al., A microfluidic liquid phase nucleic acid purification chip to selectively isolate DNA or RNA from low copy/single bacterial cells in minute sample volume followed by direct on-chip quantitative PCR assay. *Analytical Chemistry*, 2013. **85**(3): 1484–1491.

45. Morales, M.C. and J.D. Zahn, Droplet enhanced microfluidic-based DNA purification from bacterial lysates via phenol extraction. *Microfluidics and Nanofluidics*, 2010. **9**(6): 1041–1049.

46. Meagher, R.J., Y.K. Light, and A.K. Singh, Rapid, continuous purification of proteins in a microfluidic device using genetically-engineered partition tags. *Lab on a Chip*, 2008. **8**(4): 527–532.

47. Mao, X., S. Yang, and J.D. Zahn. Experimental demonstration and numerical simulation of organic-aqueous liquid extraction enhanced by droplet formation in a microfluidic channel. in *ASME International Mechanical Engineering Congress and Exposition*. 2006.

48. Reddy, V. and J.D. Zahn, Interfacial stabilization of organic–aqueous two-phase micro-flows for a miniaturized DNA extraction module. *Journal of Colloid and Interface Science*, 2005. **286**(1): 158–165.

49. Hawkins, T.L., et al., DNA purification and isolation using a solid-phase. *Nucleic Acids Research*, 1994. **22**(21): 4543.

50. Buszewski, B. and M. Szultka, Past, present, and future of solid phase extraction: A review. *Critical Reviews in Analytical Chemistry*, 2012. **42**(3): 198–213.

51. Smith, K., M. Diggle, and S. Clarke, Comparison of commercial DNA extraction kits for extraction of bacterial genomic DNA from whole-blood samples. *Journal of Clinical Microbiology*, 2003. **41**(6): 2440–2443.

52. Zhong, R., et al., Fabrication of two-weir structure-based packed columns for on-chip solid-phase extraction of DNA. *Electrophoresis*, 2007. **28**(16): 2920–2926.

53. Hagan, K.A., et al., Microchip-based solid-phase purification of RNA from biological samples. *Analytical Chemistry*, 2008. **80**(22): 8453–8460.

54. Reinholt, S.J. and A.J. Baeumner, Microfluidic isolation of nucleic acids. *Angewandte Chemie International Edition*, 2014. **53**(51): 13988–14001.

55. Breadmore, M.C., et al., Microchip-based purification of DNA from biological samples. *Analytical Chemistry*, 2003. **75**(8): 1880–1886.

56. Bhattacharyya, A. and C.M. Klapperich, Microfluidics-based extraction of viral RNA from infected mammalian cells for disposable molecular diagnostics. *Sensors and Actuators B: Chemical*, 2008. **129**(2): 693–698.

57. Wu, Q., et al., Microchip-based macroporous silica sol–gel monolith for efficient isolation of DNA from clinical samples. *Analytical Chemistry*, 2006. **78**(16): 5704–5710.

58. Wen, J., et al., DNA extraction using a tetramethyl orthosilicate-grafted photopolymerized monolithic solid phase. *Analytical Chemistry*, 2006. **78**(5): 1673–1681.

59. Cady, N.C., S. Stelick, and C.A. Batt, Nucleic acid purification using microfabricated silicon structures. *Biosensors and Bioelectronics*, 2003. **19**(1): 59–66.

60. Ji, H.M., et al., DNA purification silicon chip. *Sensors and Actuators A: Physical*, 2007. **139**(1–2): 139–144.

61. Wu, Q., et al., Integrated glass microdevice for nucleic acid purification, loop-mediated isothermal amplification, and online detection. *Analytical Chemistry*, 2011. **83**(9): 3336–3342.

62. Yamaguchi, A., et al., A novel automated device for rapid nucleic acid extraction utilizing a zigzag motion of magnetic silica beads. *Analytica Chimica Acta*, 2016. **906**: 1–6.

63. Bordelon, H., et al., Development of a low-resource RNA extraction cassette based on surface tension valves. *ACS Applied Materials & Interfaces*, 2011. **3**(6): 2161–2168.

64. Liu, C.-J., et al., Magnetic-bead-based microfluidic system for ribonucleic acid extraction and reverse transcription processes. *Biomedical Microdevices*, 2009. **11**(2): 339–350.

65. Lien, K.-Y., et al., Extraction of genomic DNA and detection of single nucleotide polymorphism genotyping utilizing an integrated magnetic bead-based microfluidic platform. *Microfluidics and Nanofluidics*, 2009. **6**(4): 539–555.

66. Hagan, K.A., et al., An integrated, valveless system for microfluidic purification and reverse transcription-PCR amplification of RNA for detection of infectious agents. *Lab on a Chip*, 2011. **11**(5): 957–961.

67. Wang, J., et al., Microfluidic platform for isolating nucleic acid targets using sequence specific hybridization. *Biomicrofluidics*, 2013. **7**(4): 044107.

68. Wang, C.-H., et al., A magnetic bead-based assay for the rapid detection of methicillin-resistant Staphylococcus aureus by using a microfluidic system with integrated loop-mediated isothermal amplification. *Lab on a Chip*, 2011. **11**(8): 1521–1531.

69. Root, B.E., et al., Purification of HIV RNA from serum using a polymer capture matrix in a microfluidic device. *Analytical Chemistry*, 2011. **83**(3): 982–988.

70. Dhaliwal, A., DNA extraction and purification. *Mater Methods*, 2013. **3**: 191.

71. Robin, J.D., et al., Comparison of DNA quantification methods for next generation sequencing. *Scientific Reports*, 2016. **6**(1): 1–10.

72. Ashton, R., C. Padala, and R.S. Kane, Microfluidic separation of DNA. *Current Opinion in Biotechnology*, 2003. **14**(5): 497–504.

73. Minc, N., et al., Quantitative microfluidic separation of DNA in self-assembled magnetic matrixes. *Analytical Chemistry*, 2004. **76**(13): 3770–3776.

74. Huang, F.C., C.S. Liao, and G.B. Lee, An integrated microfluidic chip for DNA/RNA amplification, electrophoresis separation and on-line optical detection. *Electrophoresis*, 2006. **27**(16): 3297–3305.

75. Li, Y., et al., Ultrahigh-throughput approach for analyzing single-cell genomic damage with an agarose-based microfluidic comet array. *Analytical Chemistry*, 2013. **85**(8): 4066–4073.

76. Duarte, G.R., et al., Disposable polyester–toner electrophoresis microchips for DNA analysis. *Analyst*, 2012. **137**(11): 2692–2698.

77. Vanderschaeghe, D., et al., High-throughput profiling of the serum N-glycome on capillary electrophoresis microfluidics systems: Toward clinical implementation of GlycoHepatoTest. *Analytical Chemistry*, 2010. **82**(17): 7408–7415.

78. Durney, B.C., C.L. Crihfield, and L.A. Holland, Capillary electrophoresis applied to DNA: determining and harnessing sequence and structure to advance bioanalyses (2009–2014). *Analytical and Bioanalytical Chemistry*, 2015. **407**(23): 6923–6938.

79. Datinská, V., et al., Recent progress in nucleic acids isotachophoresis. *Journal of Separation Science*, 2018. **41**(1): 236–247.

80. Rogacs, A., L.A. Marshall, and J.G. Santiago, Purification of nucleic acids using iso-tachophoresis. *Journal of Chromatography A*, 2014. **1335**: 105–120.

81. Wu, R., Y. Seah, and Z. Wang, Microfluidic chip for stacking, separation and extraction of multiple DNA fragments. *Journal of Chromatography A*, 2016. **1437**: 219–225.

82. Lao, A.I.K. and I.-M. Hsing, Flow-based and sieving matrix-free DNA differentiation by a miniaturized field flow fractionation device. *Lab on a Chip*, 2005. **5**(6): 687–690.

83. Jones, P.V., G.L. Salmon, and A. Ros, Continuous separation of DNA molecules by size using insulator-based dielectrophoresis. *Analytical Chemistry*, 2017. **89**(3): 1531–1539.

84. Täuber, S., et al., Reaching for the limits in continuous-flow dielectrophoretic DNA analysis. *Analyst*, 2017. **142**(24): 4670–4677.

85. Regtmeier, J., et al., Dielectrophoretic trapping and polarizability of DNA: The role of spatial conformation. *Analytical Chemistry*, 2010. **82**(17): 7141–7149.

86. Huang, L.R., et al., Tilted Brownian ratchet for DNA analysis. *Analytical Chemistry*, 2003. **75**(24): 6963–6967.

87. McGrath, J., M. Jimenez, and H. Bridle, Deterministic lateral displacement for particle separation: A review. *Lab on a Chip*, 2014. **14**(21): 4139–4158.

88. Huang, L.R., et al., A DNA prism for high-speed continuous fractionation of large DNA molecules. *Nature Biotechnology*, 2002. **20**(10): 1048–1051.

89. Wunsch, B.H., et al., Gel-on-a-chip: Continuous, velocity-dependent DNA separation using nanoscale lateral displacement. *Lab on a Chip*, 2019. **19**(9): 1567–1578.

90. Nussbaum, A.L., G. Scheuerbrandt, and A.M. Duffield, Stepwise synthesis of certain deoxyribotrinucleotides. *Journal of the American Chemical Society*, 1964. **86**(1): 102–106.

91. Hughes, R.A. and A.D. Ellington, Synthetic DNA synthesis and assembly: Putting the synthetic in synthetic biology. *Cold Spring Harbor Perspectives in Biology*, 2017. **9**(1): a023812.

92. Song, L.-F., et al., Large-scale de novo oligonucleotide synthesis for whole-genome synthesis and data storage: Challenges and opportunities. *Frontiers in Bioengineering and Biotechnology*, 2021. **9**(526): 689797.

93. Miller, M. and Y.-W. Tang, Basic concepts of microarrays and potential applications in clinical microbiology. *Clinical Microbiology Reviews*, 2009. **22**: 611–33.

94. Oh, S.J., et al., Fully automated and colorimetric foodborne pathogen detection on an integrated centrifugal microfluidic device. *Lab Chip*, 2016. **16**(10): 1917–1926.

95. Alhasan, A.H., et al., Scanometric MicroRNA array profiling of prostate cancer markers using spherical nucleic acid–gold nanoparticle conjugates. *Analytical Chemistry*, 2012. **84**(9): 4153–4160.

96. Zhao, Z., et al., Microfluidic bead trap as a visual bar for quantitative detection of oligonucleotides. *Lab Chip*, 2017. **17**(19): 3240–3245.

97. Chabinyc, M.L., et al., An integrated fluorescence detection system in poly (dimethylsiloxane) for microfluidic applications. *Analytical Chemistry*, 2001. **73**(18): 4491–4498.

98. Li, Z., et al., A rapid microfluidic platform with real-time fluorescence detection system for molecular diagnosis. *Biotechnology, Biotechnological Equipment*, 2019. **33**(1): 223–230.

99. Nguyen, H.T., et al., Microfluidics-assisted fluorescence in situ hybridization for advantageous human epidermal growth factor receptor 2 assessment in breast cancer. *Laboratory Investigation*, 2017. **97**(1): 93–103.

100. Peng, H.-I., C.M. Strohsahl, and B.L. Miller, Microfluidic nanoplasmonic-enabled device for multiplex DNA detection. *Lab Chip*, 2012. **12**(6): 1089–1093.

101. Luo, M., et al., Chemiluminescence biosensors for DNA detection using graphene oxide and a horseradish peroxidase-mimicking DNAzymeElectronic supplementary information (ESI) available: Experimental details and additional figures. 2012. **48**(8): 1126–1128. doi: 10.1039/c2cc16868e.

102. Špringer, T., M. Piliarik, and J. Homola, Surface plasmon resonance sensor with dispersionless microfluidics for direct detection of nucleic acids at the low femtomole level. Sensors and actuators. *B, Chemical*, 2010. **145**(1): 588–591.

103. Soares, L., et al., Localized surface plasmon resonance (LSPR) biosensing using gold nanotriangles: Detection of DNA hybridization events at room temperature. *Analyst*, 2014. **139**(19): 4964–4973.

104. Tian, S., et al., Aluminum nanocrystals: A sustainable substrate for quantitative SERS-based DNA detection. *Nano Letters*, 2017. **17**(8): 5071–5077.

105. He, Y., et al., Silicon nanowires-based highly-efficient SERS-active platform for ultra-sensitive DNA detection. *Nano Today*, 2011. **6**(2): 122–130.

106. Zhou, J., et al., Fabrication of a microfluidic Ag/AgCl reference electrode and its application for portable and disposable electrochemical microchips. *Electrophoresis*, 2010. **31**: 3083–3089.

107. Yeung, S.-W., et al., A DNA biochip for on-the-spot multiplexed pathogen identification. *Nucleic Acids Research*, 2006. **34**(18): e118.

108. Liu, R.H., et al., Self-contained, fully integrated biochip for sample preparation, polymerase chain reaction amplification, and DNA microarray detection. *Analytical Chemistry*, 2004. **76**(7): 1824–1831.

109. Ferguson, B.S., et al., Integrated microfluidic electrochemical DNA sensor. *Analytical Chemistry*, 2009. **81**(15): 6503–6508.

110. Hong, S.R., H.D. Jeong, and S. Hong, QCM DNA biosensor for the diagnosis of a fish pathogenic virus VHSV. *Talanta*, 2010. **82**(3): 899–903.

111. Zhang, Y., et al., A surface acoustic wave biosensor synergizing DNA-mediated in situ silver nanoparticle growth for a highly specific and signal-amplified nucleic acid assay. *Analyst*, 2017. **142**(18): 3468–3476.

112. Mullis, K., et al., Specific enzymatic amplification of DNA in vitro: the polymerase chain reaction. *Cold Spring Harbor Symposia on Quantitative Biology*, 1986. **51**(Pt 1): 263–273.

113. Malorny, B., et al., Standardization of diagnostic PCR for the detection of foodborne pathogens. *International Journal of Food Microbiology*, 2003. **83**(1): 39–48.

114. Jeffery, K.J.M., et al., Diagnosis of viral infections of the central nervous system: clinical interpretation of PCR results. *The Lancet*, 1997. **349**(9048): 313–317.

115. De Medici, D., et al., Rapid methods for quality assurance of foods: the next decade with polymerase chain reaction (PCR)-based food monitoring. *Food Analytical Methods*, 2015. **8**(2): 255–271.

116. Fierer, N., et al., Assessment of soil microbial community structure by use of taxon-specific quantitative PCR assays. *Applied and Environmental Microbiology*, 2005. **71**(7): 4117–4120.

117. Chen, J., et al., Microfluidic PCR chips. *Nano Biomedicine and Engineering*, 2011. **3**.doi: 10.5101/nbe.v3i4.p203-210.

118. Northrup, M.A., et al., DNA amplification with a microfabricated reaction chamber. Proceedings of International Conference Solid-State Sensors and Actuators (TRANSDUCERS' 93), Yokohama, June, 1993, pp. 924–926.

119. Kopp, M.U., A.J. Mello, and A. Manz, Chemical amplification: Continuous-flow PCR on a chip. *Science*, 1998. **280**(5366): 1046–1048.

120. Li, Z., et al., All-in-one microfluidic device for on-site diagnosis of pathogens based on an integrated continuous flow PCR and electrophoresis biochip. *Lab on a Chip*, 2019. **19**(16): 2663–2668.

121. Chung, K.H., Y.H. Choi, and M.Y. Jung. Natural convection PCR in a disposable polymer chip. in *SENSORS, 2009*. IEEE. 2009.
122. Wiedmann, M., et al., Ligase chain reaction (LCR)--overview and applications. *PCR Methods and Applications*, 1994. **3**(4): S51–S64.
123. Marshall, R.L., et al., Detection of HCV RNA by the asymmetric gap ligase chain reaction. *PCR Methods and Applications*, 1994. **4**(2): 80–84.
124. Peng, Z., et al., Ligase detection reaction generation of reverse molecular beacons for near real-time analysis of bacterial pathogens using single-pair fluorescence resonance energy transfer and a cyclic olefin copolymer microfluidic chip. *Analytical Chemistry*, 2010. **82**(23): 9727–9735.
125. Xu, L., et al., Recent advances in rolling circle amplification-based biosensing strategies-A review. *Analytica Chimica Acta*, 2021. **1148**: 238187.
126. Li, J. and J. Macdonald, Advances in isothermal amplification: Novel strategies inspired by biological processes. *Biosensors and Bioelectronics*, 2015. **64**: 196–211.
127. Hernández-Neuta, I., et al., Microfluidic magnetic fluidized bed for DNA analysis in continuous flow mode. *Biosensors and Bioelectronics*, 2018. **102**: 531–539.
128. Cao, H., X. Zhou, and Y. Zeng, Microfluidic exponential rolling circle amplification for sensitive microRNA detection directly from biological samples. *Sensors and Actuators B: Chemical*, 2019. **279**: 447–457.
129. Walker, G.T., et al., Strand displacement amplification—an isothermal, in vitro DNA amplification technique. *Nucleic Acids Research*, 1992. **20**(7): 1691–1696.
130. Yan, L., et al., Isothermal amplified detection of DNA and RNA. *Molecular BioSystems*, 2014. **10**(5): 970–1003.
131. Giuffrida, M.C., et al., Isothermal circular-strand-displacement polymerization of DNA and microRNA in digital microfluidic devices. *Analytical and Bioanalytical Chemistry*, 2015. **407**(6): 1533–1543.
132. Vincent, M., Y. Xu, and H. Kong, Helicase-dependent isothermal DNA amplification. *EMBO Reports*, 2004. **5**(8): 795–800.
133. Ramalingam, N., et al., Microfluidic devices harboring unsealed reactors for real-time isothermal helicase-dependent amplification. *Microfluidics and Nanofluidics*, 2009. **7**(3): 325.
134. Mahalanabis, M., et al., An integrated disposable device for DNA extraction and helicase dependent amplification. *Biomed Microdevices*, 2010. **12**(2): 353–359.
135. Van Ness, J., L.K. Van Ness, and D.J. Galas, Isothermal reactions for the amplification of oligonucleotides. *Proceedings of the National Academy of Sciences of the United States of America*, 2003. **100**(8): 4504–4509.
136. Reid, M.S., X.C. Le, and H. Zhang, Exponential isothermal amplification of nucleic acids and assays for proteins, cells, small molecules, and enzyme activities: An EXPAR example. *Angewandte Chemie (International ed. in English)*, 2018. **57**(37): 11856–11866.
137. Deng, H., et al., Paperfluidic chip device for small RNA extraction, amplification, and multiplexed analysis. *ACS Applied Materials & Interfaces*, 2017. **9**(47): 41151–41158.
138. Zhang, K., et al., Digital quantification of miRNA directly in plasma using integrated comprehensive droplet digital detection. *Lab on a Chip*, 2015. **15**: 4217–4226.
139. Piepenburg, O., et al., DNA detection using recombination proteins: e204. *PLoS Biology*, 2006. **4**(7): e204.
140. Symington, L.S., Focus on recombinational DNA repair. *EMBO Reports*, 2005. **6**(6): 512–517.
141. Lobato, I.M. and C.K. O'Sullivan, Recombinase polymerase amplification: Basics, applications and recent advances. *TrAC Trends in Analytical Chemistry*, 2018. **98**: 19–35.

142. Liu, D., et al., A microfluidic-integrated lateral flow recombinase polymerase amplification (MI-IF-RPA) assay for rapid COVID-19 detection. *Lab on a Chip*, 2021. **21**(10): 2019–2026.

143. Bai, Y., et al., Recombinase polymerase amplification integrated with microfluidics for nucleic acid testing at point of care. *Talanta*, 2022. **240**: 123209.

144. Notomi, T., et al., Loop-mediated isothermal amplification of DNA. *Nucleic Acids Research*, 2000. **28**(12): E63–E63.

145. Gansen, A., et al., Digital LAMP in a sample self-digitization (SD) chip. *Lab on a Chip*, 2012. **12**(12): 2247–2254.

146. Zhang, L., et al., Integrated microcapillary for sample-to-answer nucleic acid pretreatment, amplification, and detection. *Analytical Chemistry*, 2014. **86**(20): 10461–10466.

147. Zhang, H., et al., LAMP-on-a-chip: Revising microfluidic platforms for loop-mediated DNA amplification. *TrAC Trends in Analytical Chemistry*, 2019. **113**: 44–53.

148. Dirks, R.M. and N.A. Pierce, Triggered amplification by hybridization chain reaction. *Proceedings of the National Academy of Sciences of the United States of America*, 2004. **101**(43): 15275.

149. Yuan, D., et al., Hybridization chain reaction coupled with gold nanoparticles for the allergen gene detection of peanut, soybean and sesame. *The Analyst*, 2019. **144**: 3886–3891.

150. Li, Z., et al., Hybridization chain reaction coupled with the fluorescence quenching of gold nanoparticles for sensitive cancer protein detection. *Sensors and Actuators B: Chemical*, 2017. **243**: 731–737.

151. Guo, Q., et al., Electrochemical sensing of exosomal MicroRNA based on hybridization chain reaction signal amplification with reduced false-positive signals. *Analytical Chemistry*, 2020. **92**(7): 5302–5310.

152. Wang, X., et al., Enzyme-free and label-free fluorescence aptasensing strategy for highly sensitive detection of protein based on target-triggered hybridization chain reaction amplification. *Biosensors and Bioelectronics*, 2015. **70**: 324–329.

153. Si, H., et al., Highly sensitive fluorescence imaging of Zn^{2+} and Cu^{2+} in living cells with signal amplification based on functional DNA self-assembly. *Analytical Chemistry*, 2018. **90**(15): 8785–8792.

154. Bi, S., S. Yue, and S. Zhang, Hybridization chain reaction: A versatile molecular tool for biosensing, bioimaging, and biomedicine. *Chemical Society Reviews*, 2017. **46**(14): 4281–4298.

155. Yin, P., et al., Programming biomolecular self-assembly pathways. *Nature*, 2008. **451**(7176): 318–322.

156. He, H., et al., Self-assembly of DNA nanoparticles through multiple catalyzed hairpin assembly for enzyme-free nucleic acid amplified detection. *Talanta*, 2018. **179**: 641–645.

157. Wu, M., et al., Cascade-amplified microfluidic particle accumulation enabling quantification of lead ions through visual inspection. *Sensors and Actuators B: Chemical*, 2020. **324**: 128727.

158. Zheng, A.-X., et al., Enzyme-free fluorescence aptasensor for amplification detection of human thrombin via target-catalyzed hairpin assembly. *Biosensors and Bioelectronics*, 2012. **36**(1): 217–221.

159. Qian, C., et al., Rapid exosomes concentration and in situ detection of exosomal microRNA on agarose-based microfluidic chip. *Sensors and Actuators B: Chemical*, 2021. **333**: 129559.

2 Single-Molecule Detection by Solid-State Nanopores

Chun-Yen Lee
The University of Tokyo

Amer Alizadeh
University of Birmingham

Hirofumi Daiguji and Wei-Lun Hsu
The University of Tokyo

CONTENTS

2.1 Nanopore Technologies 50
    2.1.1 Introduction 50
    2.1.2 Types of Solid-State Nanopores 51
    2.1.3 Fabrication of Solid-State Nanopores...................................52
    2.1.4 Single-Molecule Detection by Solid-State Nanopores 55
2.2 Fundamental Electrokinetic Transport Phenomena in Solid-State
    Nanopores 56
    2.2.1 Electrostatics...56
    2.2.2 Electric Double Layer 58
    2.2.3 Electroosmosis in a Cylindrical Pore and Pore Conductance 61
    2.2.4 Electrophoresis 62
    2.2.5 Steric Effects, Dielectric Saturation, and Viscoelectric Effects.........64
    2.2.6 Computational Simulation 66
2.3 Nanopore Electrokinetics in the Presence of a Concentration Gradient........68
    2.3.1 Salt Concentration Gradients in Nanopores 68
    2.3.2 Nonuniform Electroosmotic Flow (EOF)...................................70
    2.3.3 Transport-Induced-Charge Electroosmosis................................71
    2.3.4 Diffusiophoresis and Electrodiffusiophoresis 74
2.4 Conclusions and Future Outlook 76
Acknowledgement 77
References..77

2.1 NANOPORE TECHNOLOGIES

2.1.1 INTRODUCTION

In our DNAs, massive genomic information, which plays a critical role in the development of next-generation diagnostic systems for precision medicine technologies, is recorded. The state-of-the-art technology used to perform DNA sequencing involves a polymerase chain reaction (PCR) method; however, the standard PCR procedure is time-consuming since it requires complex steps, involving reverse transcription, denaturation, annealing, polymerization and heat denaturation followed by cooling to obtain fluorescence signals (Bustin, 2002). In the post-COVID-19 era, exploring simple and fast molecular sequencing alternatives to labor-intensive PCR testing is important for determining preventive measures for the spread of infection. Moreover, the rapid detection of diseases may accelerate the development procedures of treatments and vaccines.

Continuous structural information on DNA molecules has been obtained using mutants of α-hemolysin pores with unwinding enzymes (Clarke et al., 2009), as illustrated in Figure 2.1. Hence, using biological nanopores embedded in a lipid bilayer has emerged as a promising method for DNA sequencing. The molecule detection mechanism is based on the principle proposed by Coulter in 1953 (USA, 1953), which was originally developed for cell counting using a small insulating orifice. In a process known as resistive pulse sensing, cells are detected based on the temporary resistance change that results in current blockage pulses. For DNA sequencing, α-hemolysin pores on a lipid bilayer membrane are submerged in an electrolyte solution (*e.g.*, a potassium chloride aqueous solution). Two electrodes are inserted into the solution on

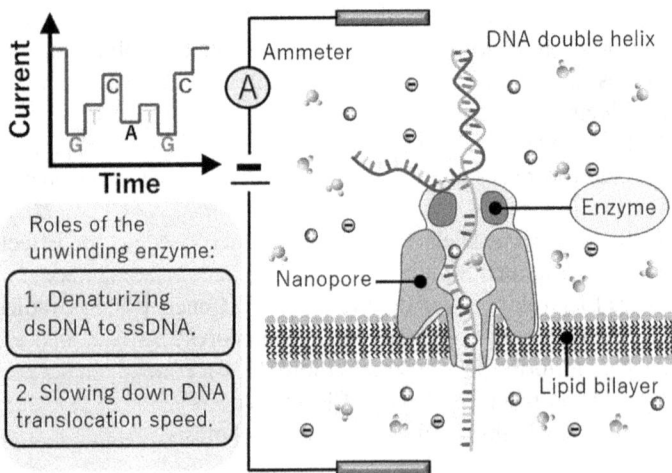

FIGURE 2.1 Biological nanopore-based DNA sequencing uses unwinding enzymes to concurrently denature double-stranded DNA into single-stranded DNA molecules and decrease the DNA translocation speed for high-resolution sequencing. However, these protein nanopores have poor mechanical and unstable chemical properties, and inflexible pore dimensions.

each side of the nanopores, separated by a membrane. The electrodes are connected to a voltage supplier and current meter to establish a closed circuit with the solution. Once a voltage bias is applied, the double-stranded DNA (dsDNA) molecules near the membrane are driven from the *cis* reservoir toward the nanopores by electrophoresis, which involves the motion of charged particles under the influence of an applied electric field. The direction of the applied electric field is opposite to that of the negatively charged dsDNAs. When an electrophoretically driven dsDNA molecule encounters an enzyme mounted at the entrance of the mutated α-hemolysin pores, the molecule is denatured into two single-stranded DNA (ssDNA) molecules, one of which is slowly ratcheted into the nanopore. The presence of the ssDNA molecule hinders ion transport through the nanopore, resulting in a salient decrease in the current measured at the external circuit. Due to the difference in the structure between the nucleotides, the nucleotide sequence of the ssDNA can be identified based on the ionic current variation.

Despite the success in acquiring genomic information from current signals using machine learning-based post-processing, α-hemolysin nanopores have the disadvantages of weak mechanical strength and insufficient spatial resolution. The bottleneck thickness of α-hemolysin nanopores is greater than the distance between nucleotides in ssDNA; consequently, during the sequencing process, approximately 12 nucleotides simultaneously occupy the nanopore (Meller et al., 2001; Branton et al., 2008). The size of the α-hemolysin pores bottleneck apertures is approximately 2 nm (Hall et al., 2010), which may not be compatible with the detection of other biomarkers for medical diagnosis; for example, antibodies against COVID-19, such as IgG and IgA, which are approximately 15 nm in diameter (Bachmann et al., 2021; Chen et al., 2004). Thus, flexible pore sizes for a range of target molecules are desirable for diagnostic applications.

2.1.2 Types of Solid-State Nanopores

In comparison with biological nanopores, artificial nanopores using solid-state materials possess the advantages of superior mechanical strength, flexible geometric capability, and stable chemical properties over biological nanopores and are therefore more favorable for diverse biomolecules. The first solid-state nanopore drilling of silicon nitride was conducted at the beginning of the current century (Li et al., 2001). Silicon nitride (Figure 2.2a) was selected because of its hydrophilicity; stable thermal properties, with a melting point of 1900°C; and compatibility with semiconductor manufacturing processes (Webster et al., 2012; Rumble, 2021). Regardless of the advanced fabrication methods for nanopores on silicon nitride membranes, the resistive pulse sensing of DNA molecules using these pores still has limitations, in terms of both spatial and temporal resolution limits (Iqbal et al., 2007), largely because of the absence of unwinding enzymes in solid-state nanopores. These enzymes not only denature dsDNA to ssDNA for sequencing but also slow down the translocation speed, resulting in high sequencing resolution. Compared with the 0.63 nm gap between each nucleotide on ssDNA (Ambia-Garrido et al., 2010), the thinnest silicon nitride membranes are nearly one order of magnitude thicker. Thus,

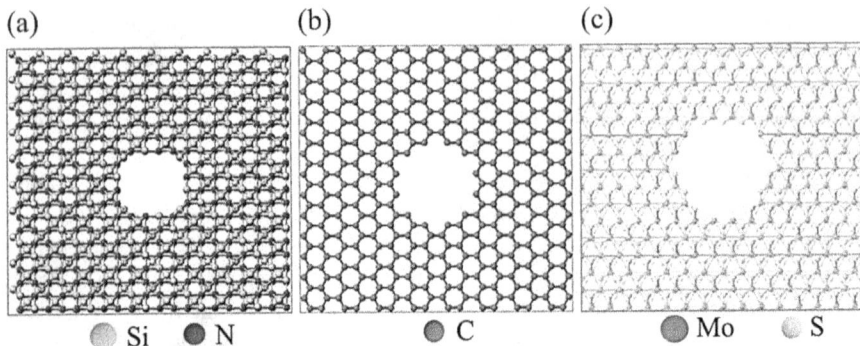

FIGURE 2.2 Atomic structure of a nanopore on different solid-state materials: (a) silicon nitride, (b) graphene and (c) monolayer molybdenum disulfide (MoS_2).

it remains extremely challenging to obtain a single-base-level resolution during DNA translocation (Bellini et al., 2012).

In this regard, nanopores made of two-dimensional (2D) materials have been adopted to enhance the spatial resolution whose thicknesses coincide with the distance between each nucleotide. Graphene, with a thickness of 3.4 Å, has been widely employed for this purpose (Schneider et al., 2010; Merchant et al., 2010), as shown in Figure 2.2b. Nevertheless, owing to the hydrophobic nature of carbon atoms, the attractive interactions between the molecules and outer surface of graphene can negatively affect the translocation efficiency. Hence, monolayer molybdenum disulfide (MoS_2), being slightly thicker (~6.5 Å) than graphene (Splendiani et al., 2010), was employed for nanopore sensing (Feng et al., 2015), as shown in Figure 2.2c. Similar to silica and silicon nitride, the hydrophilic surface results in the deprotonation of functional groups, leading to a negatively charged surface in basic solutions. This results in an electrostatic repulsive force between the DNAs and material, thereby avoiding the sticking problem. In addition to graphene and MoS_2, other 2D materials may also be suitable for molecule-sensing applications. Using a controllable low-power laser beam on a monolayer tungsten disulfide (WS_2) nanopore, Danda et al. (2017) demonstrated that the pore size could be tuned using a laser. Monolayer hexagonal boron nitride (h-BN), with a typical thickness of 5 Å, could be another candidate for high-resolution molecular sensing (Chen et al., 2020).

2.1.3 FABRICATION OF SOLID-STATE NANOPORES

The previous section described the advantages of solid-state nanopores over biological nanopores, including enhanced stability and flexible pore dimensions for different biomolecules. In this section, we summarize the fabrication processes for solid-state nanopores. With recent advances in nanotechnology, a range of methods for nanopore fabrication have been developed for different scales, as shown in Figure 2.3. Motivated by biological pores such as α-hemolysin (Branton et al., 2008; Schneider and Dekker, 2012) and MspA (Schneider and Dekker, 2012; Derrington et al., 2010), Li et al. (2001) proposed an approach to fabricate solid-state nanopores on a free-standing silicon nitride membrane using focused ion beam (FIB) sculpting.

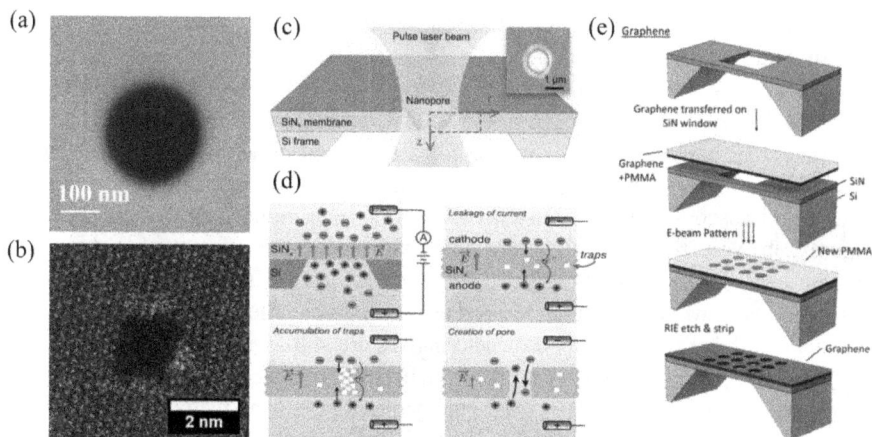

FIGURE 2.3 (a) A scanning electron microscopy (SEM) image of a silicon nitride nanopores drilled by focused ion beam (FIB) sculpting. (Reprinted with permission from Paul, 2022). (b) A scanning transmission electron microscopy (STEM) image of a 2 nm pore on MoS_2 prepared by STEM drilling. (Reprinted with permission from Hsu, 2019). (c) Laser pulse drilling: A nanosecond laser pulse is adopted to evaporate unwanted molecules. (Reprinted with permission from Yuan et al., 2018.) (d) Dielectric breakdown: Intrinsic defects in nitride thin films induce leakage current and bond-breaking processes under an applied electric field; the bond-breaking process further generates defects inside the film, eventually leading to nanopore formation. (Reprinted with permission from Kwok et al., 2014.) (e) Lithography and dry etching: Graphene is dry-transferred utilizing poly methyl methacrylate (PMMA), which is further used as an electron beam (e-beam) resist. Adopting an e-beam to define the pattern on PMMA, and subsequently dry etching the unwanted materials. (Reprinted with permission from Verschueren et al., 2018.)

First, a 60 nm nanopore was drilled using Argon ions (Ar^+), followed by irradiation with an Ar^+ ion beam to trigger lateral diffusion from the silicon nitride thin film, thus shrinking the nanopore to 2 nm. The targeted nanopore dimensions are controlled by lateral diffusion of the material, subject to the flux of ion beams and temperature effects. A scanning electron microscopy image of a 230 nm pore on a silicon nitride membrane is shown in Figure 2.3a. FIB sculpting using other ion sources, such as gallium ions (Ga^+) or helium ions (He^+), is conducted. Previous studies have suggested that the limitation of the pore size for a Ga^+ ion source is close to 10 nm (Balčytis et al., 2017), whereas the He^+ ion source can obtain sub-5 nm pores (Yang et al., 2011). Because He^+ is lighter than Ga^+, it undergoes a smaller lateral scattering process during ionic sculpturing (Allen et al., 2019); therefore, adopting an He^+ ion source with a smaller focal spot enables the processing of finer structures (Hlawacek and Gölzhäuser, 2018). In this regard, Emmrich et al. (2016) demonstrated a nanopore down to 1.3 nm in diameter on a carbon nanomembrane substrate via He^+ microscopy. However, owing to high energy of the ion beam (~keV), ions are implanted into the structure of the nanopore, which may locally alter the surface properties of the nanopores (Melngailis, 1987; Stein et al., 2004).

Both transmission electron microscopy (TEM) drilling and scanning transmission electron microscopy (STEM) drilling use electrons that are much smaller than Ga^+

and He$^+$. The electromagnetic lens focuses the electron beam under a high-vacuum system with an acceleration voltage of up to several hundred kV to penetrate the suspended samples (Williams and Carter, 2009). Storm et al. (2003) controlled silicon oxide nanopores based on the surface tension effects of the material to open a 3 nm nanopore. FIB and TEM/STEM methods are processed in a single-vacuum chamber that is below 10^{-6}Torr, reducing the risk of sample pollution. A scanning transmission electron microscopy (STEM) image of a 2 nm pore on monolayer MoS$_2$ is shown in Figure 2.3b. Nevertheless, drilling nanopores with a controlled diameter less than 2 nm remains challenging (Graf et al., 2019). Furthermore, both TEM and STEM are costly and require advanced operational skills. Therefore, despite the high resolution required for nanopore fabrication, inexpensive alternatives are still required. Material sculpting that uses laser pulses (Yuan et al., 2018), which can be operated without a high-vacuum system during the drilling process, has emerged as a promising method. The nanopore size can be adjusted using experimental parameters such as the laser power, laser pulse width, and wavelength of the laser source. Nevertheless, previous studies have suggested that while using similar operating conditions, the pore size can vary, yielding low reproducibility (Bian et al., 2011). Further investigation of the local heat transfer during the laser–matter interaction needs to be conducted to control the pore dimensions.

Apart from direct nanopore sculpting, Kwok et al. (2014) proposed a dielectric breakdown method for rapid nanopore fabrication in highly concentrated electrolyte solutions. The basic principle of dielectric breakdown involves the application of an electric field across a dielectric membrane to induce a leakage current, triggering the bond-breaking process inside the dielectric material, where the magnitude of the leakage current is proportional to the number of defect-induced traps. Once the bond-breaking proceeds under the applied voltage bias, more defects or vacancies are generated, and defect-induced traps accumulate, eventually leading to the formation of an aperture (Kwok et al., 2014). Dielectric breakdown nanopore fabrication is simple and inexpensive, which does not require high-vacuum conditions, thereby significantly reducing the operation time and complexity. However, care must be taken since the pore size is highly associated with the impurities and defects inside the thin-film system. Thus, it is challenging to precisely control the pore dimension.

Electron beam (e-beam) lithography technology, combined with dry etching using high-energy ions for nanoscale patterning (Nojiri, 2014), can be used for nanopore array fabrication, paving the way for various applications. Nevertheless, the nanopore size is limited by electron interaction with the substrate and the corresponding resist on the substrate (Broers, 1988). Verschueren et al. (2018) suggested that the smallest nanopore size obtained on graphene using e-beam lithography is approximately 16 nm. Similarly, optical lithography uses a single wavelength light source to create nanoscale patterns, whose critical dimensions (CD), being the minimum pattern of optical lithography, can be defined via the Rayleigh criterion (Lin, 2021):

$$CD = \frac{k_1 \lambda}{NA} \tag{2.1}$$

where k_1 is a constant related to the manufacturing process, λ the wavelength in vacuum, and NA the numerical aperture of the imaging lens. The shorter the wavelength, the finer the feature size that can be defined. Notably, optical lithography technology reduces its wavelength from the extreme ultraviolet (EUV) scale to 13.5 nm, enabling the definition of a finer pattern compared with the deep ultraviolet (DUV) light source having wavelengths of 248 and 193 nm (Wagner and Harned, 2010). Considering the light diffraction limits, Mojarad et al. (2015) suggested that a further combination of the EUV with atomic etching (Kanarik et al., 2018) might achieve ultra-high resolutions of up to 3.5 nm.

2.1.4 SINGLE-MOLECULE DETECTION BY SOLID-STATE NANOPORES

Attributed to the controllable pore size of solid-state nanopores depending on the fabrication process, the application of solid-state nanopore-resistive pulse sensing is not limited to DNA molecules, but can also be applied to the investigation of various biomolecules (Talaga and Li, 2009; van der Hout et al., 2010; Yang and Yamamoto, 2016; Darvish et al., 2019; Cai et al., 2021), as shown in Figure 2.4. For instance, silicon nitride nanopores have been used to identify the folding states of proteins (Talaga and Li, 2009). In addition, novel technologies combined with mass spectrometry and deep learning algorithms may facilitate protein sequencing and sensing in the near future (Alfaro et al., 2021). To characterize cell walls, Cai et al. (2021) utilized a 4 nm free-standing silicon nitride nanopore to distinguish hyperacetylated

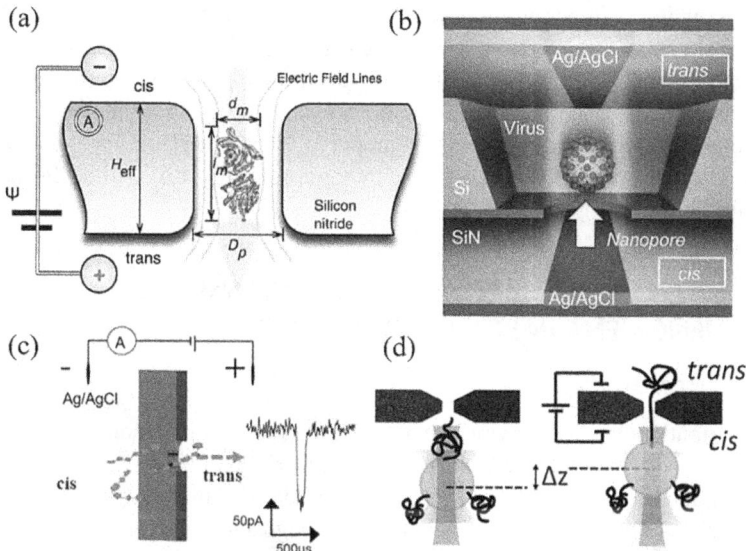

FIGURE 2.4 Using nanopores on a silicon nitride membrane to characterize (a) conformational states of proteins, where d_m, D_p, H_{eff} and ψ denote the diameter of the molecule, average pore diameter, effective pore thickness and applied voltage, respectively (Talaga and Li, 2009), (b) coronaviruses. (Reprinted with permission from Taniguchi et al., 2021.) (c) Cell walls, i.e., plant polysaccharides. (Reprinted with permission from Cai et al., 2021.) (d) dsRNAs with optical tweezers, where Δz denotes the position displacement. (Reprinted with permission from van der Hout et al., 2010.)

and unacetylated modifications in plant cells, improving our understanding of how polysaccharides assist plants to become accustomed to the environment.

One of the primary advantages of biomolecule detection using nanopores is its capability of characterizing unknown species, without the prior design of PCR primers or antibodies required in conventional methods. This renders solid-state nanopore technology a possible candidate for the investigation of new variants of pandemic diseases, such as coronaviruses (Taniguchi et al., 2021). Yang and Yamamoto (2016) characterized the binding of antibodies to virus particles and compared the results with unbonded virus particles. They observed that the antibody-bonded viruses have a larger volume exclusion effect than their counterparts, giving rise to more salient current blockage signals. Furthermore, the application of solid-state nanopores can be extended to specify the level of infection of viruses by identifying their deformation state. Darvish et al. (2019) found that non-infective viruses have more robust mechanical properties than infective viruses. Direct virus characterization can be important because some existing sequencing tools are limited to short-read sequencing. For example, Ion Torrent and Illumina Solexa (Hölzer and Marz, 2017) are designed for short-chain oligomers, making the characterization of viral mutations by mRNA sequence challenging. On the other hand, by integrating with optical tweezers, van der Hout et al. (2010) directly measured intramolecular force on dsRNA in nanopores to identify the secondary structure. It is found the force on DNA and RNA molecules is different under the same bias voltage. In addition to biological sample detection and characterization, studies have been conducted on the characterization of nanoparticles such as polystyrene (Lan and White, 2012). Although these nanopores conceptually promise high sensing resolution for different biomolecules, understanding the kinetic behavior of these charged entities through nanopores is indispensable for the control of translocation behavior enabling accurate nanopore sensing; this will be discussed in detail in the following sections.

2.2 FUNDAMENTAL ELECTROKINETIC TRANSPORT PHENOMENA IN SOLID-STATE NANOPORES

2.2.1 ELECTROSTATICS

To understand transport phenomena and molecule behavior in nanopores, it is crucial to investigate electrokinetic transport phenomena originating from the charged nanopore wall and molecule surface. In this section, the fundamental principles governing the fluid and molecule transport behavior through a nanopore are summarized for typical molecule sensing in nanopores. During the transport of molecules through an electric potential-biased nanopore, the (electro-)osmotic flow and molecular migration are subjected to the electric force acting on dissolved ions in electrolytes and immobile charges on the molecule, respectively. Hereafter, we introduce the concept of an electric potential to analyze electrostatics in a charged solution.

When considering an electric field **E** around a charge q separated by a distance r, the electric force acting on a unit positive charge is inversely proportional to the square of the distance, according to Coulomb's law (Schey, 2005):

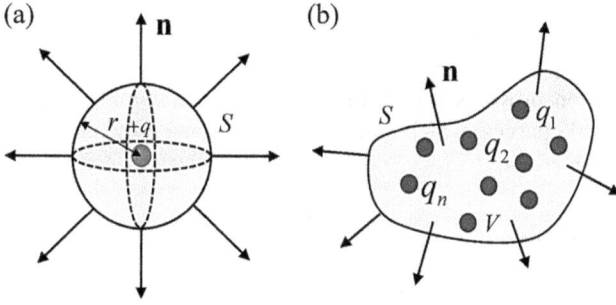

FIGURE 2.5 Electrostatics of charges in a finite volume. (a) One single charge q in a spherical shell of radius r; and (b) multiple charges q_i, where $1 \le i \le n$ and n is the total number of charges, in an arbitrary geometry. S, V and $\mathbf{n}$ denote the surface, control volume and unit normal vector, respectively.

$$E = k_e \frac{q}{r^2} \mathbf{i} \tag{2.2}$$

where $k_e = 8.99 \times 10^9$ Nm^2C^{-2} is Coulomb's constant, and $\mathbf{i}$ refers to the unit vector of the unit charge pointing away from q. Integrating the electric field in the normal direction on a spherical surface S of radius r centered at the charge location, as illustrated in Figure 2.5a, the following relation can be derived:

$$\iint_S \mathbf{E} \cdot \mathbf{n} \, dA = \iint_S k_e \frac{q}{r^2} (\mathbf{i} \cdot \mathbf{n}) \, dA = 4\pi k_e q \tag{2.3}$$

where $\mathbf{n}$ is the unit normal vector on S. Similarly, for multiple charges in an arbitrary geometry, as shown in Figure 2.5b, the electric field flux on the surface can be estimated by the summation of each charge q_i, where $1 \le i \le n$ and n is the total number of charges in the control volume. According to Gauss' law, which converts the surface integral to volume integral, one can derive the following relation for charges in an arbitrary medium after substituting Coulomb's constant with $\dfrac{1}{4\pi\varepsilon_0\varepsilon}$, where $\varepsilon_0 = 8.85 \times 10^{-12}$ F/m is the permittivity in vacuum and ε denotes the dielectric constant of the medium, which is $\cong 78.5$ for water at room temperature.

$$\iint_S \mathbf{E} \cdot \mathbf{n} \, dA = \iiint_V \nabla \cdot \mathbf{E} \, dV = \frac{\sum\limits_{i=1}^n q_i}{\varepsilon_0\varepsilon} = \frac{1}{\varepsilon_0\varepsilon} \iiint_V \rho_e \, dV \tag{2.4}$$

Here $\sum\limits_{i=1}^n q_i$ represents the total charge inside the control volume V and ρ_e is the charge density. By definition, the electric field can be expressed in the opposite direction of the electric potential gradient as $\mathbf{E} = -\nabla\psi$. Accordingly, one can derive Poisson's equation for the variation of electric potential as a function of the local charge density when ε is uniform in the system:

$$\nabla^2 \psi = -\frac{\rho_e}{\varepsilon_0 \varepsilon} \qquad (2.5)$$

Poisson's equation provides information of electric potential variation due to the presence of local charges, which can be used to estimate the electric field strength near a charged interface.

2.2.2 ELECTRIC DOUBLE LAYER

When a charged surface is in contact with an electrolyte solution, co-ions possessing the same kind of charge as the surface are electrostatically repelled, while a large number of counterions accumulate near the interface subjected to the electrostatic attractive force. At equilibrium, as the electrical drift is offset by ionic diffusion, a layer of asymmetric ion distribution develops near the charged surface on the liquid side. The immobile charges on the solid surface along with the inversely charged solution layer are called the "electric double layers (EDLs)." The ion distribution within the charged solution layer, known as the diffuse layer, can be mathematically depicted by the Gouy–Chapman model (Masliyah and Bhattacharjee, 2006), which states that the conductive flux of ions $\mathbf{J}_{cond}$ normal to a charged wall is balanced by the diffusive flux $\mathbf{J}_{diff}$ (namely, there is no net ionic flux into the ion-impermeable surface) as follows:

$$\mathbf{J}_{cond} = \frac{\mp D_i e n_i}{k_B T} \nabla \psi = -\left(-D_i \nabla n_i\right) = -\mathbf{J}_{diff} \qquad (2.6)$$

where D_i is the diffusion coefficient of ionic species i (for binary electrolytes, $i = 1$ for cations and $i = 2$ for anions), z_i the valence of ionic species i, $e = 1.6 \times 10^{-19}$ C the elementary charge, n_i the concentration of ionic species i, $k_B = 1.38 \times 10^{-23}$ J/K the Boltzmann constant and the T the absolute temperature, and $\mathbf{J}_{diff}$ the diffusive flux of ions. In the bulk, where the electrostatic force from the charged surface becomes insignificant, it can be assumed that the ionic concentration n_i equals the solute concentration n_0 for ideal electrolytes. From Eq. 2.6, one can derive a Boltzmann distribution of ions as a function of the electric potential for monovalent electrolytes:

$$n_i = n_0 \exp\left(m \frac{e \psi}{k_B T} \right) \qquad (2.7)$$

Substituting Eq. 2.7 into Eq. 2.5 gives the well-known Poisson–Boltzmann equation for binary systems as follows:

$$\nabla^2 \psi = -\frac{\rho_e}{\varepsilon_0 \varepsilon} = -\frac{\sum_{i=1}^{2} z_i e n_i}{\varepsilon_0 \varepsilon} = \frac{2 e n_0}{\varepsilon_0 \varepsilon} \sinh\left(\frac{e \psi}{k_B T} \right) \qquad (2.8)$$

For the low surface potential regime at $|\psi| \ll \dfrac{k_B T}{e} \cong 25.7$ mV under the Debye–Hückel approximation, a simplified expression of the Poisson–Boltzmann equation can be derived as $\nabla^2 \psi \cong \kappa^2 \psi$. Here, κ is the reciprocal of a characteristic length of the electric double-layer λ_D, which is referred to as the Debye length.

$$\kappa^{-1} = \lambda_D = \left(\frac{\varepsilon_0 \varepsilon k_B T}{2e^2 n_0} \right)^{1/2} \tag{2.9}$$

For monovalent electrolytes at 298 K, the Debye length can be described as the following equation, which indicates that the thickness of the diffusion layer is inversely proportional to the square root of the bulk electrolyte concentration.

$$\lambda_D [nm] = \frac{0.304}{\sqrt{C_0 [M]}} \tag{2.10}$$

where C_0 is the solution molar concentration.

It is worth noting that ions are regarded as point charges in the Gouy–Chapman model whose volume is ignored. This assumption is generally valid in the bulk until it approaches a surface. The discretization of ion concentrations in the vicinity of an interface can appear due to significant ionic steric effects, in which the minimum distance between the ion center and surface is limited by the ion radius. To take this effect into account, a modified EDL was model proposed by Stern (1924) stating that there exists an atomically thin layer on the surface with a thickness equal to the ion radius in which no charge is enclosed. This yields a capacitance structure that two charged layers possessing opposite charges exist in parallel, separated by a short distance. As illustrated in Figure 2.6a, the capacitance C_{Stern} within the so-called Stern layer can be expressed as:

$$C_{\text{Stern}} = \frac{\sigma_S}{\psi_S - \psi_D} \tag{2.11}$$

where σ_S, ψ_S, and ψ_D are the surface charge density, surface potential, and electric potential at the junction of the Stern and diffuse layers, respectively. In the basic Stern model, it is considered that water molecules are immobile in the Stern layer, and thus the outer boundary of the Stern layer is regarded as the shear layer of the surface; and the potential at which is defined as the zeta potential ζ, viz. $\psi_D = \zeta$. By solving the Poisson–Boltzmann equation within the diffuse layer using the inner potential boundary condition of ζ, the relationship between σ_S and ζ can be captured by the Grahame equation as follows (Grahame, 1947):

$$\sigma_S = \sqrt{8 \varepsilon_0 \varepsilon k_B T n_0} \sinh \left(\frac{e\zeta}{2k_B T} \right) \tag{2.12}$$

When exposed to an aqueous solution, the surface charge on hydrophilic materials originates from protonation/deprotonation of the surface functional groups. For instance, the charge condition of the silica/water interface is governed by silanol groups (Hsu et al., 2016). Based on a site-binding model, the relationship between ζ and σ_S with a known bulk pH at equilibrium can be acquired (Behrens and Grier, 2001; van der Heyden et al., 2005):

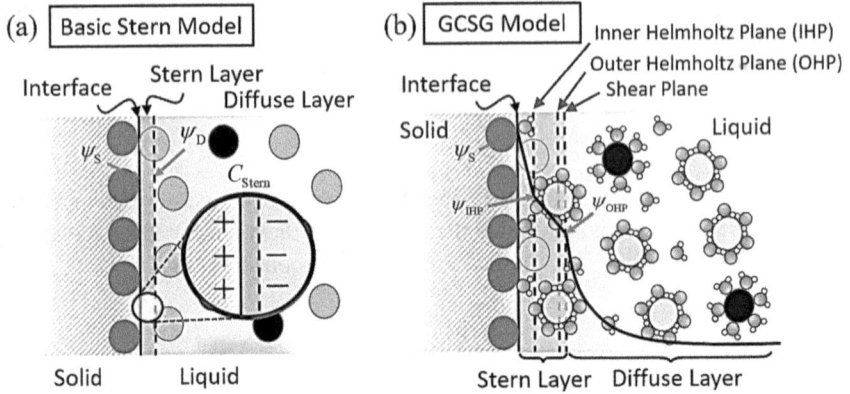

FIGURE 2.6 (a) The basic Stern model: A capacitance structure is considered to account for the charge discretization at the interface, where C_{Stern} ψ_S and ψ_D are the Stern layer capacitance, surface potential and electric potential at the inner boundary of diffuse layer, respectively. (b) The Gouy–Chapman–Stern–Grahame (GCSC) model: An inner Helmholtz plane (IHP), and an outer Helmholtz plane (OHP) are defined, for the adsorption of non-hydrated and hydrated ions, respectively. ψ_{IHP} and ψ_{OHP} are the electric potential on the IHP and OHP, respectively.

$$\zeta = \frac{k_B T}{e} \ln \frac{-\sigma_S}{e\Gamma + \sigma_S} - \left(pH - pK_a\right)\frac{k_B T}{e} \ln 10 - \frac{\sigma_S}{C_{Stern}} \quad (2.13)$$

where Γ and K_a refer to the number density and equilibrium constant of surface functional groups, respectively. Notably, the basic Stern model neglects different adsorption locations of non-hydrated and hydrated ions on the surface, which are further considered by the Gouy–Chapman–Stern–Grahame (GCSG) model (Grahame, 1947), as illustrated in Figure 2.6b. Similar to the basic Stern model, the interface is split into an immobile (Stern) layer and a diffuse layer. The locations where the non-hydrated and hydrated ions reside are referred to as the inner Helmholtz plane (IHP) and outer Helmholtz plane (OHP), respectively. ψ_{IHP} and ψ_{OHP} are the electric potential on the IHP and OHP, respectively. Based on the GCSC model, Yates et al. (1974) estimated surface charge properties at the silica/water interface. It is indicated that the charge conditions on oxide surfaces depend upon the local concentration of proton (H^+) and hydroxyl ions (OH^-).

It is worth emphasizing that although the shear plane at the interface is not necessarily coincide with the OHP, they are considered to be at the same location in the GCSG model for simplicity; namely ζ on the shear plane is approximated as ψ_{OHP}. On this account, an electrical-quad layer (EQL) model considering the electric potential difference between the OHP and shear plane was suggested by Alizadeh and Wang (2019). An inactive electrokinetic layer, so-called buffer layer, was inserted between the OHP and shear plane in the EQL model. Moreover, the EQL model was employed to investigate temperature effects on zeta potential (Alizadeh and Wang, 2020), being crucial for electrokinetic phenomena.

2.2.3 ELECTROOSMOSIS IN A CYLINDRICAL PORE AND PORE CONDUCTANCE

When an electric field is applied in the tangential direction to the surface, the charged solution in the diffuse layer is electrostatically driven in the direction of the electric field (Alizadeh et al. 2021). By considering the momentum balance between the viscous stress and the electric body force, one can arrive at the classical Helmholtz–Smoluchowski equation of electroosmosis for thin Debye lengths (Masliyah and Bhattacharjee, 2006):

$$v_{EOF} = -\frac{\varepsilon_0 \varepsilon \zeta E_\infty}{\eta} \qquad (2.14)$$

where v_{EOF} and E_∞ are the electroosmotic velocity and imposed external electric field strength, respectively. According to Eq. 2.10, the length scale of the diffuse layer in nanopores could become similar to the pore size, particularly at low concentrations. This implies that the variation in electroosmotic velocity within the diffuse layer is non-trivial. Using the Debye–Hückel approximation in cylindrical coordinates, the radial potential distribution in the diffuse layer can be derived, which is expressed by the modified zero-order Bessel function as $\zeta \frac{I_0(\kappa r)}{I_0(\kappa R)}$. Substituting this potential profile in the momentum balance equation to estimate the local electric body force, the electroosmotic velocity distribution in an infinitely long cylindrical pore v_{pore} can be obtained as follows (Rice and Whitehead, 1965):

$$v_{pore}(r) = -\frac{\varepsilon_0 \varepsilon \zeta E_{pore}}{\eta} \left[1 - \frac{I_0(\kappa r)}{I_0(\kappa R)} \right] \qquad (2.15)$$

where E_{pore} is the axial electric field strength in the cylindrical pore. For thin double layers, when $\kappa R \gg 1$, this equation converges to the classical Helmholtz–Smoluchowski equation, viz., Eq. 2.14.

It is worth mentioning that Eq. 2.15 is derived for infinitely long tubes, where the entrance effect is not considered. For thin nanopores connecting two large reservoirs, the effective electric potential field inside the nanopore largely depends on the aspect ratio of the pore radius and pore length for a given electric potential bias. In long pores, the bulk resistance is primarily contributed by the inner resistance of the pore R_{inner}, which is proportional to the channel length L but inversely proportional to the pore cross-sectional area (Hsu et al., 2021a):

$$R_{inner} = \frac{L}{\sigma_b \pi R^2} \qquad (2.16)$$

Here, σ_b is the bulk conductivity and R the pore radius. Conversely, for short nanopores, the access resistance R_{access} at the junction of the reservoir and pore entrance becomes significant; hence, the total pore resistance R_{pore} can be estimated as (Hall, 1975):

$$R_{pore} = R_{inner} + R_{access} = \frac{L}{\sigma_b \pi R^2} + \frac{1}{2\sigma_b R} \quad (2.17)$$

For ultrathin nanopores, the latter term becomes predominant, implying that the resistance is inversely proportional to pore radius. Note that Eq. 2.17 only includes the contribution of the bulk conductance, where the additional conductance attributed to the presence of EDLs is excluded. Kowalczyk et al. (2011) investigated the effect of the electrolyte concentration on the conductance of nanofluidic systems. At the nanoscale, when the pore size coincides with the EDL thickness, the surface conductance becomes significant, leading to a modified expression of pore conductance G:

$$G = \sigma_b \left(\frac{L}{\pi R^2} \frac{1}{1+Du} + \frac{1}{2R} \right)^{-1} \quad (2.18)$$

In this equation, Du denotes the Dukhin number, which is the ratio of the surface conductivity σ_s, attributed by the inner surface charge of the nanopore (Lyklema and Minor, 1998), to the product of σ_b and R_2. Given that Eq. 2.18 neglects the conductance contributed by the charges on the outer pore surface, Lee et al. (2012) proposed a modified model to estimate the overall conductance for thin nanopores as follows:

$$G = \sigma_b \left(\frac{L}{\pi R^2} \frac{1}{1+Du} + \frac{2}{2\alpha R + \beta l_{Du}} \right)^{-1} \quad (2.19)$$

where α, β, and $l_{Du} = \dfrac{\sigma_s}{\sigma_b}$ are a geometrical prefactor, a numerical constant, and the Dukhin length, respectively. According to the resistance model, the electric potential drop and thus E_{pore} in Eq. 2.15 can be estimated. Notably, although Eq. 2.15 is derived based on the assumption of an infinitely long tube, neglecting the inertia stress at the pore/reservoir junctions, the low Reynolds number in nanopores (typically in the order of 10^{-4}) yields assumption valid for most nanopore systems.

2.2.4 ELECTROPHORESIS

When a charged biomolecule migrates toward a nanopore, the electric field in the access region of the nanopore activates the electrophoretic motion of the molecule. In the presence of an electric field, the charged solution surrounding the molecule surface gives rise to an electro-retardation flow that pushes the molecule in the opposite direction. Meanwhile, the existence of an electric field can distort the counterion cloud for thick EDLs (O'Brien and White, 1978), as illustrated in Figure 2.7a.

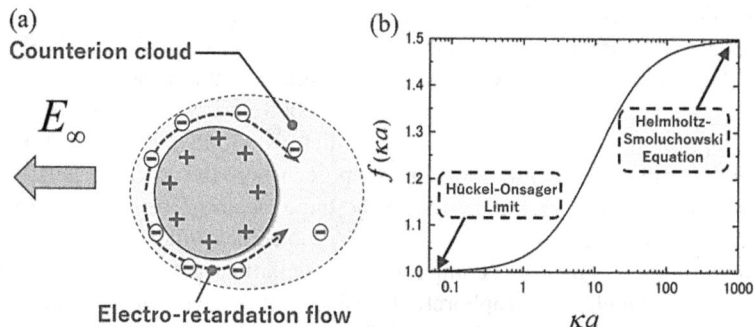

FIGURE 2.7 (a) Schematic of the electrophoresis of a charged spherical particle. (b) Variation in the Henry function f as a function of κa approximated by Eq. 2.23.

For thin EDLs (i.e., $\kappa a \gg 1$), the distortion of the EDL and curvature effects on the surface are negligible. In this manner, the electrophoretic mobility μ can be described by the Helmholtz–Smoluchowski equation of electrophoresis:

$$\mu\big|_{\kappa a \gg 1} = \frac{U}{E_\infty} = \frac{\varepsilon_0 \varepsilon \zeta}{\eta} \tag{2.20}$$

where U is the electrophoretic velocity. At the other extreme, in a dilute system, Hückel (1924) estimated both the electric and hydrodynamic forces acting on a spherical particle to derive μ at steady state at which the net force vanishes, which gives:

$$\mu\big|_{\kappa a \ll 1} = \frac{U}{E_\infty} = \frac{2}{3}\frac{\varepsilon_0 \varepsilon \zeta}{\eta} \tag{2.21}$$

In Hückel's model, it was assumed that the EDL distortion is insignificant, which is valid when the applied electric field is much weaker than that within the EDL; therefore, the diffuse layer thickness can be regarded as uniform. In a similar vein, Henry (1931) utilized a perturbation method splitting the electric potential into two terms: (i) the equilibrium electric potential from the EDL in the absence of the external electric field and (ii) the perturbed electric potential due to the applied electric field. When the imposed electric field magnitude is much smaller than the inherent electric potential within the EDL (i.e., $E_\infty \ll \kappa \zeta$), one can linearly decompose these two components. Accordingly, Henry derived a general expression of U for arbitrary concentrations expressed in function f as follows:

$$U = \left(\frac{2}{3}\frac{\varepsilon_0 \varepsilon \zeta}{\eta} E_\infty \right) f(\kappa a) \tag{2.22}$$

$$f(\kappa a) \cong \frac{3}{2} - \frac{1}{2\left[1 + 0.072(\kappa a)^{1.13}\right]} \tag{2.23}$$

The variation in $f(\kappa a)$ as a function of κa is shown in Figure 2.7b. Note that Eq. 2.22 is consistent with previous models in that f converges to 1.5 when $\kappa a \gg 1$, yielding the Helmholtz–Smoluchowski equation, whereas it approaches unity at $\kappa a \ll 1$ for the Hückel and Onsager limit (Schoch et al. 2008). It is worth emphasizing that, the assumption of the weak external electric field is not necessarily guaranteed in nanopore systems, specifically for thin nanopores where the applied electric potential is focused in the pore region due to the huge electrical resistance. Moreover, molecules with different shapes, such as DNA molecules, other than spherical particles are of interest. Therefore, computational simulations are extensively employed to analyze the complex electrophoretic translocation behavior of molecules, which we will provide more details in Section 2.2.6.

2.2.5 STERIC EFFECTS, DIELECTRIC SATURATION, AND VISCOELECTRIC EFFECTS

In addition to the distortion of EDLs, in this section, we summarize complex but important effects in nanopore systems that cannot be easily resolved analytically. First, in the aforementioned EDL models based on the mean-field theory, ions in the diffuse layer are regarded as point charges concentrated at the center point without considering the ionic steric effect. This assumption could be invalid for nanopores, particularly when both the concentration and surface charge are high. On this account, Bikerman considered that the work required to insert ions in the solution is comprised of two terms: (i) the addition of the charge to the solution and (ii) the dipoles displaced by ions (Bazant et al., 2009). Similarly, Borukhov et al. (1997) evaluated ion steric effects and proposed a modified Boltzmann equation based on a thermodynamic approach:

$$n_i = \frac{n_{i,0}e^{\mp z_i e \psi / k_B T}}{1 + 2v \sinh^2\left(z_i e \psi / k_B T\right)} \tag{2.24}$$

where $n_{i,0}$ is the bulk concentration of ionic species i and $v = 2a^3 n_{i,0}$ is the bulk volume fraction of solvated ions. The results of Eq. 2.24 indicate that, due to the ion steric effect, the classical Poisson–Boltzmann equation (i.e., Eq. 2.8) can overestimate the counterion concentration in the vicinity of the surface for high surface charge density.

On the other hand, the presence of an intense electric field and nanoconfinement effects can give rise to dissimilar physical properties of molecules from their bulk values. For instance, the electrostatic permittivity of polar solvents is remarkably influenced by the local electric field strength. Consequently, the effective dielectric permittivity of water molecules reduces at the interface. The theoretical model that accounts for the dielectric saturation was first attempted by Debye (1929) using a virtual spherical model:

$$\frac{\varepsilon - 1}{\varepsilon + 2} = \frac{4\pi}{3} \sum N \left(\xi + \frac{\chi^2}{3 k_B T} \right) \tag{2.25}$$

where N, ξ, and χ denote the number concentration, polarizability, and dipole moment of the polarized molecule, respectively. Nevertheless, the Debye model showed little agreement with the experimental measurements. A probable reason is that the Lorentz field for non-polar dielectrics is adopted for both the internal field and directing field, where the former refers to the difference between the external electric field and the field due to the polarization of the molecule, and the latter is the electric field that directs the permanent dipoles. Onsager (1936) pointed out that this assumption is not permissible for polar solvents, and that the internal field and directing field differ. Kirkwood (1939) further extended Onsager's theory by including the effect of the surrounding molecules on electric moments. Because both the Onsager and Kirkwood models are limited to weak electric field magnitudes, a more general form for arbitrary field strengths was obtained by Booth (2004) and later simplified by Hunter (1966) as follows:

$$\varepsilon = \tilde{\varepsilon}\left(1 - B | \mathbf{E} |^2\right) \tag{2.26}$$

where $\tilde{\varepsilon}$ is the relative permittivity in the absence of $\mathbf{E}$, and the constant B is estimated to be 4×10^{-18} m^2/V^2 for water at room temperature (Hunter, 1988).

Not only does the water molecule polarization affect the solution property, but also the orientation of water molecules near a charged surface favoring interactions between water molecules results in a higher viscosity for the flow in the tangential direction to the charged surface. From the perspective of activation energy, the intense electric field within the EDL decreases the vibration frequency of molecules, providing enhanced activity energy E_a, according to Reynolds' equation:

$$\eta = A \exp\left(\frac{E_a}{k_B T}\right) \tag{2.27}$$

where the coefficient A is a constant. Andrade and Dodd experimentally demonstrated the influence of electric fields on the viscosity of polar solvents by measuring the flow rate of a pressure-driven flow through a slit channel with an applied electric field normal to the flow direction (Andrade and Dodd, 1939). In polar solvents, the apparent viscosity increased considerably with the applied electric field magnitude. In stark contrast, it remained constant for non-polar solvents. In subsequent studies, the phenomenon was quantitatively investigated revealing that the increase in viscosity is proportional to the square of the applied electric field magnitude at low applied electric fields (Andrade and Dodd, 1946, 1951):

$$\eta = \tilde{\eta}\left(1 + f_v |\mathbf{E}^2|\right) \tag{2.28}$$

where $\tilde{\eta}$ is the viscosity in the absence of $\mathbf{E}$ and $f_v = \Delta E_a / (k_B T)$ is referred to as the viscoelectric coefficient, in which ΔE_a is the activation energy variation due to the presence of the electric field. The low electric field assumption $\left(f_v |\mathbf{E}|^2 \ll 1\right)$ breaks down owing to the large electric field within EDLs. This makes the high-order terms in the Maclaurin series of the exponential function in Eq. 2.27 non-negligible and thus an exponential-type viscoelectric equation has to be considered:

$$\eta = \tilde{\eta} \exp\left(f_v |\mathbf{E}^2|\right) \tag{2.29}$$

Based on Debye and Onsager's theories of polarization, f_v is proportional to the square of χ (Lyklema and Overbeek 1961):

$$f_v \propto \frac{\chi^2}{3k_\mathrm{B}T}\left(\frac{3\varepsilon}{2\varepsilon+1}\right)^2 \tag{2.30}$$

Lyklema and Overbeek provided a detailed explanation of f_v based on the theory of the polarization of spherical molecules. By assuming that the structural coefficient does not vary significantly for different polar molecules, an estimate of f_v was obtained for water as $10.2\times10^{-16}\,\mathrm{m}^2/\mathrm{V}^2$ using an average value from previous data of organic polar solvents. Later, Hunter and Leyendekkers (1978) experimentally obtained an estimate of $f_v = (5-10)\times10^{-16}\,\mathrm{m}^2/\mathrm{V}^2$ for water between two parallel sheets, which is in close agreement with recent measurements obtained by the surface force balance at the nanoscale (Jin et al., 2022).

Molecular dynamics simulations have also predicted an increased viscosity in nanochannels. Qiao and Aluru (2005) pointed out that when the interaction between the surface and water molecules is strong, the increase in viscosity is evident, which is consistent with the viscoelectric theory. In addition, the enhancement of viscosity suppresses the ionic diffusivity of an ion of radius a_{ion} based on the Stokes–Einstein equation:

$$D_i\eta = \tilde{D}_i\tilde{\eta} = \frac{k_\mathrm{B}T}{6\pi a_{\mathrm{ion}}} \tag{2.31}$$

where $\tilde{D}_i$ is the local ionic diffusivity when an electric field is present, respectively, in the absence of an electric field. Kaji et al. (2006) observed an increase in the viscosity and a decrease in DNA diffusivity in nanospaces, in agreement with the viscoelectric theory. By comparing the continuum model with molecular dynamics, Hsu et al. (2017) indicated that viscoelectric effects become dominant in nanofluidic systems when the surface charge density is greater than 40 mC/m².

2.2.6 Computational Simulation

To incorporate the aforementioned nonlinear terms for nanopore systems, numerical simulations have been extensively employed in the literature to estimate the molecule translocating speed and current blockage. In continuum models, the electric potential and ionic concentration distributions can be obtained by solving the Poisson–Nernst–Planck equation:

$$\begin{cases} \nabla\cdot\varepsilon_0\varepsilon\nabla\psi = -\rho_e; \\ \dfrac{\partial n_i}{\partial t} + \nabla\cdot\left(-D_i\nabla n_i \mp \dfrac{D_i e n_i}{k_\mathrm{B}T}\nabla\psi + n_i\mathbf{v}\right) = 0. \end{cases} \tag{2.32}$$

Here t is the time and $\mathbf{v}$ is the velocity vector of the liquid. The velocity coupled in the advection term $n_i\mathbf{v}$ in the Nernst–Planck equation can be derived by simultaneously resolving the continuity and Navier–Stokes equations for incompressible fluids.

$$\begin{cases} \nabla \cdot \mathbf{v} = 0; \\ \rho \dfrac{\partial \mathbf{v}}{\partial t} + \mathbf{v} \cdot \nabla \mathbf{v} = -\nabla p + \nabla \cdot \eta \left[\nabla \mathbf{v} + (\nabla \mathbf{v})^T \right] - \rho_e \nabla \psi. \end{cases} \qquad (2.33)$$

In these expressions, ρ is the solution density and p is the pressure. For non-polarizable materials, the surface charge density on the surfaces can be derived as a function of the local electric potential gradient as $\sigma = -\varepsilon_0 \varepsilon \mathbf{n} \cdot \nabla \phi$ based on the electroneutrality of the EDL. Conversely, non-slip and ion-impermeable boundary conditions are generally employed on the surface for the velocity and ion concentrations, respectively. From the obtained electric field and velocity distributions, the electric force $\mathbf{F}_E$ and hydrodynamic force $\mathbf{F}_H$ exerted on the molecule can be estimated by integrating the Maxwell stress and hydrodynamic stress on the molecule surface, respectively.

$$\mathbf{F}_E = \varepsilon_0 \varepsilon \iint_S \tau_E \cdot \mathbf{n} \, dA = \varepsilon_0 \varepsilon \iint_S \left[\mathbf{EE} - \frac{1}{2} (\mathbf{E} \cdot \mathbf{E}) \mathbf{I} \right] \cdot \mathbf{n} \, dA \qquad (2.34)$$

$$\mathbf{F}_H = \iint_S \tau_H \cdot \mathbf{n} \, dA = \iint_S \left[-p\mathbf{I} + \eta \left[\nabla \mathbf{v} + (\nabla \mathbf{v})^T \right] \right] \cdot \mathbf{n} \, dA \qquad (2.35)$$

Using a pseudo-steady-state assumption, the translocation velocity at each position can be estimated. Moreover, the instant current variation can be traced for ionic fluxes in the axial direction. Zhang et al. (2012) numerically investigated the effects of the local permittivity environment (LPE) surrounding DNA molecules. In the model, an ion-penetrable layer on DNA molecules was considered, in which the relative permittivity ε_i was 1.76, being much lower than that of bulk water owing to EDL formation and hydrogen bonding interactions between the nucleotides. The estimated translocation velocity of DNA molecules became slower due to the presence of the LPE effect. The LPE effect yields more cations attached to the surface of the DNA molecules, leading to a lower effective charge density and reduced local electric field in the nanopore. Importantly, the scaled current blockage χ_0 estimated from the LPE-modified model is in good agreement with previous experimental measurements (Smeets et al., 2006), as shown in Figure 2.8a.

In addition to continuum simulations, molecular dynamic simulations incorporating both molecular properties and their steric effects have been employed to study nonlinear transport behavior of molecules through nanopores. In addition to the electric force, the van der Waals force is captured by the Lennard–Jones potentials for short-range interactions. The water molecule orientation, along with hydrogen bonding and its influence on electroosmotic transport properties, can be investigated as well. Farimani et al. (2014) revealed the nonlinear responses of DNA molecule translocation speed to the applied voltage through a monolayer molybdenum disulfide nanopore, that a threshold applied voltage is required to active a translocation event, as shown in Figure 2.9b. Compared to graphene, monolayer MoS_2 was found to be more suitable for ionic current measurements. Furthermore, in MoS_2, Mo exhibits hydrophilic properties, whereas S is hydrophobic. Therefore, when hydrophobic DNA molecules pass through a Mo-terminated MoS_2 nanopore, the DNA molecules

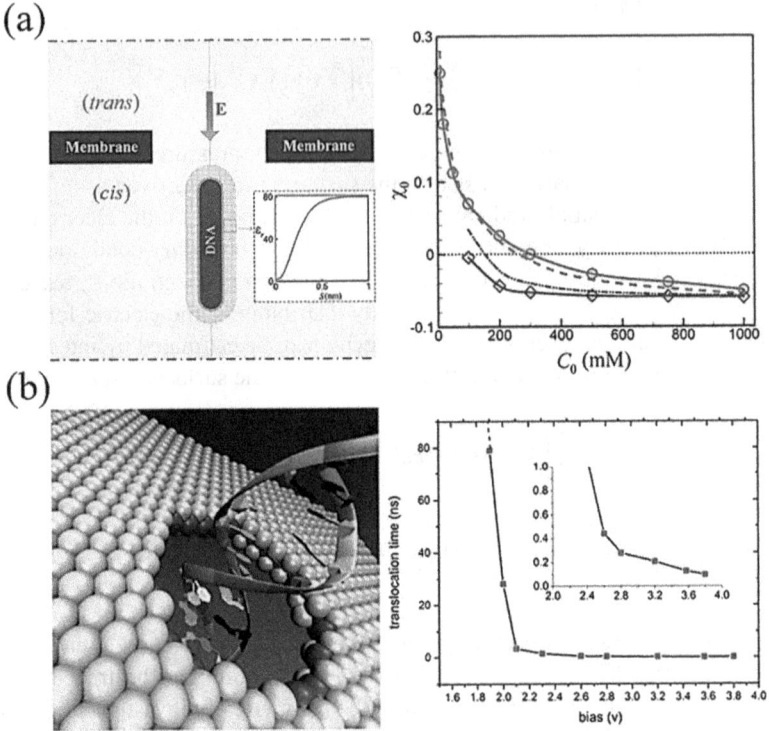

FIGURE 2.8 Computational approaches of dsDNAs through a solid-state nanopore based on the (a) continuum simulation. (Reprinted with permission from Zhang et al., 2012.) and (b) molecular dynamics simulation (Farimani et al., 2014). In (a): Variation of the normalized blockage current χ_0 as a function of the bulk salt concentration C_0. Symbols are experimental observations (Reprinted with permission from Smeets et al., 2006.) Solid curves are theoretical predictions considering the local permittivity environment (LPE) effect and dashed curves are theoretical results without the LPE effect. In (b): Variation of translocation time of dsDNA in a MoS$_2$ nanopores as a function of the applied voltage bias.

do not adhere to the MoS$_2$ membrane, which benefits DNA translocation. To improve the signal-to-noise ratio (SNR), it was found that two main factors enhance the SNR. One is to increase the applied bias as it can stretch DNA to avoid the formation of DNA knots. The other is to decrease the operation temperature to reduce the thermal vibration of DNA molecules for a smoother translocation.

2.3 NANOPORE ELECTROKINETICS IN THE PRESENCE OF A CONCENTRATION GRADIENT

2.3.1 SALT CONCENTRATION GRADIENTS IN NANOPORES

Nonuniform salt concentrations are ubiquitous in nanopores, leading to nonlinear electrokinetic behavior. Owing to the ion-selective characteristic of a charged

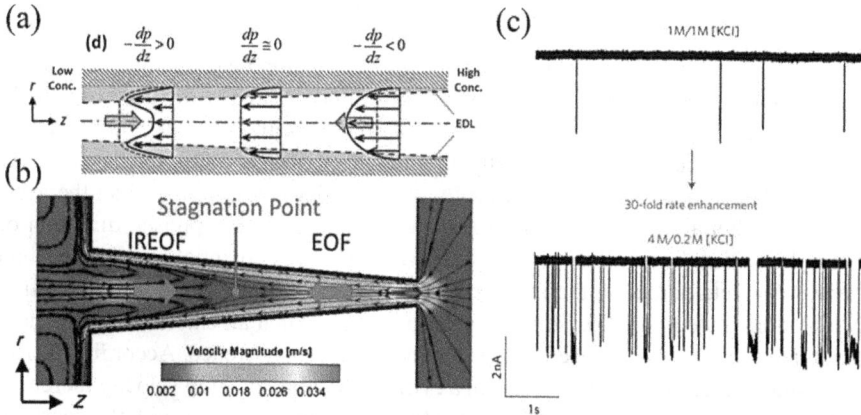

FIGURE 2.9 (a) Schematic of nonuniform electroosmotic flow (EOF) in a cylindrical nanopore. (Reprinted with permission from Hsu and Daiguji, 2016.) (b) Simulation results of induced reverse electroosmotic flow (IREOF) in a conical nanopore due to the presence of a concentration gradient, resulting in an induced pressure gradient. (Reprinted with permission from Hsu and Daiguji, 2016.) (c) Experimental results of simultaneous enhancement of DNA molecule capture rate and translocation duration in a salt-concentration-biased nanochannel. (Reprinted with permission from Wanunu et al., 2010.)

nanopore, a simultaneous local concentration gradient appears as an electric potential difference is imposed. For instance, when an electric field is applied across a negatively charged cation-selective membrane, both cations and anions accumulate on the cathode side of the membrane, whereas they are depleted at the other end of the nanopore junction. The former region is called the ion enrichment zone, and the latter the ion depletion zone. This ion concentration polarization (ICP) phenomenon leads to a salt concentration gradient within the nanopore parallel to the applied electric field (Li and Anand, 2016; Berzina and Anand, 2020). Not limited to ions, ICP can be used to concentrate charged molecules, resolving the detection issue for biosensing dealing with extremely low sample concentrations in the analyte (Gholinejad et al., 2021). In addition, the ion-depleted nature of ICP has been applied to the desalination of seawater as well as for other purification purposes (Kim et al., 2010b). Importantly, other than the nonuniform concentration induced by the ICP phenomenon, externally applied salt concentration gradients are widely used in nanopore systems to enhance sensing performance. (Wanunu et al., 2010). In the following parts of this section, we discuss the effect of nanopore electrokinetics on the presence of a concentration gradient. Understanding the physics behind the concentration gradient effects on electroosmosis in nanopores, such as nonuniform electroosmotic flow (EOF) in nanopores (Section 2.3.2), as well as EOF in ultrathin nanopores made on 2D materials (Section 2.3.3), will guide us to design proper nanopore systems for biosensing. It is equally important to understand the complex behavior of biomolecules under the concentration gradient, which will be introduced in Section 2.3.4.

2.3.2 Nonuniform Electroosmotic Flow (EOF)

The existence of a concentration difference across a nanopore results in nonuniform EOF in nanopores owing to variations in the zeta potential and diffuse layer thickness of EDLs. Herr et al. (2000) used two capillaries with different zeta potentials connected in series to experimentally investigate nonlinear EOF via fluorescence imaging. For the region of the capillary with a smaller zeta potential than the average zeta potential, a pressure-driven flow was induced in the opposite direction of the electric field. In contrast, in the region where the zeta potential was greater than the average zeta potential, the induced pressure-driven flow had the same direction as the applied electric field. Similar behavior occurs when an external electric field is applied to a nanopore biased by a salt concentration difference. According to Eq. 2.15, which describes electroosmosis in a cylindrical tube, the EOF velocity increases with the magnitude of the zeta potential, which generally increases with the decrease in salt concentration according to the Grahame equation, i.e., Eq. 2.12. Considering the case where the surface charge density does not change significantly with the variation in the salt concentration, the zeta potential and therefore electroosmotic velocity decrease in the vicinity of the nanopore inner surface along the concentration gradient. Consequently, a pressure gradient is developed in the axial direction, suppressing the flow at the low-concentration end while accelerating it near the high concentration end, thereby resulting in a nonuniform EOF. Similar to the analysis by Herr et al. for the variation in EOF in two channels with asymmetric zeta potentials, the local EOF velocity can be estimated by considering the superposition of the electric-force-driven flow and pressure-driven flow as follows:

$$U_z(r,z) = \frac{-\varepsilon_0 \varepsilon \zeta_z E_z}{\eta}\left[1 - \frac{I_0(\kappa r)}{I_0(\kappa R)}\right] - \left(\frac{\partial p}{\partial z}\right)_z \frac{R^2}{4\eta}\left(1 - \frac{r^2}{R^2}\right) \qquad (2.36)$$

where $U_z(r,z)$, ζ_z, E_z, and $\left(\dfrac{\partial p}{\partial z}\right)_z$ are the axial electroosmotic velocity, zeta potential, electric field magnitude, and induced pressure gradient at a specific location z, respectively. According to the mass conservation by which the integral of the velocity profile in the radial direction over each cross-sectional area must be constant, the following relation can be derived for induced pressure gradient along the z-axis:

$$\left(\frac{\partial p}{\partial z}\right)_z = \frac{8\varepsilon_0 \varepsilon E_z}{R}\frac{I_1(\kappa R)}{I_0(\kappa R)}\left(\bar{\zeta} - \zeta_z\right) \qquad (2.37)$$

where $\bar{\zeta}$ is the average zeta potential at which the induced pressure gradient vanishes, as ζ_z is larger than $\bar{\zeta}$. In large channels filled with higher concentration solutions at $\kappa R \gg 1$, Eq. 2.37 can be simplified as follows:

$$\left(\frac{dp}{dz}\right)_z = \frac{8\varepsilon_0 \varepsilon E_z}{R^2}\left(\bar{\zeta} - \zeta_z\right) \qquad (2.38)$$

Once the local zeta potential deviates from the average zeta potential, the pressure force balances the flow rate (Hsu and Daiguji, 2016). As illustrated in Figure 2.9a, when $\zeta_z > \bar{\zeta}$ is on the low-concentration side of the channel, the induced pressure gradient pushes against the EOF, resulting in a concave flow pattern. In contrast, a convex shape is formed on the other end at $\zeta_z < \bar{\zeta}$ when the concentration is high. Substituting Eq. 2.37 into Eq. 2.36, we obtain a general form of the flow profile in terms of the local zeta potential as follows:

$$U(r,z) = \frac{-\varepsilon_0 \varepsilon \zeta_z E_z}{\eta}\left[1 - \frac{I_0(\kappa r)}{I_0(\kappa R)}\right] - \frac{8\varepsilon_0 \varepsilon E_z}{R}\frac{I_1(\kappa R)}{I_0(\kappa R)}\left(\bar{\zeta} - \zeta_z\right)\frac{R^2}{4\eta}\left(1 - \frac{r^2}{R^2}\right) \quad (2.39)$$

For thin EDLs, we have

$$U(r,z) = \frac{-\varepsilon_0 \varepsilon E_z}{\eta}\left[\zeta_z + 2\left(\bar{\zeta} - \zeta_z\right)\left(1 - \frac{r^2}{R^2}\right)\right] \quad (2.40)$$

The nonuniformity of electroosmosis can be further amplified by employing a nonuniform pore shape to enlarge the induced pressure gradient. Hsu and Daiguji (2016) considered a concentration-biased conical pore and reported that the flow can reverse when the induced pressure gradient becomes dominant near the pore centerline, as shown in Figure 2.9b, which is identified as induced reverse electroosmotic flow (IREOF). IREOF results in local vortices facilitating molecule capturing for resistive pulse sensing, which can be utilized to enhance the throughput of biomolecule detection.

The application of a salt concentration to nanopore systems was found to be useful for resistive pulse sensing by Wanunu et al. (2010), who demonstrated that by imposing a concentration gradient between the cis and trans chambers of the nanofluidic system, both the DNA capture rate and molecule translocation rate are significantly enhanced. The aforementioned nonuniform velocity profile may help to explain the experimental observations of Wanunu et al. (Figure 2.9c) that the concave shape of the opposing EOF at the low concentration end in the cis chamber near the silicon nitride nanopore, where the molecules enter, enhances the capture rate. On contrary, the strengthened flow near the outlet at the high concentration end yields a longer translocation time as the electric field direction is against the flow direction. Other possible mechanisms for the enhancement of capture rate, including osmotic flow and local space charge, have also been proposed (Hatlo et al., 2011; He et al., 2013). He et al. (2013) suggested that the salt concentration gradient gives rise to K^+ ions accumulation in the nanopore close to the cis side. This elevates the local electric field and consequently may promote the capture rate. Hatlo et al. (2011) suggested that the improved capture rate is due to the contributions of electrophoresis and diffusioosmosis. Therefore, the capture rate increases with the increment of the concentration gradient.

2.3.3 Transport-Induced-Charge Electroosmosis

As previously discussed, the adoption of ultrathin membranes, including 2D materials, for nanopore systems has been extensively investigated owing to their capability to elevate the spatial resolution for biosensing (Schneider et al. 2010; Feng et al. 2015). For such high-aspect-ratio (radius to thickness) nanopores, the influence of the zeta potential distribution becomes less significant because of the narrow inner surface area in the nanopore. However, the decrease in the pore thickness amplifies the electric field. With the presence of a concentration gradient inside the nanopore, local ion separation can be initiated, generating free ionic charges. In the case where a concentration gradient exists, induced charges are generated by the nonuniform electric field due to the varying solution conductivity. Based on the Nernst–Planck equation, one can obtain the following equation at steady state when the conduction current is dominant at high concentrations, such that the ionic current flux $\mathbf{J}$ is approximated by the product of σ_b and $\mathbf{E}$:

$$\nabla \cdot \mathbf{J} \cong \nabla \cdot (\sigma_b \mathbf{E}) = 0 \tag{2.41}$$

By substituting Poisson's equation into Eq. 2.41, we obtain the following formula for the local space charge density:

$$\rho_e = \varepsilon_0 \varepsilon \mathbf{E} \cdot \left(\frac{\nabla \varepsilon}{\varepsilon} - \frac{\nabla \sigma_b}{\sigma_b} \right) \tag{2.42}$$

This extra space charge in addition to the charges from EDLs is known as salt-gradient-induced charge or, more generally, transport-induced charge (TIC) for non-equilibrium systems simultaneously involving an electric field and a conductivity or permittivity gradient (He et al., 2013; Hsu et al., 2018). Considering that the dielectric constant barely changes with salt concentration, the first term in Eq. 2.42 becomes negligible (Wang et al., 2020) for isothermal systems. Accordingly, we derived a simple expression for TIC density $\rho_{e,\text{TIC}}$ in a nanopore owing to the coexistence of an electric field and a salt concentration gradient.

$$\rho_{e,\text{TIC}} = -\frac{\varepsilon_0 \varepsilon E_{\text{pore}}}{L} \left(\frac{\Delta n_0}{n_0} \right) \tag{2.43}$$

where E_{pore} is the axial electric field strength in the nanopore.

The local space charges result in a body force exerting on the solution due to the external electric field, leading to an electroosmotic motion of the solution. To distinguish this phenomenon from the conventional electroosmosis caused by EDLs, it is referred to as TIC electroosmosis. The TIC velocity $v_{z,\text{TIC}}(r)$ can be estimated by the following equation (Hsu et al., 2021b):

$$v_{z,\text{TIC}}(r) = -\frac{\varepsilon_0 \varepsilon R^2 E_{\text{pore}}^2}{4\eta L} \left(\frac{\Delta n_{0,\text{pore}}}{n_0} \right) \left[1 - \left(\frac{r}{R} \right)^2 \right] \tag{2.44}$$

Intriguingly, the parabolic velocity profile of TIC electroosmosis is proportional to the square of E_{pore}. This leads to a flow in a consistent direction from the high-concentration end to the dilute end, regardless of the electric field direction. Figure 2.10a and b demonstrate the numerical simulation results of TIC distribution

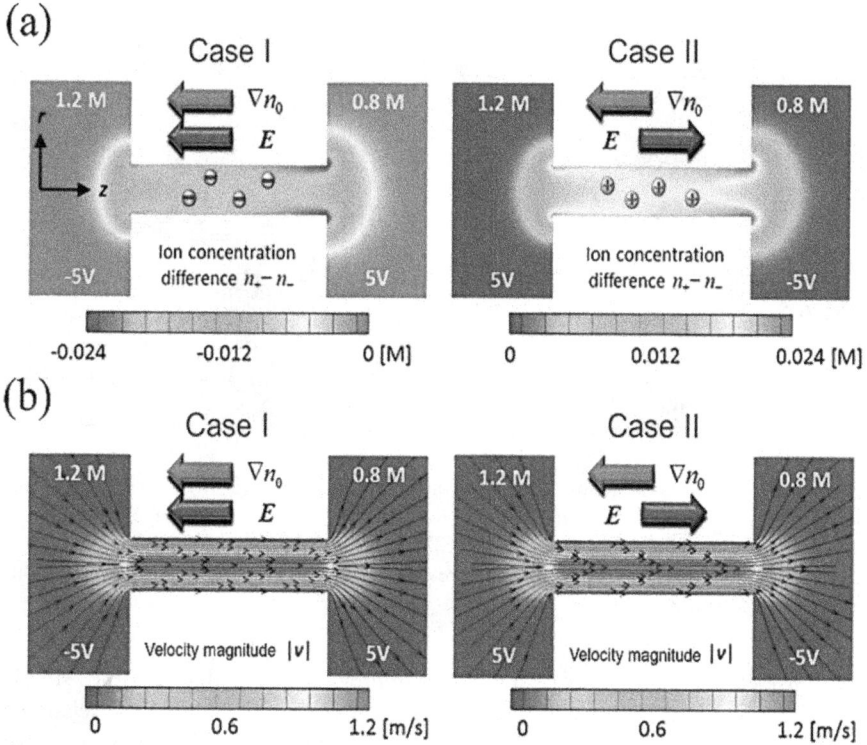

FIGURE 2.10 Transport-induced-charge phenomena in an ultrathin nanopore: contours of (a) charge and (b) electroosmotic velocity distributions. Case I: both the external electric field and concentration gradient point toward the same direction; Case II: the external electric field is in the opposite direction to the concentration gradient. (Reprinted with permission from Hsu et al., 2018.)

and TIC electroosmosis without the effect of EDLs, respectively. It is indicated that although the sign of TIC is inversed owing to the reverse electric field direction, the resultant TIC electroosmosis remains the same, consistent with Eq. 2.44.

Based on the contribution of electroosmosis from EDLs, we obtain a general expression of EOF profile in thin nanopores biased by a salt concentration difference.

$$U(r) = \frac{-\varepsilon_0\varepsilon\zeta E_{pore}}{\eta}\left[1 - \frac{I_0(\kappa r)}{I_0(\kappa R)}\right] - \frac{\varepsilon_0\varepsilon R^2 E_{pore}^2}{4\eta L}\left(\frac{\Delta n_{0,pore}}{n_0}\right)\left[1 - \left(\frac{r}{R}\right)^2\right] \quad (2.45)$$

When the charges in the EDL are opposite to the induced charges, the direction of the net EOF depends on the axial electric field strength in the nanopore, leading to the complex translocation behavior of molecules. Given that the former term in Eq. 2.45 is dominant at lower electric field strengths and the influence of the latter significantly becomes remarkable when increasing the electric field strength due to the second order relation. This nonlinearity may explain the observed threshold voltage

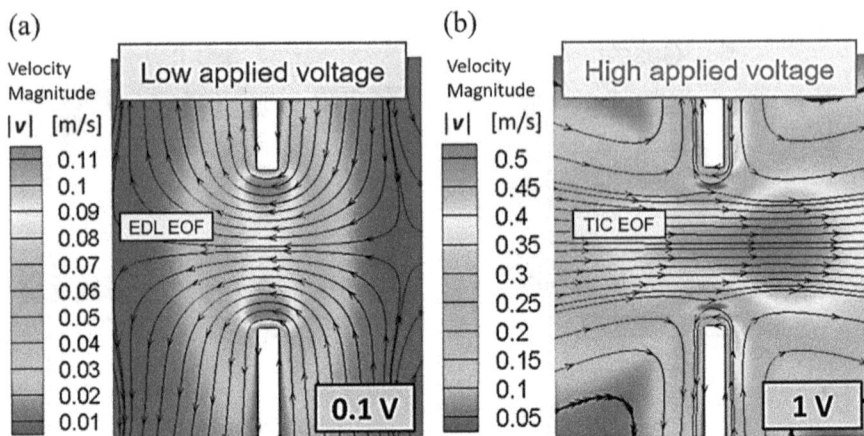

FIGURE 2.11 Simulation results for the electroosmotic flow (EOF) distribution around a monolayer molybdenum disulfide nanopore at the applied electric voltage: (a) 0.1 V and (b) 1 V. At low applied voltages in (a), the EDL EOF is dominant, while the flow direction reverses at high voltages in (b) due to the presence of TIC EOF. (Reprinted with permission from Hsu et al., 2018.)

for translocation events of DNA molecules in previous molecular dynamics simulation (Farimani et al., 2014; Zhang et al., 2012). As shown in Figure 2.11, regardless of the absence of an external solute concentration bias, the ICP phenomenon gives rise to a local concentration gradient in the nanopore, yielding TIC EOF, which is predominant at high voltages. The EOF at low applied voltages is subject to surface charge within the EDL, opposite to the molecule migration direction. On the other hand, the reversal of the electroosmosis direction could largely facilitate the translocation of molecules at larger applied potential bias (Hsu et al., 2018).

2.3.4 DIFFUSIOPHORESIS AND ELECTRODIFFUSIOPHORESIS

The presence of a concentration gradient generates a tangential pressure gradient and an electric field along charged molecule surfaces. The migration of molecules due to these effects is known as chemiphoresis. For open circuit systems, where the net current vanishes, the difference in ionic diffusivity between cations and anions yields an induced electric field in the solution. The motion of the molecule driven by this induced electric field in combination with chemiphoresis is called diffusiophoresis. Chemiphoresis occurs in a solution containing ions with an identical diffusivity, whereas diffusiophoresis includes the effect of the difference in ionic diffusivity between cations and anions. By neglecting the influence of the tangential electric field, Prieve et al. (1984) derived an analytical solution for diffusiophoretic behavior of particles. For thin double layers (i.e., $\kappa a \gg 1$), the flow on a spherical particle can be estimated by considering it as a flat surface in Cartesian coordinates, similar to the treatment of the Helmholtz–Smoluchowski model for electrophoresis. Within the EDL, the electric force acting on the charged solution in the direction vertical to the particle surface is balanced by the opposing pressure. When a subtle concentration variation is present in the x-direction parallel to the surface, the variation in pressure $p(x, y)$ as a function of the local electric potential is derived by solving the momentum balance equation in the wall normal direction y:

$$p(x,y) - p_\infty = 2k_BT \left\{ \cosh\left[\frac{e(\psi(x,y) - \psi_\infty(x))}{k_BT} \right] - 1 \right\} n_\infty(x) \tag{2.46}$$

where p_∞ is the bulk pressure at which the salt concentration is n_∞. Accordingly, the x-component of the Navier–Stokes equation can be expressed as follows:

$$\eta \frac{d^2 v_x}{dy^2} = e(n_+ - n_-) \frac{d\psi_\infty}{dx} + 2k_BT \left\{ \cosh\left[\frac{e(\psi - \psi_\infty)}{k_BT} \right] - 1 \right\} \frac{dn_\infty}{dx} \tag{2.47}$$

where v_x denotes the flow velocity in the x-direction and ψ_∞ is the electric potential far from the surface. In the above expression, the electric potential distribution can be expressed as:

$$\tanh \frac{e(\psi - \psi_\infty)}{4k_BT} = \tanh \frac{e\zeta}{4k_BT} e^{-\kappa y} \tag{2.48}$$

By substituting Eq. 2.48 into Eq. 2.47 and integrating twice with the boundary conditions $v_x = dv_x / dy = 0$ at $y \to \infty$, one can estimate the particle diffusiophoretic velocity U_{Diff} as follows:

$$U_{\text{Diff}} = v_x(0) = \frac{\varepsilon_0 \varepsilon}{\eta} \frac{k_BT}{e} \left[\beta\zeta - \frac{2k_BT}{e} \ln(1 - \gamma^2) \right] \frac{d \ln n_\infty}{dx} \tag{2.49}$$

Here $\beta = \dfrac{D_+ - D_-}{D_+ + D_-}$ and $\gamma = \tanh \dfrac{e\zeta}{4k_BT}$. The former term in the square brackets arises from the induced electric field $\mathbf{E}_{\text{induced}}$ due to the difference in ionic diffusivity when the ionic current vanishes, which can be expressed in terms of β:

$$\mathbf{E}_{\text{induced}} = -\nabla \psi_\infty = \frac{k_BT\beta}{e} \frac{d \ln n_\infty}{dx} \tag{2.50}$$

Prieve et al.'s model neglects the tangential electric field caused by the deformation of EDL owing to the nonuniform salt concentration. Using computational simulations, Hsu et al. (2009) demonstrated two types of EDL polarization effects in the presence of a salt concentration gradient owing to the imbalance of both counterions and co-ions. The former, along with the tangential pressure gradient, drives the molecule toward the high concentration end, whereas the latter forces it in the opposite direction, leading to a local maximum mobility when increasing the zeta potential.

The classical diffusiophoretic system investigated by Prieve et al. does not include an external circuit for current measurements. For biomolecule sensing using a salt concentration-biased nanopore, an electric potential difference across the nanopores

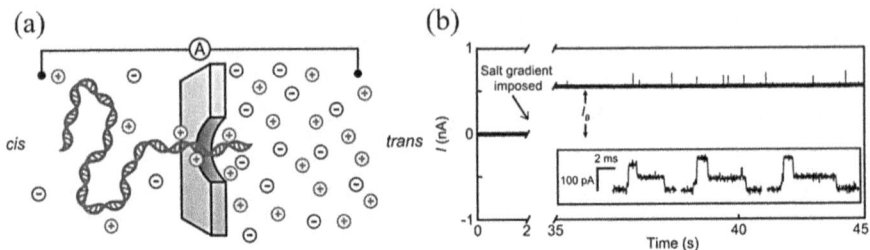

FIGURE 2.12 (a) Schematic of (electro-)diffusiophoretic motion of a DNA molecule through a nanopore. (b) Current variation of DNA (electro-)diffusiophoresis in a silicon nitride nanopore. (Reprinted with permission from McMullen et al., 2019.)

must be considered because of the asymmetric redox reactions at the electrodes immersed in electrolyte solutions of different concentrations (McMullen et al., 2019), as illustrated in Figure 2.12a, This includes both the Nernst potential V_{redox} and externally applied voltage V_{ext}. The system then shares the same circuit model as reverse electrodialysis systems that harvest electricity for salinity power (Kim et al., 2010a). Accordingly, the electric potential drop across nanopore V_{pore} can be expressed as follows:

$$V_{pore} = V_{diff} - IR_{pore} = V_{ext} - V_{redox} = V_{ext} - \frac{k_B T}{e} \ln \frac{\gamma_{n_H} n_H}{\gamma_{m_L} n_L} \qquad (2.51)$$

Here $n_H, n_L, V_{diff}, I, \gamma_{n_H}$, and γ_{m_L} represent the high salt concentration, low salt concentration, diffusion potential, ionic current, activity coefficient of the high concentration solution, and activity coefficient of the low concentration solution, respectively. In addition to diffusiophoresis, the term "electrodiffusiophoresis" has been used in the literature to distinguish it from the classical diffusiophoresis system without the presence of an ionic current. Yalcin et al. (2010) suggested that particle transport speed can be controlled by the magnitude and direction of the concentration gradient. As shown in Figure 2.12b, McMullen et al. (2019) experimentally demonstrated the DNA molecule detection using the (electro-)diffusiophoretic method in silicon nitride nanopores, allowing the slow translocation of molecules for improved temporal resolution.

2.4 CONCLUSIONS AND FUTURE OUTLOOK

Resistive pulse sensing using solid-state nanopores has emerged as a direct method for single-molecule detection that provides instantaneous molecule information. The mechanical strength, stable chemical properties, and loose constraints on pore geometry have made solid-state nanopores competitive over protein nanopores for practical applications in medical diagnosis. Single-molecule detection using solid-state nanopores not only pave the way for precision medicine but also offers a promising platform for the characterization of unknown virus molecules that may help to combat global pandemics in the future. Although the employment of 2D materials significantly enhances the spatial resolution, the absence of motor enzymes in solid-state nanopores results in poor temporal resolution of molecular detection due to the excessive molecule translocation speed. The precise control of molecular behavior remains a critical issue to be resolved for the development of high-resolution solid-state nanopore technology. It is, therefore, of crucial importance to understand nonlinear electrokinetics in ultrathin nanopores for the investigation of molecule transport phenomena. This chapter gives an overview of fundamental electrokinetic phenomena in nanopore systems.

Furthermore, it has been indicated that asymmetric solute concentrations allow for an effective method to manipulate the molecule translocation behavior. We have reviewed the pertinent mechanisms of electrokinetic phenomena attributed to the presence of a salt concentration gradient along the nanopore axis. The complex electrokinetic behavior of molecules offers various strategies to regulate molecule translocation, enabling us to explore new possibilities for nanopore sensing. In this regard, surface charge information of nanopores when exposed to an electrolyte solution is crucial, but relevant studies are still lacking, particularly for 2D materials. Fundamental investigation of the interfacial properties using electrokinetic methods or direct probing techniques for emerging materials will benefit the development of the next generation of resistive pulse sensing technology using solid-state nanopores.

ACKNOWLEDGEMENT

The authors would like to thank Haoyu Wang and Soumyadeep Paul at the University of Tokyo for providing Figures 2.2 and 2.3a.

REFERENCES

Alfaro, J. A., P. Bohländer, M. Dai, et al. 2021. The emerging landscape of single-molecule protein sequencing technologies. *Nature Methods* 18: 604–17.

Alizadeh, A., and M. Wang. 2019. Flexibility of inactive electrokinetic layer at charged solid-liquid interface in response to bulk ion concentration. *Journal of Colloid and Interface Science* 534: 195–204.

Alizadeh, A., and M. Wang. 2020. Temperature effects on electrical double layer at solid-aqueous solution interface. *Electrophoresis* 41: 1067–72.

Alizadeh, A., W.-L. Hsu, M. Wang, and H. Daiguji. 2021. Electroosmotic flow: From microfluidics to nanofluidics. *Electrophoresis* 42: 834–68.

Allen, F. I., N. R. Velez, R. C. Thayer, et al. 2019. Gallium, neon and helium focused ion beam milling of thin films demonstrated for polymeric materials: Study of implantation artifacts. *Nanoscale* 11: 1403–9.

Ambia-Garrido, J., A. Vainrub, and B. M. Pettitt. 2010. A model for structure and thermodynamics of ssDNA and dsDNA near a surface: A course grained approach. *Computer Physics Communications* 181: 2001–7.

Andrade, E. N. da C., and C. Dodd. 1939. Effect of an electric field on the viscosity of liquids. *Nature* 143: 26–7.

Andrade, E. N. da C., and C. Dodd. 1946. The effect of an electric field on the viscosity of liquids. *Proceedings of the Royal Society A* 187: 296–337.

Andrade, E. N. da C., and C. Dodd. 1951. The effect of an electric field on the viscosity of liquids. II. *Proceedings of the Royal Society A* 204: 449–64.

Bachmann, M. F., M. O. Mohsen, L. Zha, M. Vogel, and D. E. Speiser. 2021. SARS-CoV-2 structural features may explain limited neutralizing-antibody responses. *NPJ Vaccines* 6: 1–5.

Balčytis, A., C. Briosne-Fréjaville, A. Mau, X. Li, and S. Juodkazis 2017. Rescalable solid-state nanopores. *AIP Conference Proceedings* 1874: 1–5.

Bazant, M. Z., M. SabriKilic, B. D. Storey, and A. Ajdari. 2009. Towards an understanding of induced-charge electrokinetics at large applied voltages in concentrated solutions. *Advances in Colloid and Interface Science* 152: 48–88.

Behrens, S. H. and D. G. Grier. 2001. The charge of glass and silica surfaces. *The Journal of Chemical Physics* 115: 6716–21.

Bellini, T., R. Cerbino, and G. Zanchetta. 2012. DNA-based soft phases. In *Liquid Crystals: Materials Design and Self-Assembly*, ed. C. Tschierske, 225–79. Springer: Heidelberg.

Berzina, B., and R. K. Anand. 2020. Tutorial review: Enrichment and separation of neutral and charged species by ion concentration polarization focusing. *Analytica Chimica Acta* 1128: 149–73.

Bian, F., Y. C. Tian, R. Wang, H. X. Yang, H. Xu, S. Meng, and J. Zhao. 2011. Ultrasmall silver nanopores fabricated by femtosecond laser pulses. *Nano Letters* 11: 3251–7.

Booth, F. 2004. The dielectric constant of water and the saturation effect. *The Journal of Chemical Physics* 19: 391–4.

Borukhov, I., D. Andelman, and H. Orland. 1997. Steric effects in electrolytes: A modified Poisson-Boltzmann equation. *Physical Review Letters* 79: 435–8.

Branton, D., D. W. Deamer, A. Marziali, et al. 2008. The potential and challenges of nanopore sequencing. *Nature Biotechnology* 26: 1146–53.

Broers, A. N. 1988. Resolution limits for electron-beam lithography. *IBM Journal of Research and Development* 32: 502–13.

Bustin, S. A. 2002. Quantification of MRNA using real-time reverse transcription PCR (RT-PCR): Trends and problems. *Journal of Molecular Endocrinology* 29: 23–39.

Cai, Y., B. Zhang, L. Liang, et al. 2021. A solid-state nanopore-based single-molecule approach for label-free characterization of plant polysaccharides. *Plant Communications* 2: 1–9.

Chen, T. A., C. P. Chuu, C. C. Tseng, et al. 2020. Wafer-scale single-crystal hexagonal boron nitride monolayers on Cu (111). *Nature* 579: 219–23.

Chen, Y., J. Cai, Q. Xu, and Z. W. Chen. 2004. Atomic force bio-analytics of polymerization and aggregation of phycoerythrin-conjugated immunoglobulin G molecules. *Molecular Immunology* 41: 1247–52.

Clarke, J., H. C. Wu, L. Jayasinghe, et al. 2009. Continuous base identification for single-molecule nanopore DNA sequencing. *Nature Nanotechnology* 4: 265–70.

Coulter, W. H. 1953. Means for counting particles suspended in a fluid. *US Patent 2,656,508*, filed August 27, 1949, and issued October 20, 1953.

Danda, G., P. M. Das, Y.-C. Chou, et al. 2017. Monolayer WS_2 nanopores for DNA translocation with light-adjustable sizes. *ACS Nano* 11: 1937–45.

Darvish, A., J. S. Lee, B. Peng, J. Saharia, et al. 2019. Mechanical characterization of HIV-1 with a solid-state nanopore sensor. *Electrophoresis* 40: 776–83.

Debye, P. 1929. Polar molecules. *Zeitschrift Für Angewandte Chemie* 42: 995–995.

Derrington, I. M., T. Z. Butler, M. D. Collins, et al. 2010. Nanopore DNA sequencing with MspA. *Proceedings of the National Academy of Sciences of the United States of America* 107: 16060–5.

Emmrich, D., A. Beyer, A. Nadzeyka, et al. 2016. Nanopore fabrication and characterization by helium ion microscopy. *Applied Physics Letters* 108: 163103-1–4.

Farimani, A. B., K. Min, and N. R. Aluru. 2014. DNA base detection using a single-layer MoS_2. *ACS Nano* 8: 7914–22.

Feng, J., K. Liu, R. D. Bulushev, et al. 2015. Identification of single nucleotides in MoS_2 Nanopores. *Nature Nanotechnology* 10: 1070–6.

Gholinejad, M., A. J. Moghadam, D. T. Phan, A. K. Miri, and S. A. M. Shaegh. 2021. Design and application of ion concentration polarization for preconcentrating charged analytes. *Physics of Fluids* 33: 051301-1–20.

Graf, M., M. Lihter, M. Thakur, et al. 2019. Fabrication and practical applications of molybdenum disulfide nanopores. *Nature Protocols* 14: 1130–68.

Grahame, D. C. 1947. The electrical double layer and the theory of electrocapillarity. *Chemical Reviews* 41: 441–501.

Hall, A. R., A. Scott, D. Rotem, K. K. Mehta, H. Bayley, and C. Dekker. 2010. Hybrid pore formation by directed insertion of α-haemolysin into solid-state nanopores. *Nature Nanotechnology* 5: 874–7.

Hall, J. E. 1975. Access resistance of a small circular pore. *Journal of General Physiology* 66: 531–2.

Hatlo, M. M., D. Panja, and R. V. Roij. 2011. Translocation of DNA molecules through nanopores with salt gradients: the role of osmotic flow. *Physical Review Letters* 107: 068101-1–5.

He, Y., M. Tsutsui, R. H. Scheicher, C. Fan, M. Taniguchi, and T. Kawai. 2013. Mechanism of how salt-gradient-induced charges affect the translocation of DNA molecules through a nanopore. *Biophysical Journal* 105: 776–82.

Henry, D. C. 1931. The cataphoresis of suspended particles. Part I.—The equation of cataphoresis. *Proceedings of the Royal Society of London. Series A* 133: 106–29.

Herr, A. E., J. I. Molho, J. G. Santiago, M. G. Mungal, T. W. Kenny, and M. G. Garguilo. 2000. Electroosmotic capillary flow with non-uniform zeta potential. *Analytical Chemistry* 72: 1053–7.

Hlawacek, G. and A. Gölzhäuser. 2018. *Helium Ion Microscopy*. Springer Tracts in Modern Physics. Springer, Heidelberg.

Hölzer, M., and M. Marz. 2017. Software dedicated to virus sequence analysis "Bioinformatics Goes Viral". In *Advances in Virus Research*, ed. M. Beer, and D. Höper, 233–57. Elsevier, Amsterdam.

Hsu, C., C. Y. Lin, A. Alizadeh, H. Daiguji, and W.-L. Hsu. 2021a. Investigation of entrance effects on particle electrophoretic behavior near a nanopore for resistive pulse sensing. *Electrophoresis*, 42: 2206–14.

Hsu, J.-P., W.-L. Hsu, and Z.-S. Chen. 2009. Boundary effect on diffusiophoresis: Spherical particle in a spherical cavity. *Langmuir* 25: 1772–84.

Hsu, W.-L., and H. Daiguji. 2016. Manipulation of protein translocation through nanopores by flow field control and application to nanopore sensors. *Analytical Chemistry* 88: 9251–8.

Hsu, W.-L., H. Daiguji, D. E. Dunstan, M. R. Davidson, and D. J. E. Harvie. 2016. Electrokinetics of the silica and aqueous electrolyte solution interface: Viscoelectric effects. *Advances in Colloid and Interface Science*, 234: 108–31.

Hsu, W.-L., D. J. E. Harvie, M. R. Davidson, D. E. Dunstan, J. Hwang, and H. Daiguji. 2017. Viscoelectric effects in nanochannel electrokinetics. *Journal of Physical Chemistry C* 121: 20517–23.

Hsu, W.-L., J. Hwang, and H. Daiguji. 2018. Theory of transport-induced-charge electroosmotic pumping toward alternating current resistive pulse sensing. *ACS Sensors* 3: 2320–6.

Hsu, W.-L., S. Paul., Z. Gu, et al. 2019. Isothermal nanopore DNA sensing using diffusion current. *The 23rd International Conference on Miniaturized Systems for Chemistry and Life Sciences*. Basel, Switzerland.

Hsu, W.-L., Z. Wang, and H. Daiguji. 2021b. Transport-induced-charge electroosmosis. arXiv 2018.02564:1–5.

Hückel, E. 1924. Die kataphorese der kugel. *Physikalische Zeitschrift* 25: 204–10.

Hunter, R. J. 1966. The interpretation of electrokinetic potentials. *Journal of Colloid and Interface Science* 22: 231–9.

Hunter, R. J. 1988. *Zeta Potential in Colloidal Science*. Academic Press, Cambridge, MA.

Hunter, R. J., and J. V. Leyendekkers. 1978. Viscoelectric coefficient for water. *Journal of the Chemical Society, Faraday Transactions 1* 74: 450–5.

Iqbal, S. M., D. Akin and R. Bashir 2007. Solid-state nanopore channels with DNA selectivity. *Nature Nanotechnology* 2: 243–8.

Jin, D., Y. Hwang, L. Chai, N. Kampf, and J. Klein. 2022. Direct measurement of the viscoelectric effect in water. *Proceedings of the National Academy of Sciences* 119: 1–7.

Kaji, N., R. Ogawa, A. Oki, Y. Horiike, M. Tokeshi, and Y. Baba. 2006. Study of water properties in nanospace. *Analytical and Bioanalytical Chemistry* 386: 759–64.

Kanarik, K. J., S. Tan, and R. A. Gottscho. 2018. Atomic layer etching: Rethinking the art of etch. *The Journal of Physical Chemistry Letters* 9: 4814–21.

Kim, D. K., C. Duan, Y. F. Chen, and A. Majumdar. 2010a. Power generation from concentration gradient by reverse electrodialysis in ion-selective nanochannels. *Microfluidics and Nanofluidics* 9: 1215–24.

Kim, S. J., S. H. Ko, K. H. Kang, and J. Han. 2010b. Direct seawater desalination by ion concentration polarization. *Nature Nanotechnology* 5: 297–301.

Kirkwood, J. G. 1939. The dielectric polarization of polar liquids. *The Journal of Chemical Physics* 7: 911–9.

Kowalczyk, S. W., A. Y. Grosberg, Y. Rabin and C. Dekker. 2011. Modeling the conductance and DNA blockade of solid-state nanopores. *Nanotechnology* 22: 1–5.

Kwok, H., K. Briggs, and V. Tabard-Cossa. 2014. Nanopore fabrication by controlled dielectric breakdown. *PLoS ONE* 9: 1–6.

Lan, W. J., and H. S. White. 2012. Diffusional motion of a particle translocating through a nanopore. *ACS Nano* 6: 1757–65.

Lee, C., L. Joly, A. Siria, A. L. Biance, R. Fulcrand, and L. Bocquet. 2012. Large apparent electric size of solid-state nanopores due to spatially extended surface conduction. *Nano Letters* 12: 4037–44.

Li, J., D. Stein, C. McMullan, D. Branton, M. J. Aziz, and J. A. Golovchenko. 2001. Ion-beam sculpting at nanometre length scales. *Nature* 412: 166–9.

Li, M., and R. K. Anand. 2016. Recent advancements in ion concentration polarization. *Analyst* 141: 3496–510.

Lin, B. J. 2021. *Optical Lithography: Here is Why.* 2nd ed. SPIE: Bellingham.

Lyklema, J., and M. Minor. 1998. On surface conduction and its role in electrokinetics. *Colloids and Surfaces A* 140: 33–41.

Lyklema, J., and J. Th G. Overbeek. 1961. On the interpretation of electrokinetic potentials. *Journal of Colloid Science* 16: 501–12.

Masliyah, J. H., and S. Bhattacharjee. 2006. *Electrokinetic and Colloid Transport Phenomena.* Wiley, New York.

McMullen, A., G. Araujo, M. Winter, and D. Stein. 2019. Osmotically driven and detected DNA translocations. *Scientific Reports* 9: 1–10.

Meller, A., L. Nivon, and D. Branton. 2001. Voltage-driven DNA translocations through a nanopore. *Physical Review Letters* 86: 3435–8.

Melngailis, J. 1987. Focused ion beam technology and applications. *Journal of Vacuum Science & Technology B* 5: 469–95.

Merchant, C. A., K. Healy, M. Wanunu, et al. 2010. DNA translocation through graphene nanopores. *Nano Letters* 10: 2915–21.

Mojarad, N., M. Hojeij, L. Wang, J. Gobrecht, and Y. Ekinci. 2015. Single-digit-resolution nanopatterning with extreme ultraviolet light for the 2.5 nm technology node and beyond. *Nanoscale* 7: 4031–7.

Nojiri, K. 2014. *Dry Etching Technology for Semiconductors.* Springer, Cham.

O'Brien, R. W., and L. R. White. 1978. Electrophoretic mobility of a spherical colloidal particle. *Journal of the Chemical Society, Faraday Transactions 2* 74: 1607–26.

Onsager, L. 1936. Electric moments of molecules in liquids. *Journal of the American Chemical Society* 58: 1486–93.

Paul, S. 2022. E1-200nm-2_20190603.bmp. figshare. *Figure.* https://doi.org/10.6084/m9.figshare.19732114.v1

Prieve, D. C., J. L. Anderson, J. P. Ebel, and M. E. Lowell. 1984. Motion of a particle generated by chemical gradients. Part 2. electrolytes. *Journal of Fluid Mechanics* 148: 247–69.

Qiao, R., and N. R. Aluru. 2005. Scaling of electrokinetic transport in nanometer channels. *Langmuir* 21: 8972–7.

Rice, C. L., and R. Whitehead. 1965. Electrokinetic flow in a narrow cylindrical capillary. *Journal of Physical Chemistry* 69: 4017–24.

Rumble, J. 2021. *Handbook of Chemistry and Physics.* 102nd ed. CRC Press, Boca Raton, FL.

Schey, H. M. 2005. *Div, Grad, Curl, and All that An Informal Text on Vector Calculus.* 4th ed. W.W. Norton & Company, New York.

Schneider, G. F., and C. Dekker. 2012. DNA sequencing with nanopores. *Nature Biotechnology* 30: 326–8.

Schneider, G. F., S. W. Kowalczyk, V. E. Calado, et al. 2010. DNA translocation through graphene nanopores. *Nano Letters* 10: 3163–7.

Schoch, R. B., J. Han, and P. Renaud. 2008. Transport phenomena in nanofluidics. *Reviews of Modern Physics* 80: 839–83.

Smeets, R. M. M., U. F. Keyser, D. Krapf, M. Y. Wu, N. H. Dekker, and C. Dekker. 2006. Salt dependence of ion transport and dna translocation through solid-state nanopores. *Nano Letters* 6: 89–95.

Splendiani, A., L. Sun, Y. Zhang, et al. 2010. Emerging photoluminescence in monolayer MoS_2. *Nano Letters* 10: 1271–5.

Stern, O. 1924. Zur theorie der elektrolytischen doppelschicht. *Zeitschrift für Elektrochemie* 30: 508–16.

Stein, D., M. Kruithof, and C. Dekker. 2004. Surface-charge-governed ion transport in nanofluidic channels. *Physical Review Letters* 93: 035901-1–4.

Storm, A. J., J. H. Chen, X. S. Ling, H. W. Zandbergen, and C. Dekker. 2003. Fabrication of solid-state nanopores with single-nanometre precision. *Nature Materials* 2: 537–40.

Talaga, D. S., and J. Li. 2009. Single-molecule protein unfolding in solid state nanopores. *Journal of the American Chemical Society* 131: 9287–97.

Taniguchi, M., S. Minami, C. Ono, et al. 2021. Combining machine learning and nanopore construction creates an artificial intelligence nanopore for coronavirus detection. *Nature Communications* 12: 3726-1–8.

van der Heyden, F. H. J., D. Stein, and C. Dekker. 2005. Streaming currents in a single nanofluidic channel. *Physical Review Letters* 95: 116104-1–4.

van der Hout, M., I. D. Vilfan, S. Hage, and N. H. Dekker. 2010. Direct force measurements on double-stranded RNA in solid-state nanopores. *Nano Letters* 10: 701–7.

Verschueren, D. V., W. Yang, and C. Dekker. 2018. Lithography-based fabrication of nanopore arrays in freestanding SiN and graphene membranes. *Nanotechnology* 29: 1–11.

Wagner, C., and N. Harned. 2010. EUV lithography: Lithography gets extreme. *Nature Photonics* 4: 24–6.

Wang, Z., W.-L. Hsu, S. Tsuchiya, S. Paul, A. Alizadeh, and H. Daiguji. 2020. Joule heating effects on transport-induced-charge phenomena in ultrathin nanopores. *Micromachines* 11: 1041-1–12.

Wanunu, M., W. Morrison, Y. Rabin, A. Y. Grosberg, and A. Meller. 2010. Electrostatic focusing of unlabelled DNA into nanoscale pores using a salt gradient. *Nature Nanotechnology* 5: 160–5.

Webster, T. J., A. A. Patel, M. N. Rahaman, and B. S. Bal. 2012. Anti-infective and osteointegration properties of silicon nitride, poly(ether ether ketone), and titanium implants. *Acta Biomaterialia* 8: 4447–54.

Williams, D. B., and C. B. Carter. 2009. *Transmission Electron Microscopy: A Textbook for Materials Science.* Springer, New York.

Yalcin, S. E., A. Sharma, S. Qian, S. W. Joo, and O. Baysal. 2010. Manipulating particles in microfluidicsby floating electrodes. *Electrophoresis* 31: 3711–8.

Yang, J., D. C. Ferranti, L. A. Stern, et al. 2011. Rapid and precise scanning helium ion microscope milling of solid-state nanopores for biomolecule detection. *Nanotechnology* 22: 1–6.

Yang, L., and T. Yamamoto. 2016. Quantification of virus particles using nanopore-based resistive-pulse sensing techniques. *Frontiers in Microbiology* 7: 1–7.

Yates, D. E., S. Levine., and T. W. Healy. 1974. Site-binding model of the electrical double layer at the oxide/water interface. *Journal of the Chemical Society, Faraday Transactions 1* 70: 1807–18.

Yuan, Y., G. Li, H. Zaribafzadeh, et al. 2018. Sub-20 nm nanopores sculptured by a single nanosecond laser pulse. http://arxiv.org/abs/1806.08172.

Zhang, M., L. H. Yeh, S. Qian, J. P. Hsu, and S. W. Joo. 2012. DNA electrokinetic translocation through a nanopore: Local permittivity environment effect. *The Journal of Physical Chemistry C* 116: 4793–801.

3 Confocal Microscope-Based Detection

Xue Wang, Xinchao Lu, and Chengjun Huang
Institute of Microelectronics of the
Chinese Academy of Sciences

CONTENTS

DOI: 10.1201/9781003409472-3

3.1   INTRODUCTION

In this chapter, we present a comprehensive depiction of scanning confocal microscope (SCM) and applications based on its detection. In Figure 3.1, the subordination
relation of various microscopes based on SCM is shown. Two main types including laser scanning confocal microscope (LSCM) and fluorescence resonance energy
transfer (FRET) are introduced. As for LSCM, fluorescent laser scanning confocal
microscope (FLSCM) and label-free laser scanning confocal microscope (LLSCM),
as well as their application on single-molecule detection are illustrated. Furthermore,
three subtypes of FLSCM, i.e., two-photon and multi-photon FLSCM, fiber-based
and miniaturized FLSCM, as well as super-resolution FLSCM are described. Two
subtypes of LLSCM including laser scanning coherent Raman scattering microscope
(LSCRSM), which consists of coherent anti-Stokes Raman scattering microscope
(CARSM) and stimulated Raman scattering microscope (SRSM), and second harmonic generation microscope (SHGM) are indicated. In addition, we elaborate the
technique of FRET, including single-molecule fluorescence resonance energy transfer (smFRET), alternating laser excitation (ALEX) smFRET and high-order FRET.

3.2   LASER SCANNING CONFOCAL MICROSCOPE (LSCM)

With the higher requirements for imaging quality in biology and medical diagnostics,
the traditional optical microscope severely limits the imaging contrast and spatial
resolution. In 1957, the first scanning confocal microscope (SCM) was proposed by
M. Minsky for improving the imaging blur and contrast through point illuminate and

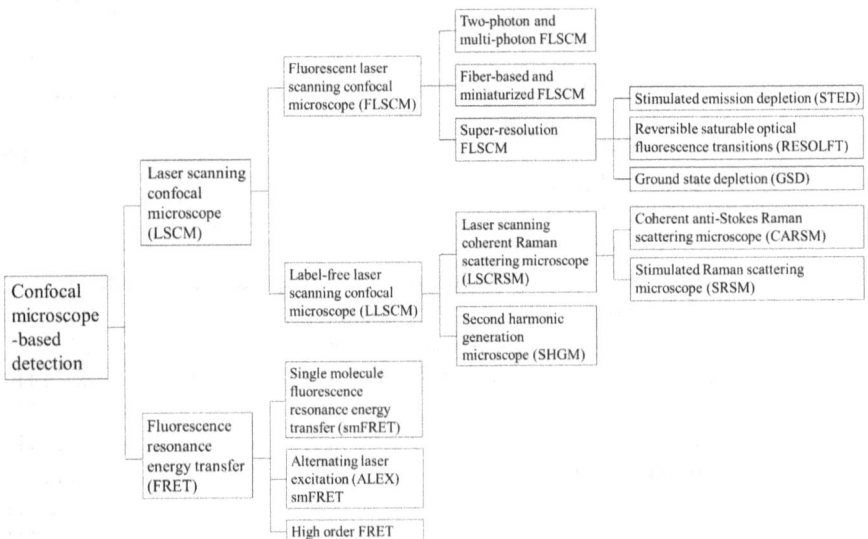

FIGURE 3.1   The subordination of various microscopes based on scanning confocal
microscope.

point detect (Minsky 1957, 1988). Since the evolvement of laser technology (Maiman 1967), laser scanning confocal microscope (LSCM) got rapid development, and the imaging techniques derived from LSCM have been a routine method for analyzing biological specimens (Paddock 1999, 2000, Hell 2007, Bayguinov et al. 2018). With the ability to approach diffraction-unlimited imaging, achieving higher and higher resolution is one of the hotspots in LSCM field (Peron et al. 2015, Liu et al. 2019b, Bayguinov et al. 2018, Barkauskas et al. 2020). LSCM has experienced rapid development in 3D imaging. Wilson et al. introduced the optical sectioning with depth selectivity (Wilson and Sheppard 1984), which illustrates the 3D imaging of LSCM.

The essential principle of LSCM was first proposed in 1955 for imaging neural networks (Minsky 1988). The basic schematic of the LSCM system is shown in Figure 3.2 (Wang et al. 2015). A source pinhole near the zirconium arc source produces a point illumination. Reflected by a dichroic mirror, the incident light is focused on the specimen through an objective lens. The reflective light from the specimen passes through the objective lens and the dichroic mirror, then is detected by a photodetector, with detector pinhole near the detector for filtering the stray light from the non-focus plane. Because of point illumination and point detection, LSCM possesses higher resolution imaging than routine optical microscope.

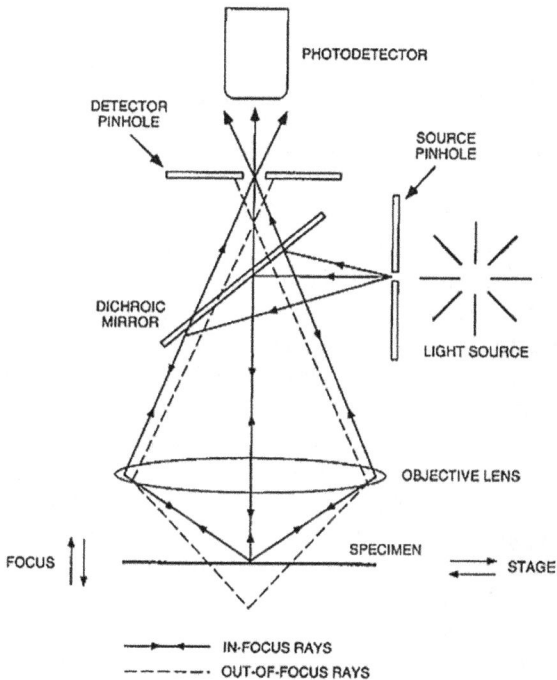

FIGURE 3.2 The schematic of typical LSCM system. With a source pinhole placed near the light source and detector pinhole near the photodetector for point illumination and point detection, respectively. (Reprinted with permission from Wang et al. 2015.)

After the 1980s, the LSCM got rapid development that relies on the scanning, laser, fluorescent staining and photoelectric detection technologies. The LSCM uses beam scanning for rapid imaging whose scanning speed can meet video rate (~ms) (Paddock 2000, Chun et al. 2009) instead of using the specimen scanning, which is slow and damages the biological specimen by apace mechanical movement. By using two high-speed galvanometers, the illumination focused point scans the specimen in the XY optical section with scanning frequency of galvanometers achieving 12 kHz. The point illumination and point detection of LSCM make it suitable for 3D reconstruction with a series of optical sections at different depth (Allgeier et al. 2018, Liu et al. 2014). Through changing the focal plane of LSCM, optical sections along different z positions are collected, and 3D image is reconstructed by processing a series of images along the z direction. When Z-series images are collected over time with the combination of multicolor images, a 4D or 5D reconstruction is possible by combining 3D stacks with time or wavelength (Bayguinov et al. 2018). The photodetector used in the early stage of LSCM is photomultiplier tube (PMT), which produces current signal with the photons arriving through the detector pinhole. Data collected by PMT are used to reconstruct the images of specimen. Benefitting from the development of CCD and CMOS technologies, modern LSCM manifests a fast and simple imaging process, as well as better sensitivity and higher contrast. Combined with high-speed spinning Nipkow disk or micro-lens array, multi-points detection has been achieved that accelerates the imaging speed reaching up to hundreds of frames per second.

3.2.1 FLUORESCENT LASER SCANNING CONFOCAL MICROSCOPE

In order to improve the imaging quality of conventional wide filed fluorescence microscope (Born et al. 1959), fluorescent laser scanning confocal microscope (FLSCM), which combines the LSCM with fluorescent labels, has been one of the most powerful and versatile methods for studying cells, tissues and organisms (Hell 2007, Bayguinov et al. 2018). With the development of various fluorescent labels, e. g. fluorescent proteins, organic dyes and nano-luminescent particles, FLSCM has achieved high sensitivity, high-resolution, molecular specificity, prominent contrast, multi-label ability, and 3D imaging. Besides, FLSCM is also a vital tool for super-resolution biological imaging that breaks through the optical diffraction limitation (Hell 2007). Various super-resolution imaging methods based on FLSCM have been invented, such as stimulated emission depletion (STED) (Hell and Wichmann 1994, Vicidomini et al. 2018), reversible saturable optical fluorescence transitions (RESOLFT) and ground state depletion (GSD) (Fölling et al. 2008, Hell and Kroug 1995), which is introduced in Section 3.2.2.3. Solving the bottle-lock of photobleaching and small imaging depth has prompted researchers to develop more stable and high-efficient fluorescent labels and novel fluorescence excitation mode. Combining with other imaging technologies for comprehensively investigating specimen (Alhede et al. 2012, Waharte et al. 2010) has been development trends of FLSCM. A combination of fluorescence correlation spectroscopy (FCS) and STED has been used for investigating molecular diffusion in sub-diffraction limitation volume (Kastrup et al. 2005).

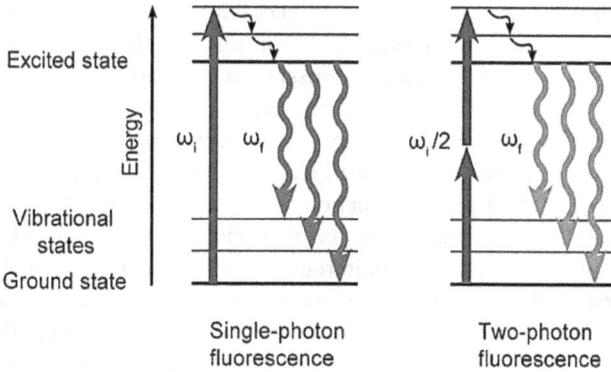

FIGURE 3.3 The process of photon absorption and emission. The absorption of one or more photons promote the molecule from ground state to excited state. Fluorescence is emitted when the molecule returns to the ground state (Ebersbach et al. 2018).

3.2.1.1 Two-Photon and Multi-Photon FLSCM

In 1931, two-photon absorption phenomenon was first predicted by M. G. Mayer (Göppert-Mayer 1931, Pawlicki et al. 2009). In 1986, the presence of mode-locked laser made two-photon excitation fluorescence possible (Valdmanis and Mourou 1986). The first two-photon FLSCM was reported by W. Denk et al. (1990). The process of photon absorption and emission is shown in Figure 3.3 (Ebersbach et al. 2018). For one-photon excited fluorescence, the fluorescence absorbs one photon and excites one photon. For multiphoton excited fluorescence, two or more photons are absorbed simultaneously to accumulate their energies for achieving transition energy of the fluorophore.

As single-photon FLSCM needs one photon with short wavelength to excite the fluorescence, photodamage and photobleaching caused by UV laser is inevitable. Two-photon FLSCM is a non-linear optical process with two long-wavelength photons exciting fluorescence, which avoids the photodamage and photobleaching induced by short wavelength photons. By using high energy mode-locked pulsed laser with red or infrared light with pulse width less than 100 fs at repetition rates 100 MHz, the high peak and low average energy brings low damage to cells. Also, with focusing the pulsed laser by high NA objective, the two-photon FLSCM excites fluorescence at the focal point, and detector pinhole is unnecessary. Moreover, compared with UV laser used in one-photon FLSCM, long wavelength excitation has less scattering and deeper penetration depth (~1 mm), which makes two-photon FLSCM more suitable for 3D and living cells imaging.

In addition, multi-photon FLSCM not only inherited the low damage, low photobleaching, and deep penetration depth of two-photon FLSCM but also makes multiwavelength fluorescence emission simultaneously owing to different fluorophores absorbing different photons (Abdeladim et al. 2019, Lanin et al. 2020). For example, two photons of red light are absorbed by rhodamine for 550 nm wavelength emission, and three photons of red light are absorbed by DAPI (4′, 6-diamidino-2-phenylindole) with a maximum emission wavelength of 461 nm. By filtering light to corresponding channels, color fluorescent imaging is obtained. Four-color imaging of rat basophilic leukemia cells was achieved by using a 700 nm wavelength laser, with exciting four different fluorochromes and collecting emissions by four different channels (Xu et al. 1996). To prevent laser damage to living tissues, an adaptive femtosecond source that illuminates only the region of interest has been used for two or three-photon FLSCM imaging to improve high-speed longitudinal neuroimaging of awake mice, which leads to a 30-fold power reduction (Li et al. 2020). Recently, multi-photon FLSCM with multi-color imaging has been developed for 3D imaging with a wide field of view. Abdeladim et al. (2019) introduced chromatic multi-photon microscope and demonstrated 3D multi-color imaging as well as brain-wide 2D multi-channel imaging. An in vivo imaging of the mouse brain labeled by quantum dots has been reported in both structural and hemodynamic three-photon microscope (Liu et al. 2019a).

3.2.1.2   Fiber-Based and Miniaturized FLSCM

The conventional FLSCM needs a bulky stage to hold the specimen, which greatly limits the flexibility of the imaging system and the applications. Through employing the fiber-optic imaging bundle, the confocal microscope can be extended to image samples that are not easily accessible (Gu et al. 2014), e.g., real-time imaging to cells within hollow or solid tissues and freely moving animals. Fiber-based FLSCM enables miniaturization, portable endoscopic microscope and long-term observation of cells or tissues in living animals (Miyamoto and Murayama 2016, Carlson et al. 2005, Yang and Yuste 2017).

By using fibers, fiber-based FLSCM allows separation between emitter and detector, providing a natural pinhole through small aperture diameter and flexible light path (Gmitro and Aziz 1993, Sung et al. 2002, Flusberg et al. 2005). As a result, fibers are the key component in fiber-based FLSCM. Initially, the single mode fiber (SMF) is used in fiber-based FLSCM, known as step-index fibers that consist of outer cladding and inner core (Peters 2011, Nagel 1984). The SMF delivers the incident light to specimen and collects the fluorescent photons, while the fiber pigtail acts as a pinhole to obtain optical sectioning. Compared with SMF, multimode fiber (MMF) has larger core diameter and greater numerical aperture (NA), which makes it more suitable for collecting specimen fluorescence. Different from step-index fiber, the refractive index of gradient refractive index fiber (GRIF) decreases quadratically from the center of core (Yariv 1976). The specific length of GRIF can be used as a lens to focus light on specimen or collect fluorescence from the focal plane.

As for two-photon FLSCM, self-phase modulation (SPM) effect existed in regular fibers distorts both pulse shape and spectrum because of the nonlinear effect from high-power short-pulse laser used for two-photon excitation. Photonic bandgap fiber (PBF), which consists of periodic arrays of air holes embedded in silica glass (Tai et al. 2004), can eliminate SPM (Flusberg et al. 2005). PBF usually transmits a pulse with group velocity dispersion vanishing during propagating in fibers. Therefore, PBF is a good choice for pulse delivery used for two-photon and multi-photon FLSCM. However, for light wavelength being tens of nanometer away from the center wavelength, the dispersion of PBF is more severe than SMF. The large mode area (LMA) fiber delivery pulses with reduced intensity and endlessly single-mode guidance

(Jing et al. 2005). In LMA fiber, the SPM is eliminated with dispersion compensation methods, and all light whose wavelength is within the transparency window propagates through certain spatial mode. Double-clad LMA fiber is an excellent alternative for two-photon FLSCM. It achieves excitation pulse transmission in inner core and fluorescence acquisition in outer cladding, simultaneously. The highly nonlinear photonic crystal fiber has been also developed, as the ultrashort excitation pulses are broadened by using its small core (Flusberg et al. 2005).

The scanning mechanism of fiber-based FLSCM includes proximal and distal scanning, depending on the usage of fiber bundles or single fiber. By using up to 100,000 individual fibers, fiber bundles acquire the pixilated intensity imaging of FLSCM and two-photon FLSCM. By using proximal scanning with fiber bundles, the separation of scanner and specimen is achieved. Proximal scanning mechanism includes point-scanning, line-scanning and micromirror array scanning. However, fiber-based FLSCM using fiber bundles indicates low lateral resolution and cross talk in adjacent fibers. The image quality can be improved by using non-ordered bundles, calibrating the intrinsic autofluorescence, precise position and optical throughput for each fiber. For the fiber-based FLSCM using single fiber, the main scanning mechanism consists of distal 2D mirror scanning, distal fiber tip scanning and distal fiber-objective scanning (Flusberg et al. 2005).

Due to the FLSCM miniaturization, scanner based on optical microelectromechanical system (MEMS) has attracted plenty of attentions for miniaturization of scanners, actuators and focus components (Zong et al. 2017, Zhen and Wibool 2015, 2019, Liu et al. 2014, 2019b, Solgaard et al. 2014, Duan et al. 2017). MEMS component is usually mounted at the distal end of fiber for better focal volume control. MEMS actuation mechanisms include bulky piezoelectric tubing, electrostatic comb drive or parallel plates, electrothermal, electromagnetic, and thin-film piezoelectric materials (Solgaard et al. 2014). According to different actuation mechanisms, single-axis with pre-objective scanning and dual-axis with post-objective scanning MEMS confocal micro-endoscope have been developed in recent years (Piyawattanametha and Wang 2010). The emergence of multifunctional active optical devices supports further miniaturization. A MEMS-in-the-lens architecture has been reported to feature an integrated MEMS device for performing biaxial scanning, axial focus adjustment, and spherical aberration control (Liu et al. 2019b). As the progress of advanced fiber and MEMS micromachining technology, miniaturized and flexible FLSCM has been greatly developed and widely applied in microendoscopy imaging for human clinical applications and tissue imaging in freely behaving animals (Zhen and Wibool 2019, Doronina-Amitonova et al. 2015). The MEMS-based two-photon FLSCM that weighed only 2.15 g has been reported for long-term real-time imaging of neuronal dendrites and spines in freely moving animals (Zong et al. 2017).

3.2.1.3 Super-Resolution FLSCM

As the spatial-resolution of an optical microscope is limited by diffraction barrier, it indicates that sample details smaller than diffraction limitation cannot be observed in conventional optical microscope. Super-resolution imaging methods breaking diffraction limitation are necessary for observing and investigating biological structures and interactions in nano scale. In FLSCM, as the spot size of incident focal light cannot be smaller than diffraction limitation, the key point of super resolution lies in controlling the size of fluorescent probe, e. g. point spread function (PSF), smaller than diffraction limitation. The methods for achieving super resolution through controlling the size of fluorescent probe include stimulated emission (Neupane et al. 2014, Blom and Widengren 2017), photoactivation (Nickerson et al. 2014), cis-trans isomerization (Pletnev et al. 2008), triplet pumping and saturated excitation (Fujita et al. 2007, Lehnhardt et al. 2011). The super-resolution technique-based FLSCM is stimulated emission depletion (STED), reversible saturable optical fluorescence transitions (RESOLFT), and ground state depletion (GSD).

STED is the first super-resolution technique used for biological imaging. In 1994, S. W. Hell and J. Wichmann developed STED, then Hell and Thomas Klar experimentally demonstrated it in 1999 (Hell and Wichmann 1994, Klar and Hell 1999). As shown in Figure 3.4, two beams, e. g. excitation and depletion beams, are used to control the saturated depletion. The excitation laser closely followed by the depletion laser with coaxial excitation and generates saturated depletion. The depletion laser is modulated by phase plate to generate doughnut-shaped intensity distribution, i.e., STED pattern. Hence, the STED beam has a zero-intensity node at the center of the beam, which guarantees normal fluorescence emission near the beam center, and the

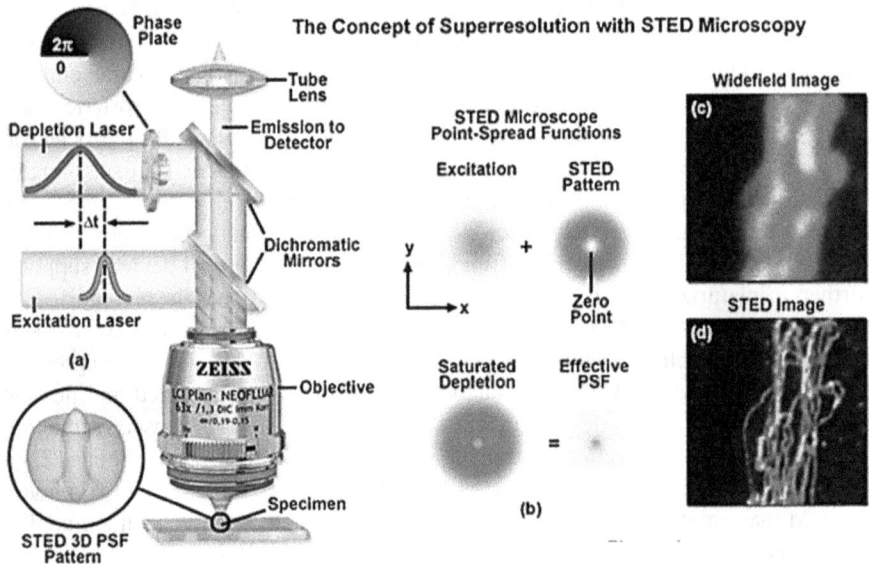

FIGURE 3.4 A typical STED optical configuration. (a) The configuration composed by the phase plate, excitation and depletion lasers, dichromatic mirrors, tube lens, and objective. A three-dimensional drawing shows the STED 3D PSF pattern. (b) The STED pattern is produced by the excitation and depletion lasers, and effective PSF with area smaller than diffraction limitation is obtained through saturated depletion. (c) Images of microtubules stained with Alexa Fluor 594 demonstrate the improved spatial resolution. (http://zeiss-campus. magnet.fsu.edu/print/superresolution/introduction-print.html).

effective area size is determined by the intensity of depletion laser. The spot size of normal fluorescence emission can be reduced with increasing intensity of depletion laser, which achieves an area smaller than diffraction limitation. The spontaneous fluorescence emission near zero-intensity node is detected after filtering out non-fluorescent signal.

A spatial resolution of 20 nm has been approached in nuclear pore complexes imaging by using fiber-based STED microscope (Gottfert et al. 2013). An ultimate spatial resolution of 6 nm has also been demonstrated for revealing colour centres in diamonds with STED microscope (Rittweger et al. 2009). 3D super resolution imaging based on STED has been demonstrated through the combination of lateral and axial incoherent STED beams (Shcherbakova and Verkhusha 2014). Also, 3D super resolution imaging is achieved by using STED microscope based on 4 pi configuration (Hell 2003) and integration of STED with selective plane illumination microscope (SPIM) (Friedrich et al. 2011).

By using high-power pulsed laser for saturated depletion to fluorescence, photodamage and photobleaching have been the bottle neck of the STED microscope. Low-power STED microscopes have been implemented by using time-gating detection (Vicidomini et al. 2011). A continuous laser source with moderated light intensity is used as depletion laser and makes lower photodamage and photobleaching possible (Vicidomini et al. 2013). It's worth noting that the continuous stimulating photons induce unnecessary fluorescence transitions and reduce the signal-noise rate of imaging. The lateral spatial resolution of 20 nm has been approached by using perovskite CsPbBr3 nanoparticles with high photoluminescence and excellent photobleaching resistance (Ye et al. 2018). Two-photon STED microscope offers both two-photon excitation and super-resolution STED, which achieves deep imaging to living cells and tissues (Polzer et al. 2019, Moneron and Hell 2009, Bethge et al. 2013). Novel two-photon STED microscope manifests the spatial resolution of sub-100 nm through imaging to presynaptic protein clusters in fixed primary cultured neurons without severe photobleaching (Ishii et al. 2019).

Similar to STED, reversible saturable optical fluorescence transition (RESOLFT) is an optical microscopy technique that breaks diffraction limitation through reversibly switching fluorescence molecular as on and off states. The fluorescence molecules have at least two distinguishable states, one state is bright with generating fluorescence signal, and the other state is dark that gives no signal. The transition between bright and dark states can be induced by light. The specimen is illuminated inhomogeneously to excite fluorescence signal to bright state with intensity decreasing to zero at one position. Another laser with threshold intensity close to zero is used for switching the molecules to dark state, so only molecules very close to the zero position can be in the bright state, which generates spot size smaller than diffraction limitation. An optical RESOLFT microscope that records raw images from living cells and tissues with low light intensity had been demonstrated in 2011 (Grotjohann et al. 2011).

Ground state depletion (GSD) microscope is a super-resolution strategy based on RESOLFT. In the bright state, the molecule can be driven between ground and the first excited state, and fluorescence can be sent out. In the dark state, the ground state of the molecule is depopulated, a transition to a long-lived excited state takes place, and no fluorescence is emitted. No cycling between ground and excited state emerges, fluorescence is hence turned off. However, the long-lived state induces slow switching speed, which is not suitable for real-time imaging to living cells or tissues. The multi-beam depletion and new fluorescence probe have fastened the switching speed and developed faster imaging speed (Grotjohann et al. 2012). Fast imaging to living cells has been achieved with spatial resolution of 77 nm, fields of view being 120 μm × 100 μm, and the imaging speed within 1 second (Chmyrov et al. 2013).

In addition, a variety of novel fluorescence probes have been developed, such as nanodiamonds (Han et al. 2009), quantum dots (QDs) (Lesoine et al. 2013) and up-conversion nanophosphors (UCNPs) (Liu et al. 2017a). UCNPs is a new fluorescence nanomaterial that possesses biocompatibility, sharp emission bandwidth, and low cytotoxicity. In addition, UCNPs are a type of lanthanide-doped nanocomposites, which absorbs multi-photons to release short wavelength photons. Different from two and multi-photon FLSCM, UCNPs are excited by continuous wave instead of mode-locked femtosecond laser. FLSCM based on UCNPs has been reported widely in recent years and is commercially available (Yu et al. 2009, Liu et al. 2017a, Wang et al. 2010), which has good photostability, strong resistance to photobleaching and high saturation intensity. The key challenge for this marker family is to establish facile and robust strategies for specific targeting (Lin et al. 2018).

3.2.2 LABEL-FREE LASER SCANNING CONFOCAL MICROSCOPE (LLSCM)

Although FLSCM has been applied extensively in biological imaging, both photo-damage and photo-bleaching induced by high-intensity laser and fluorophore hinder the imaging and analysis to biological specimen. Furthermore, functional distur-bance caused by fluorophore may also limit the applications of FLSCM. In contrast, advanced label-free laser scanning confocal microscopy (LLSCM) techniques shows advantages in label-free imaging and dynamic tracing to biological tissues and cells. This section discusses LLSCM and its label-free biological imaging applications. First, the laser scanning coherent Raman scattering microscope (LSCRSM) and its two subclasses, i.e. CARSM and stimulated Raman scattering microscope (SRSM) are introduced, and their applications in imaging of cells and issues are reviewed and compared. Then, the progress of SHGM is discussed. Performances of two LLSCM techniques, e. g. work wavelength, lateral resolution, and axial resolution are summa-rized and compared. The advantages and potential applications for biological imag-ing are discussed.

3.2.2.1 Laser Scanning Coherent Raman Scattering Microscope (LSCRSM)

The characteristic vibration of molecule chemical bonds at specific frequencies can be utilized to obtain the information of molecules. Based on Raman active vibra-tional modes of molecules, coherent Raman scattering (CRS) is used for vibrational spectroscopic imaging method, which has been an emerging technique to map the distribution of specific molecules inside specimen. The CRS has two types: coher-ent anti-Stokes Raman scattering (CARS) and stimulated Raman scattering (SRS),

which has been used for label-free imaging techniques in biology with high specificity and quantitative analysis.

3.2.2.1.1 Laser Scanning CARSM

In 1965, Terhune and Maker discovered the CARS phenomenon. When the frequency
difference of two incident lasers matches the benzene Raman scattering, a blue-shift
of Raman scattering spectrum is observed (Terhune et al. 1965). Then Duncan and
coworkers demonstrated the first laser scanning CARSM in 1982 (Duncan et al.
1982). As shown in Figure 3.5a (Potcoava et al. 2014), the CARS is a coherent nonlinear optical process. Two laser beams, with one (frequency f_p) for pump and probe
and another for Stokes laser (frequency f_s)), are impinged to target molecules within
the specimen. When the frequency difference between pump and Stokes laser satis-
fies the Raman active vibrational frequency of molecules $f_{vib} = f_p - f_s$, the CARS
occurs with frequency being $f_{CARS} = 2f_p - f_s$. Specifically, the molecule absorbs a
pump photon for a lower virtual state transition and emits a Stokes photon for return-
ing to a higher energy ground state. Then the molecule immediately absorbs a probe
photon to a higher virtual state, emits a photon with frequency of $f_p + f_{vib}$ and returns
to the ground state, as shown in Figure 3.5a.

The optical schematic of laser scanning CARSM is shown in Figure 3.5b (Li et al.
2018). Two incident lasers are collimated and tightly focused on sample to excite the
CARS signals. A time delay is used to control the temporal overlap of two excitation
beams for the effective excitation of CARS. Two objectives are used to collect the
CARS signals. The signal of backward propagation is collected by a high NA objective, which is also used to focus incident light. The signal of forward propagation is
also collected by another objective on transmission light path. The CARS signals are
detected by PMT after filtering the incident lasers.

The CARSM starts a rapid development through improving the detection speed,
bandwidth, and sensitivity. Cheng et al. (2002a) developed a high-speed CARS

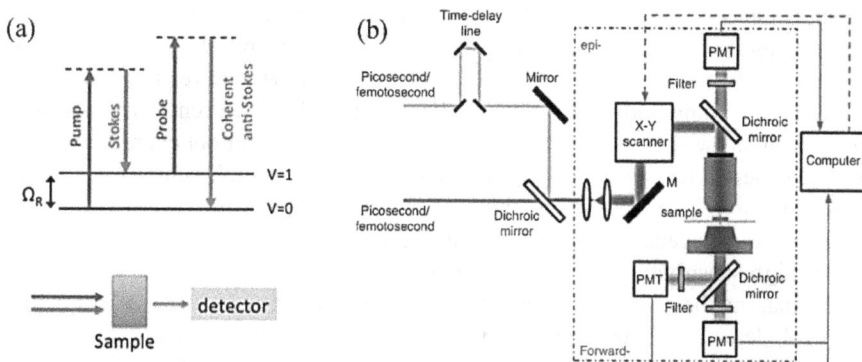

FIGURE 3.5 (a) Principle of CARS shown by Jablonski diagram (energy level diagram).
(Reprinted with permission from Potcoava et al. 2014.) (b) Diagram of a collinear laser-
scanning coherent anti-Stokes Raman scattering microscope (CARSM). (Reprinted with permission from Li et al. 2018.)

vibrational imaging method, which is two orders of magnitude faster than traditional CARSM by combining high repetition rate laser scanning and analog signal detection scheme. Then, through detecting strong backscattering of CARS signal with video-rate scanning microscope, a CARSM with imaging speed of video-rate was developed for vibrational imaging to tissues in vivo by Xie's group (Evans et al. 2005). Today, higher speed imaging has been achieved by employing Fourier-transform (Kinegawa et al. 2019) and deep-learning noise reduction (Yamato et al. 2020). The spectral bandwidth of CARSM has also showed great improvement by using multiplex (Müller and Schins 2002, Cheng et al. 2002b) and broadband CARSM techniques (Petrov et al. 2007, Kano and Hamaguchi 2006). By using a picosecond pump beam and a femtosecond Stokes beam, Xie and coworkers reported a two-pulse multiplex CARS micro-spectroscopy to demonstrate the polarization CARS imaging (Cheng et al. 2002b). Lee et al. (2007) demonstrated the three-color CARSM with continuum pulse of different frequency components as pump and Stokes, and a narrowband pulse as probe. As sensitivity of CARSM is restrained by strong non-resonant background noise, the techniques of polarization CARSM (Cheng et al. 2001), phase and polarization coherent control CARSM (Oron et al. 2003), and femtosecond adaptive spectroscopic through CARS were introduced to suppress background noise (Pestov et al. 2007). Cicerone's group developed a broadband CARS imaging platform by using intrapulse three-color excitation and exploiting the strong non-resonant background to amplify the inherently weak fingerprint signal, which provides an improvement of speed, sensitivity and spectral bandwidth, simultaneously (Camp et al. 2014).

Recently, the study on improving the spatial resolution of CARSM has attracted lot of attention. By using laser beam shaping and high-order CARS signals, spatial resolution of CARSM has been remarkably enhanced. Resolution approaching 130 nm was achieved by narrowing the microscope's excitation volume in the focal plane through combining a Toraldo-style pupil phase filter with the four-wave mixing (Kim et al. 2012). A supercritical-focused CARSM has been developed by using two optimized phase patterns with concentric rings to obtain 0 and π phase alternately, which is generated by a spatial light modulator and applied to the pump beam for minimizing the focal spot size (Gong et al. 2018). The theoretical research has indicated that diffraction limit can be broken by controlling the relative phase around the center of the pump and Stokes pulses (Wang et al. 2019a). Recently, a novel high-order CARSM was investigated by using high-order nonlinear optical processes and detected the CARS signal at new wavelength to achieve high-contrast and super-resolution vibrational imaging (Gong et al. 2019a).

Many advantageous characteristics, including high specificity, high spatial resolution, high sensitivity and label-free, make CARSM potentially ideal for noninvasively imaging the complex biological and chemical samples. In living cells imaging, CARSM detects various molecule vibrational modes, such as phosphate stretch vibration, amide I vibration (Dorosz et al. 2020, Cheng et al. 2002a), OH stretching vibration (Nuriya et al. 2019, Dufresne et al. 2003), and CH stretch vibration (Nan et al. 2004, Potma and Xie 2005), which can be applied for visualizing DNA, protein, water and lipids, respectively. Fast imaging speed and high signal level make CARSM more suitable for observing biological dynamic process in vivo (Evans et al.

2005, Jurna et al. 2007). In addition, CARSM also found applications in gas phase analysis (Roy et al. 2010) and material characterization (Koivistoinen et al. 2017). Multiplex CARSM was demonstrated and used to visualize the thermodynamic state, including the liquid crystalline and gel phase of lipid membranes (Müller and Schins 2002). In addition, CARSM has also been a powerful characterization tool for Hexagonal Boron Nitride (Ling et al. 2019), porous carbon (Li et al. 2019), and graphene (Virga et al. 2019).

3.2.2.1.2 Laser Scanning Stimulated Raman Scattering Microscope

Although CARSM has good performance on high sensitivity and high spatial resolution imaging to biological and chemical specimen, it is not suitable for quantitative chemical imaging and has to restrain the strong non-resonant background. As SRS signals depend linearly on concentration of signaling molecule inside specimen with background-free contrast, SRSM is a good candidate for quantitatively analyzing the molecule map information inside specimen. The excitation process of SRS is displayed in Figure 3.6b (Dan 2017). Similar with CARS, when the frequency difference between pump and Stokes beams is equal to Raman active vibrational frequency of molecules, amplification of the Raman signal is achieved by virtue of stimulated excitation. Consequently, the intensity of Stokes beam experiences a stimulated Raman gain, and the intensity of pump beam experiences a stimulated Raman

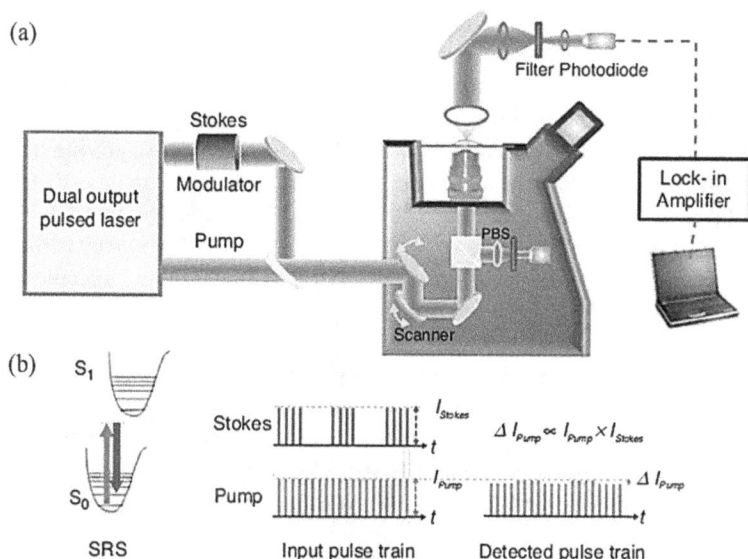

FIGURE 3.6 (a) A typical SRS microscope setup. The stokes laser is amplitude modulated with either an AOM or an EOM. The combined beams are sent into a laser scanning microscope. Either the transmitted or the back-scattered pump light is detected with a photodiode and a lock-In amplifier. (b) Illustration of the SRS process and signal generation. Detected pump pulse train has a tiny modulation at the frequency of the Stokes modulation due to the SRS of the sample. (Reprinted with permission from Dan 2017.)

loss. Therefore, unlike CARS, SRS does not exhibit a nonresonant background. For SRSM, the Stokes beam is modulated at a high frequency, and the amplitude modulation of the pump beam can be detected. As shown in Figure 3.6a, the Stokes beam of SRSM is modulated by an acoustic-optic modulator (AOM) or/and electro-optic modulator (EOM). A photodiode with a lock-in amplifier is used to detect the transmitted and reflected light after filtering. Combining with fast scanning, 3D imaging is performed. Moreover, there is linear dependence of SRL signal intensity on concentrations of target molecules.

SRS was first discovered by Woodbury and Ng in 1962 (Eckhardt et al. 1962, Hellwarth 1963). Until 2007, SRSM was first introduced by Ploetz et al. (2007) for imaging polystyrene beads. In 2008, Xie's group reported a high-speed, high sensitivity, and three-dimensional multiphoton vibrational imaging technique based on SRS, which showed powerful label-free chemical contrast capability in bio-imaging (Freudiger et al. 2008). The sensing range of typical SRSM is μM-mM of molecules concentration. Through using a microsecond resonant delay line, the detection sensitivity approaches 21.2 mM with 83 μs acquisition time (Liao et al. 2016). The super-resolution SRS imaging has been proposed and implemented by combining stimulated emission depletion (STED) microscope and femtosecond stimulated Raman spectroscopy (FSRS) (Silva et al. 2015). The far-field spatial resolution of SRSM has achieved ~130 nm with excitation wavelength of 800–1064 nm (Bi et al. 2018). A virtual sinusoidal modulation method was also proposed for retrieving super-resolution SRS images, which demonstrated a spatial resolution of 255 nm (Gong et al. 2019b). Volumetric imaging provides global understanding of 3D complex systems, Chen et al. demonstrated the Bessel-beam-based simulated Raman projection microscope and tomography for label-free volumetric imaging with spatial resolution being 0.83 μm (2017). Typically, the imaging depth of SRSM is 100 μm for high-scattering biological samples and 300–500 μm for less-scattering biological samples. Wei et al. developed a volumetric chemical imaging method that couples Raman-tailored tissue-clearing with SRSM, which achieves greater than 10-times depth compared with the standard SRS (Figure 3.7) (2019). By now, SRSM has achieved a high speed with video-rate, sensitivity down to single molecules, spatial resolution breaking the diffraction limit and volumetric imaging depth with millimeter (Hu et al. 2019).

Recently, SRSM experienced rapid development and was widely applied in various branches of life science, such as histopathology, cell biology, neuroscience, and tumor research. Histopathological diagnosis in vivo has been demonstrated by using multicolor images originating from CH_2 and CH_3 vibrations of lipids and proteins, which retrieved the subcellular resolution from fresh tissue (Freudiger et al. 2012). Moreover, SRS imaging to lipid metabolism was reported to reveal an aberrant accumulation of esterified cholesterol in lipid droplets of high-grade prostate cancer and metastases (Yue et al. 2014). Label-free SRS imaging of DNA-enabled noninvasive visualization of live cell nuclei in both human and animals. Through the distribution of DNA retrieved from the strong background of proteins and lipids by linear decomposition of SRS images at three optimally selected Raman shifts, it is possible to obtain instant histological tissue examination during surgical procedures (Lu et al. 2015). SRSM has also been proved to be able to differentiate healthy human and

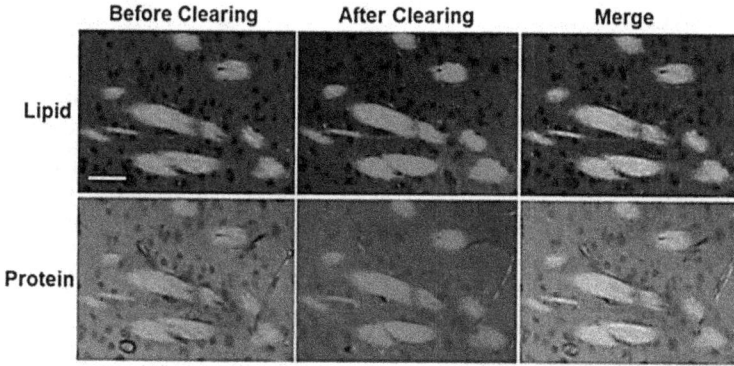

FIGURE 3.7 SRS images of regions in the same white matter region before and after tissue-clearing. (Reprinted with permission from Wei et al. 2019.)

mouse brain tissue as well as tumors from non-neoplastic tissue based on their different Raman spectra (Ji et al. 2013). Likewise, SRS imaging has shown its potential capability and application for differentiating misfolded amyloid-β on Alzheimer's disease pathology (Ji et al. 2018). These works provide new approaches for tracing dynamics and metabolism in living cells, which can be used for diagnosing and treating diseases.

SRS imaging has also been used for monitoring the breakdown of biomass to biofuels. It helps researchers to understand the conversion mechanisms of plant cell biomass to biofuels and find cost-effective industrial conversion strategy. By recognizing the Raman signal, different chemical bonding is detected. A chemical movie recording degradation of lignin and cellulose in cell wall was used to understand which parts of plant are degraded most efficiently by the treatment process (Saar et al. 2010). By comparing the two-colored SRS imaging before and after delignification reaction, another study revealed that hemicelluloses were generally resistant to acid chlorite at room temperature, and digestion of the delignified walls apparently depended on wall mass concentration but not on wall types (Ding et al. 2012). These works have greatly promoted the development of SRSM and its applications in biomedicine (Ozeki et al. 2009, 2012, Nandakumar et al. 2009).

3.2.2.2 Second Harmonic Generation Microscope

SHG process, which is also called "frequency doubling," is a special case of the sum frequency process. As shown in Figure 3.8a (Pantazis et al. 2010), by using two incident lights with same frequency, the output is twice the frequency of the incident lights. The incident lights are called fundamental frequency, and the output with frequency doubling is called second harmonic. In 1961, SHG was first confirmed with a quartz sample by Franken et al. (1961). In 1974, Hellwarth and Christensen firstly integrated SHG and microscope by imaging SHG signals from polycrystalline ZnSe (1974). A typical setup of SHGM is shown in Figure 3.8b (Sivaguru et al. 2010). The excitation wavelength was 780 nm. A galvo scanner was used to direct the beam in a

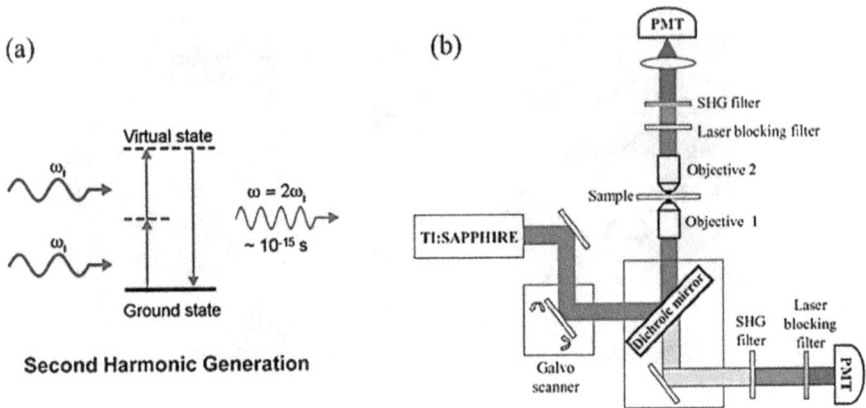

FIGURE 3.8 (a) Energy-level diagram of SHG. (Reprinted with permission from Pantazis et al. 2010.) (b) Experimental setup of SHGM with both forward and backward collection geometries. (Reprinted with permission from Sivaguru et al. 2010 © The Optical Society.)

raster-scan pattern. The beam was reflected by a dichroic mirror and focused onto the sample using a high NA water-immersion objective 1. The emitted backward SHG signal was collected by the same objective, whereas the forward signal was collected by a low NA objective 2. In both geometries the same two filters were used: one filter was used to block the fundamental laser, and the other was a band-pass filter to transmit the SHG signal at 390 nm. The PMT in both geometries were used to record the forward and backward SHG images.

In 1986, the first biological SHG imaging experiments were done by Freund and Deutsch (1986) to study the orientation of collagen fibers in rat tail tendon. In 1993, Lewis examined the second-harmonic response of biological membrane in electric fields through using chiral styryl dye (Bouevitch et al. 1993). He also demonstrated his work on live cell imaging (Campagnola et al. 1999). In 2010, P. Pantazis extended SHG to whole animal in vivo imaging. However, the spatial resolution of SHGM is vastly restricted due to the near-infrared excitation wavelength. Many methods have been carried out in order to improve the spatial resolution. Radially polarized laser illumination is an effective and easy way for dramatically enhancing the resolution of SHGM (Hashimoto et al. 2015, Yeh et al. 2018, Bautista et al. 2012). Wang et al. (2019b) demonstrated that spatial resolution was improved more than 21% by using radially polarized beam focused by elliptical mirror compared with lens. The work of Tian et al. (2015) showed that the subtractive imaging method could have been extended to SHGM, which increased the resolution to 210 nm and obtained a better contrast simultaneously. Yeh et al. (2018) combined SHGM with structured illumination based on point-scanning, which showed a resolution improvement factor of ~1.4 in the lateral and 1.56 in the axial directions. Sheppard et al. (2020) proposed pixel reassignment, in which the point detector of the traditional confocal microscope is replaced by a multi-element array detector. Combining with deblurring operation, this method showed a spatial resolution enhancement of about 1.87 compared with conventional SHGM (Wang et al. 2021).

Nowadays, SHGM has been extensively used in visualization to structures and functions of cells and tissues. As SHG is highly sensitive to collagen fibril/fiber structure (Chen et al. 2012, Freund and Deutsch 1986), SHG imaging can be used to detect collagen-associated changes occurred in various diseases (Keikhosravi et al. 2014), such as cancer, fibrosis, and connective tissue disorders. Ovarian cancer was studied by SHGM to detect the structure changes of the ovarian extracellular matrix in human normal and malignant ex vivo biopsies (Figure 3.9a) (Nadiarnykh et al. 2010). The results showed that the SHG emission distribution and bulk optical parameters are relative to the tissue structure and significantly different in the tumors. Combining with autofluorescence (AF), SHGM proves a quantitative analysis to different collagen content in the process of tumor progression and recession post-chemotherapy through calculating the ratio of SHG/AF (Figure 3.9b) (Wu et al. 2018). Polarization-dependent SHGM was developed for accurately diagnosing the tumor state of breast cancer by analyzing anisotropy and the "ratio parameter" values of SHG signal (Tsafas et al. 2020). Through further data analysis, parameters of polarization-in, polarization-out (PIPO) SHGM, i.e., susceptibility ratios and degree of linear polarization, can be used to distinguish tumor from normal tissue (Tokarz et al. 2019).

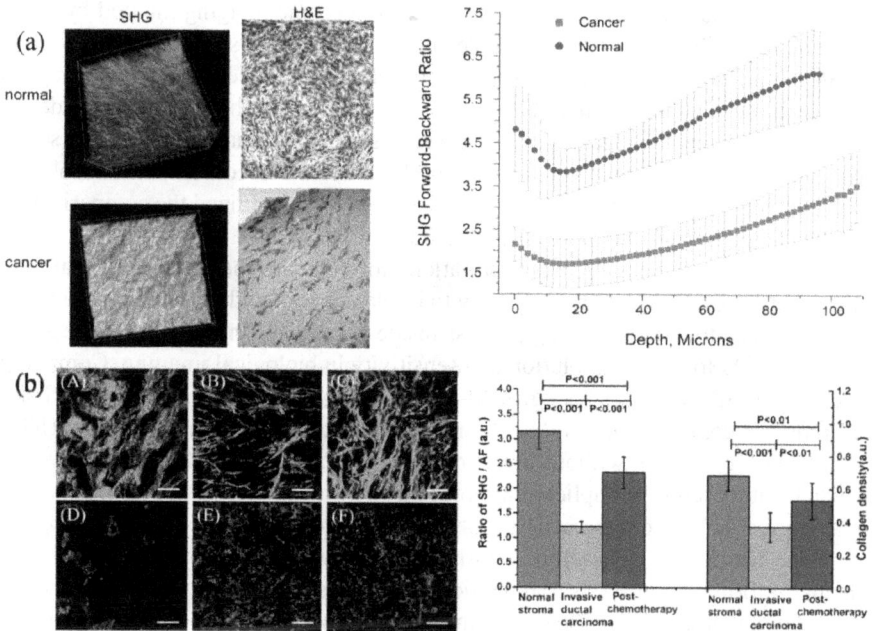

FIGURE 3.9 (a) 3D SHG renderings (left panels) and hematoxylin and eosin (H&E) staining (center panels) of normal (top) and malignant ovarian biopsies (bottom). Averaged forward/backward SHG intensities as a function of depth for normal (dark grey circles) and malignant (light grey squares) ovaries (right panels). (Reprinted with permission from Nadiarnykh et al. 2010.) (b) SHG images and correcting auto fluorescence (AF) images of the collagen in different tissues status (left panels). The ratio of SHG/AF and the collagen density of the three status of breast tissues (right panels). (Reprinted with permission from Wu et al. 2018 © The Optical Society.)

Besides, some automated or semi-automated solutions have been developed to analyze SHG images for evaluating the healthy and abnormal tissues. Early embryogenesis of zebrafish embryos was automatically reconstructed and analyzed with conformal scanning harmonic microscope scheme for better imaging quality (Olivier et al. 2010). Fast Fourier transform analysis and the gray-level co-occurrence matrix have also been applied to analyze biopsy samples of human ovarian epithelial cancer at different stages (Zeitoune et al. 2017). It showed the feasibility of identifying collagen fibril morphology based on first and second-order texture statis-tical parameters from SHG images. In addition, radiomics feature analysis, tree-based pipeline optimization tool (Wang et al. 2020), and 3D texture analysis based on k-nearest neighbor classification are also proposed for differentiating healthy and pathologic tissues (Wen et al. 2016).

3.2.2.3 Summary and Prospect

Preventing the photodamage and photobleaching induced by high-intensity laser and fluorescent molecular, label-free laser scanning microscopes shows powerful capability for imaging and dynamic tracing to biological tissues and cells. In this section, two types of label-free laser scanning microscopes are mentioned with their particular features. The LSCRSM is a vibrational spectroscopic imaging method by using Raman active vibrational modes of molecules, which has two subtypes: CARSM and SRSM. With second-harmonic response of biological samples, SHGM is achieved and has high sensitivity to the collagen fibril structure, which can be used to detect collagen-associated changes occurred in various diseases. Imaging speed and sensitivity are key issues for the label-free LSCM methods. Moreover, spatial resolution is also an important parameter for the microscopes. We organized the comparison of two label-free LSCMs, which is shown in Table 3.1.

However, there are still some limitations for current label-free laser scanning microscopes. CARSM is relatively easy to implement but with non-resonance background from a four-wave mixing process inside specimen. This impedes accessibility of CARSM to higher resolution and sensitivity in biological imaging. Compared to CARSM, SRSM shows background-free imaging, but the extraction of SRS signal requires complex lock-in detection for modulation and demodulation. SHGM is suitable for samples without inversion symmetry, such as collagen and myosin, which limit its scope of application. From the aforementioned discussion, no single imaging technology can present a global understanding to the complex biological structure and process. Therefore, multi-technique combination is a trend for imaging and information analysis to complicated samples (Sarri et al. 2019, Xiong et al. 2019, Adams et al. 2020). New imaging schemes will make label-free laser scanning microscopes to be more powerful biological imaging tool when combines with spectroscopy, laser and detection technology. Higher resolution, sensitivity and faster imaging speed are still the most important pursuits of microscope for more detailed and accurate morphology and dynamic process. With the increase of imaging speed and bandwidth, more effective and automated data methods are required to statistics and extremely large amounts of raw data. Benefitting from tremendously growth of artificial intelligence, fast and automated data processing becomes easier to implement. There's a reasonable prospect that advanced label-free

TABLE 3.1

Comparison of Three Laser Scanning Microscopic Techniques

Type		Work Principle	Work Wavelength (nm)	Lateral Resolution (nm)	Axial Resolution	References
CRS	CARSM	Supercritical focusing CARS	835 (pump) 1107 (Stokes)	295	3.32 μm	Gong et al. (2018)
		Coherent controlling of the relative phase	785 (Stokes) 727.8 (pump)	130	–	Wang et al. (2019)
		Higher-order CARS	1041 (Stokes) 680–1300 (pump)	196	–	Gong et al. (2019a)
	SRSM	Near-resonance enhancement	450	130	–	Bi et al. (2018)
		Saturated SRS	–	255	3 μm	Gong et al. (2019b)
SHGM		Structured illumination	1064	235	693 nm	Yeh et al. (2018)
		Subtractive imaging	800	210	–	Tian et al. (2015)
		Pixel reassignment	840	300	–	Wang et al. (2021)

laser scanning microscope will occupy a more dominant position in biological and biomedical imaging.

3.2.3 LASER SCANNING CONFOCAL MICROSCOPE FOR SINGLE-MOLECULE DETECTION

Single molecule is the basal function unit in all kinds of biological processes. With the development of advanced microscopes, the information, which are hidden in ensemble average of molecular population due to limitation of sensitivity and resolution, have been widely investigated for interpreting physiological phenomena at single-molecular level. Acquiring imaging and spectra of single-molecule shows a new picture never seen before and is convenient to understand complex mechanism of life activity, such as human immune response, gene expression, and cell differentiation. A lot of fundamental biological processes have been ascertained by single-molecule method, e. g. proteins folding, translocation and movement, DNA replication and remodeling, and processes relating to infection and general pathology. The primary barrier of single-molecule detection is the low signal-to-noise ratio (SNR) due to the weak intensity of the signal of single molecule in complex biological environment. Another requirement is molecular isolation to avoid crosstalk between target molecules.

LSCM is a powerful tool for observing single molecule, which has following advantages. First, due to the confocal principle, only molecules on focal plane are detected. Second, LSCM possesses little excitation volume, especially super-resolution LSCM. Third, LSCM is suitable for scanning imaging. Finally, there are low photodamage and photobleaching for non-target molecules. Therefore, with the development of laser technology, fluorescence dye and detection technology, many setups based on LSCM for studying single molecule have still been in the ascendant. For example, Dertinger et al. (2005) presented multi-focus multi-confocal setup for tracking individual molecules during diffusion in solution. Maruo et al. (2014) developed a 3D LSCM for multicolor 3D co-localization of single fluorescent molecule immobilized at a temperature of 1.5 K, whose cryogenic temperatures permit a long signal accumulation time with multi-color measurement and higher precision. Vukojevic et al. (2008) proposed an approach to quantitative single-molecule imaging by LSCM. It analyzes the molecular occurrence statistics by fluorescence intensity distribution, which is captured by digital imaging detector and enables direct determination of the number of fluorescent molecules and their diffusion rates, without resorting to temporal or spatial autocorrelation analyses. Thiele et al. (2020) validated a fluorescence lifetime super-resolution LSCM that realizes single-molecule localization-based fluorescence-lifetime super-resolution imaging and adds another dimension for fluorescence imaging.

3.3 FLUORESCENCE RESONANCE ENERGY TRANSFER BASED ON CONFOCAL MICROSCOPE

Even for super-resolution microscope, the resolution is limited to a few tens of nanometer, which is impossible for observing complex formation and dynamic of single molecules. In 1948, Theodor Förster developed the quantitative description to the FRET theory (Förster 1949, 1960). In 1967, Stryer et al. found the potential capability of FRET as the spectroscopic ruler (Stryer and Haugland 1967). First single molecule fluorescence resonance energy transfer (smFRET) measurement was proved by Ha et al. (1996). Nowadays, smFRET has been a significant biophysical tool for accurately measuring the intermolecular and intramolecular distances or conformational change of macromolecules. A pair of fluorophores (donor and acceptor) is used for signal excitation and nanoscale proximity detection, which has a typical range of 2–10 nm on single molecular level (Hohlbein et al. 2014). When both of fluorophores modify in branch of macromolecules, the conformation change of macromolecule can be speculated by the signal of smFRET. With the fluorophore modification to different molecules, the distance between molecules can be speculated by the signal from smFRET. The smFRET reveals kinetic information of single molecule that is hidden in ensemble measurement. Therefore, smFRET has been widely used in numerous single molecule studies, such as structural biology and biochemistry.

3.3.1 PRINCIPLE OF FRET

Two criteria are necessary for FRET. First, the distance between donor and accepter is within the range of 2–10 nm. Second, the overlap between emission spectrum of

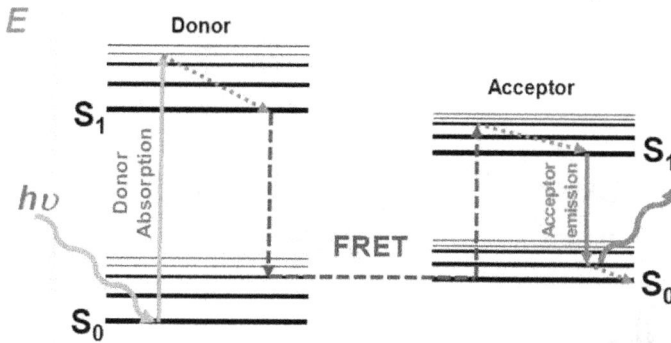

FIGURE 3.10 Energy diagram of FRET. (Reprinted with permission from Hochreiter et al. 2015.)

donor and excitation spectrum of acceptor exists. The energy transfer occurs through non-radiatively dipole–dipole coupling once the donor is close to acceptor. As shown in Figure 3.10, the electrons of donor jump to excitation state by absorbing photons, then the electrons transfer energy to acceptor by non-radiatively dipole-dipole coupling (Hochreiter et al. 2015). The electrons of acceptor on excitation state transfer to ground stage by emitting a photon. The efficiency of FRET is mainly determined by donor-acceptor distance (Figure 3.11a), which was formulated by Theodor Förster in 1940s as:

$$E_{FRET} = \frac{1}{1+(r/R_0)^6} \qquad (3.1)$$

$$R_0 = \frac{9000(\ln 10)Qk^2 J(\lambda)}{128\pi^2 N n^4} \qquad (3.2)$$

where r is the distance between donor and acceptor, R_0 is the Förster radius as E_{FRET} equal to 0.5, and it is a function of the quantum yield Q of the donor in the absence of the acceptor, the refractive index of the medium n, Avogadro's number N, the orientation factor κ^2 determined by dipole of donor and acceptor, and the spectral overlap integral $J(\lambda)$. Due to the E_{FRET} decreases sharply dependent on the sixth power of the distance between donor and accepter, the effective range of FRET detection is restricted to 10 nm. In addition, several energy processes or modes occur when the distance below 2 nm, which means the energy transfer is more complex than FRET only.

Two setups are used to measure the efficiency of smFRET according to excitation configurations: confocal microscope and total internal reflection fluorescence configurations (Figure 3.11b and c). Confocal microscope is commonly used to study freely diffusing molecules in dilute solution, which enables an observation to single molecule with sparse distribution. The FRET occurs when molecules enter into the small excitation volume, which is the confocal spot of excitation transmitting through the objective. The position of spot is stationary during the signal recording. The photons

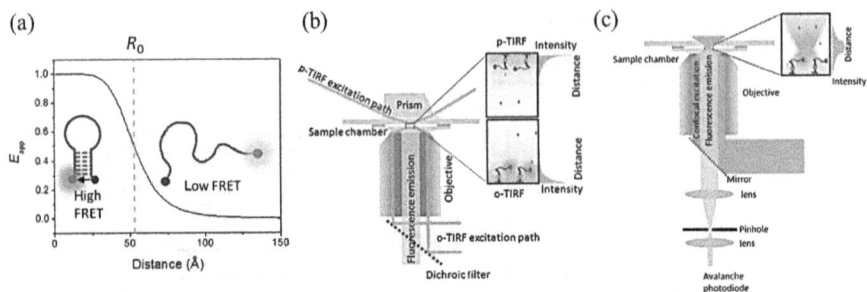

FIGURE 3.11 Principle of smFRET. (a) Efficiency of FRET as a function of interdye separation. (b) Schematic of a total internal reflection fluorescence (TIRF) setup allowing the study of smFRET on surface-immobilized molecules. (c) Schematic of a confocal microscope setup used for the acquisition of diffusion-based smFRET. (Reprinted with permission from Shaw et al. 2014.)

transferring from donor and accepter are detected by a single-photon avalanche diode (SPAD), which has high temporal resolution to photon response. The SPAD not only enables the smFRET unlimited by frame rate of CCD or CMOS but also identifies the dynamic of single-molecular on microsecond. However, due to the quick time of the molecules passing the excitation volume, the configuration is not available for the slow molecular dynamics and observation during a long period. The prism-based smFRET uses total internal reflection to produce an evanescence field for exciting donor fluorophores immobilized on prism interface. An inverted microscope is used to collect the fluorescence from the specimen. Through a dichroic filter, two images are captured by CCD or CMOS to obtain signals of donor and acceptor, respectively. The prism-based smFRET possesses low noise and long residence time because of the little penetration depth of evanescence field and molecule immobilization.

The efficiency of FRET E_{FRET} can be calculated according to the fluorescence intensities changing from donor to acceptor. The definition of E_{FRET} is the ratio of energy emission from acceptor to energy absorbed by donor. The E_{FRET} can be approximated as below:

$$E_{FRET} = I_A / (I_A + I_D) \qquad (3.3)$$

where I_A and I_D are the fluorescence intensities of donor and acceptor, respectively. The true E_{FRET} can be obtained after compensating the error between experiment and theory, which is induced by the nonradiative energy dissipation, the light loss in the optical paths and the imperfect detection efficiency of detectors. The true E_{FRET} can be described as:

$$E_{FRET} = \frac{I_A - \beta I_D}{I_A + (\gamma - \beta) I_D} \qquad (3.4)$$

where γ is the correction factor of the detection efficiency between two fluorophores, β is the compensation of the fluorescence leakage in optical path and detection.

On the other hand, the E_{FRET} can be obtained by measuring fluorescent lifetime as following:

$$E_{FRET} = 1 - \frac{\tau_{DA}}{\tau_D} \tag{3.5}$$

where τ_{DA} and τ_D are the fluorescence lifetime of the donor with appearance and absence of FRET, respectively. A pulsed laser is used as excitation source for the measurement of fluorescence lifetime.

3.3.2 SINGLE MOLECULE FLUORESCENCE RESONANCE ENERGY TRANSFER MEASUREMENT

The time-dependent FRET signals indicate that the time evolution of molecules can be obtained for studying molecular dynamics (Okamoto 2013). When the FRET occurs, the I_A and I_D show the anti-correlation relationship at same time. The peak of I_A corresponds to valley of I_D, or vice versa. The anti-correlation relationship between I_A and I_D vanishes if photochemical effects appear, i.e., blinking or photobleaching. To quantitatively describe the anti-correlation relationship, the compensated total intensity I_C can be obtained as:

$$I_C \propto \frac{I_A}{\gamma} + \left(1 - \frac{\beta}{\gamma}\right)I_D \tag{3.6}$$

If only FRET occurs, I_C is either a constant, or a large fluctuation. After the FRET occurs, the distance between donor and acceptor can be calculated by Eq. 3.1.

For diffusion-based smFRET, the intensity fluorescence from donor and acceptor changes, and the bursts indicate the signal from single molecule. The calculated statistical histogram shows the probability distribution of FRET combining two kinds of double-strand DNAs (dsDNAs), which reflects the transition rate of molecular conformational state relating to the residence time of target molecule through the observation spot. If the transition rate is faster than the residence time, only one peak in histogram emerges. When the transition rate is much slower than the residence time, two or more peaks are shown in histogram. It indicates that the multi-subpopulations can be detected, owing to the slow conformation transition. The probability distribution of FRET strongly depends on the relationship between intermediate transition rate and residence time. Fitting error functions to the cumulative distribution function (CDF) without binning is used to eliminate the influence on the binning condition in experiment (Okamoto 2013).

3.3.3 ALTERNATING LASER EXCITATION SMFRET

As mentioned above, the smFRET can plot a time-dependent E_{FRET} using single wavelength laser, and the distance between donor and acceptor can be calculated. However, its application is limited to identify distance change and kinetics of molecules. Moreover, a lot of complex corrections are necessary for accurate E_{FRET}

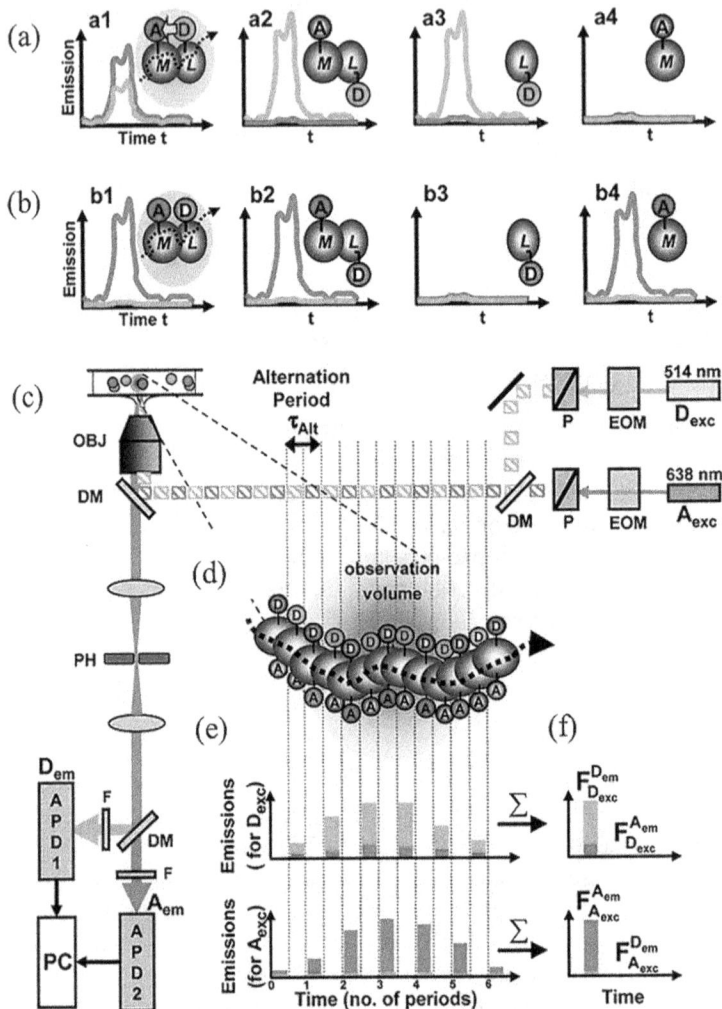

FIGURE 3.12 Fluorescent emission caused by (a) donor-excitation (grey oval) and (b) acceptor-excitation (grey oval) for different molecules in ALEX. (c) Setup schematic of ALEX FRET. (d) Probing donor-excitation and acceptor-excitation based emissions to diffusing molecules. (e) Signals of donor (light grey columns) and acceptor (dark grey columns) emission from donor excitation, respectively. (f) Single burst obtained through summing the excitation/emission streams. The E_{FRET} and S can be calculated by the series of bursts. (Reprinted with permission from Kapanidis et al. 2004 Copyright (2004) National Academy of Sciences, U.S.A.)

measurement, and the development of smFRET is limited by the cumulative effect relating to photophysics of fluorophore, photobleaching, substoichiometric labeling, and aggregation or dissociation phenomena.

ALEX is a method based on fast switch between donor and acceptor excitation (Kapanidis et al. 2004). The principle of ALEX is show as Figure 3.12c. The laser beams with two wavelengths for donor and acceptor excitation illuminate the sample with the alternation period twofold higher than diffusion time of a molecule through the observation volume. When the donor excitation laser arrives at the focal point, the time-dependent emission is shown in Figure 3.12a(a1–a4) The florescence emission of donor and acceptor appear simultaneously when distance between donor and acceptor is within the range from 2 to 10 nm (a1). The situations of large distance or donor-only indicate only florescence emission of donor (a2, a3). The case of acceptor-only has no effective detection (a4). When the acceptor excitation laser arrives at the focal point, only florescence emission of acceptor appears with molecule labeled by acceptor, as shown in Figure 3.12b (b1, b2, b4). For molecule labeled by donor, no effective emission is detected (b3).

A fluorescence burst with a series of consecutive alternating periods can be obtained when a quickly alternating excitation laser illuminates the focal point (Figure 3.12d–e). The donor-excitation-based emissions from donor and acceptor during a single burst are $F_{D_{exi}}^{D_{em}}$ and $F_{D_{exi}}^{A_{em}}$ respectively. Therefore,

$$E_{FRET} = F_{D_{exi}}^{A_{em}} / \left(F_{D_{exi}}^{D_{em}} + \gamma F_{D_{exi}}^{A_{em}} \right) \tag{3.7}$$

$$\gamma = \left(\phi_A \eta_A \right) / \left(\phi_D \eta_D \right) \tag{3.8}$$

where γ is detection correction factor, which is typically 0.5–2, depending on quantum yields of donor and acceptor ϕ_D, ϕ_A, and detection efficiencies η_A. The accurate E_{FRET} and distance-independent ratio S, which is the intensity ratio between the emission exited by only donor and only acceptor, can be used for calculating distance and determining the quantity of species, simultaneously. The accurate E_{FRET} means a more accurate measurement of distance between donor and acceptor. The acceptor-excitation-based emissions from donor and acceptor during a single burst are $F_{A_{exi}}^{D_{em}}$ and $F_{A_{exi}}^{A_{em}}$ respectively. ALEX-based ratio S that reports on donor-acceptor stoichiometry is defined as:

$$S = F_{D_{exi}} / \left(F_{D_{exi}} + F_{A_{exi}} \right) \tag{3.9}$$

$$F_{D_{exi}} = F_{D_{exi}}^{D_{em}} + F_{D_{exi}}^{A_{em}} \tag{3.10}$$

$$F_{A_{exi}} = F_{A_{exi}}^{D_{em}} + F_{A_{exi}}^{A_{em}} \tag{3.11}$$

where $F_{D_{exi}}$ and $F_{A_{exi}}$ are the sum of donor and acceptor-excitation-based emissions. As shown in Figure 3.13a, the distance-independent ratio S illustrates distinct values for species in mixtures. As $F_{D_{exi}}$ and $F_{A_{exi}}$ is distance-independent, the S and E_{FRET} can be measurement independently. Before S measurement for donor-acceptor species, it is necessary to adjust the excitation state to $F_{D_{exi}} = F_{A_{exi}}$. If the sample is donor only, $S = 1$, and S is close to 0 for acceptor only. For any specimen with donor and acceptor, the value of S is from 0.3 to 0.8. S is very sensitive to emission intensity of fluorophore, which can be used to detect changes in the local environment. The combination of E_{FRET} and S in the 2D histogram achieve FAMS (fluorescence-aided

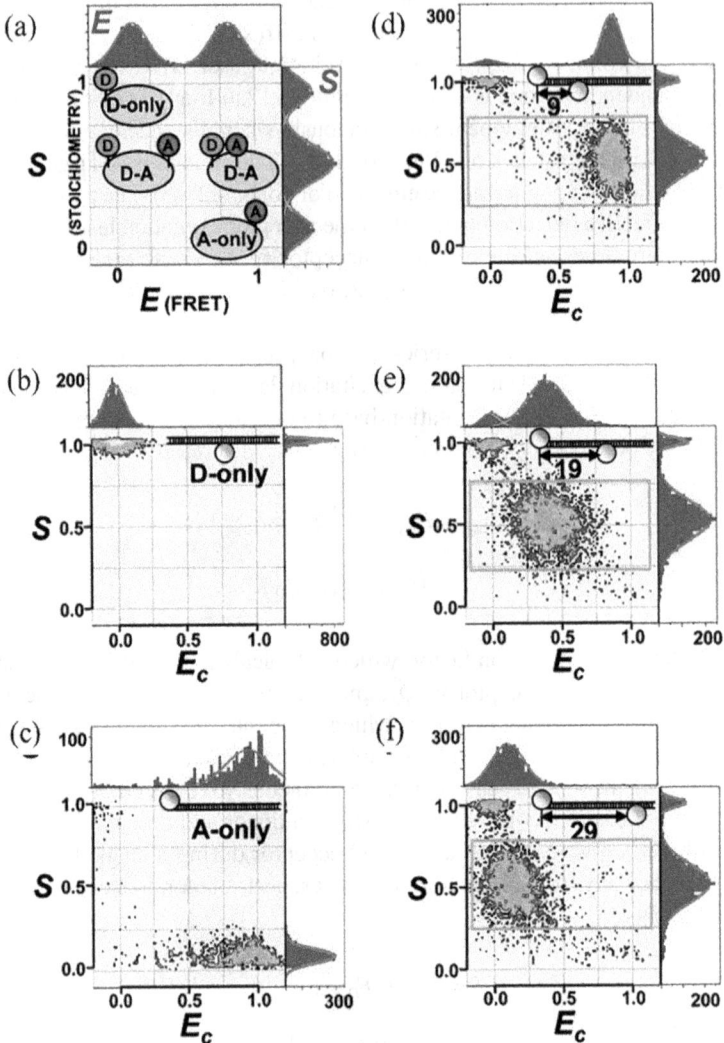

FIGURE 3.13 (a) The E_{FRET}-S histogram, which can be used to sort single molecules, for (b) donor-only DNA, (c) acceptor-only DNA, (d) High E_{FRET} DNA, (e) intermediate E_{FRET} DNA, and (f) low E_{FRET} DNA. (Reprinted with permission from Kapanidis et al. 2004 Copyright (2004) National Academy of Sciences, U.S.A.)

molecule sorting) and quantify the classified species (Figure 3.13a–f), while maintaining distance information between donor and acceptor.

3.3.4 HIGH ORDER FRET

As the conventional smFRET is insufficient for measuring complex 3D intermolecular distance and intramolecular conformation, such as protein-nucleic acid

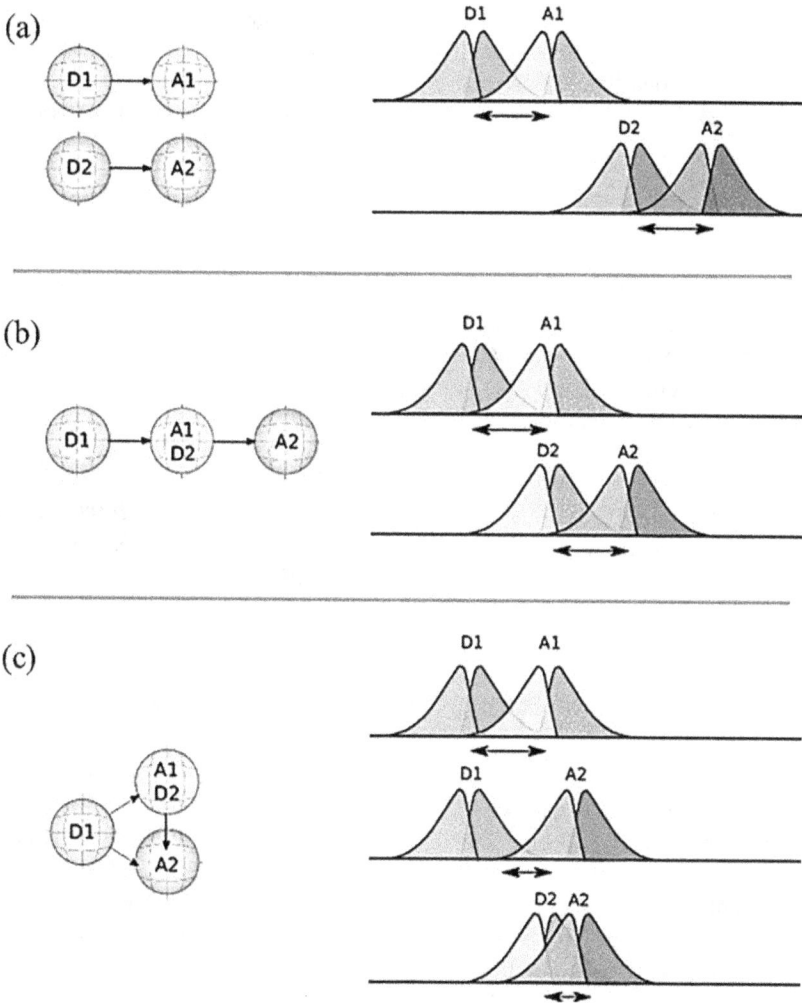

FIGURE 3.14 Types of FRET multiplexing schemes. (a) The parallel FRET schemes. (b) Three-color cascade smFRET with sequential, two-step, FRET between three fluorophores that form two consecutive FRET donor and acceptor pairs. (c) Three-fluorophore FRET used to detect multiple interactions in a complex. (Reprinted with permission from Bunt and Wouters 2017.)

interactions and protein folding, multi-color smFRET with multi-fluorophores, including parallel smFRET and three-color smFRET, had been reported for detecting multi-component interactions or spatiotemporal relationships between different conformational changes in macromolecular complexes. In Figure 3.14, the schemes show the energy transfer process of several multi-color smFRET techniques (Bunt and Wouters 2017). In parallel FRET, the two independent FRET pairs are used to detect conformational change on different part of a large macromolecule (Figure 3.14a).

In three-color cascade smFRET, the intermediate energy acceptor transfers the energy obtained from donor to the lower energy acceptor (Figure 3.14b). In three-color smFRET, one fluorophore can be used as a donor for two acceptors, and the energy transfer take place between three fluorophores without restricted limitation (Figure 3.14c).

3.3.5 APPLICATION OF SMFRET

3.3.5.1 DNA (Holliday Junction, Flip-Flop Motion, Branch Migration)

The smFRET has been extensively applied in biomolecular measurement. The first smFRET experiment was carried out in dsDNA by Deniz et al. (1999). As the distance between donor and acceptor is proportional to the number of base-pairs between donor and acceptor, dsDNA has been a standard for smFRET measurement (Okamoto 2013). Holliday junction is a special DNA structure which has four DNA strands forming a four-way junction. Dynamical DNA motions, such as flip-flop motion (Sarkar et al. 2007) and branch migration (Karymov et al. 2008), had been investigated in-depth. In addition, translation of mRNA has also been investigated through surveying the movement of tRNAs with smFRET, which involves the selective binding of tRNA, the sliding motion of the mRNA chain bound and the release of tRNA driven by the conformational dynamics of ribosome (Blanchard et al. 2004).

3.3.5.2 Protein (Protein Folding, Protein Motor (Movement of Myosin's Heads and Kinesin's Neck Linker))

The smFRET has also been extensively used to study the protein folding and unfolding, which are the most important dynamics involving numerous protein functions (Merchant et al. 2007). For example, kinesin motility has been investigated by smFRET technique and has demonstrated to be driven by the coordinated forward extension of the neck linker in one motor domain and the rearward positioning of the neck linker in the partner domain, which is accompanied by the "hand-over-hand" motion during ATP-driven motility (Tomishige et al. 2006).

3.3.5.3 Enzymatic Reaction

Also, the application of smFRET renewed people's understanding of enzymatic reactions. Traditionally, an enzyme is perceived to stay in open stage waiting for a substrate and in enclose stage catalyzing a substrate. The work of Hanson and coworkers indicates that the conformational equilibrium between open and enclose stage of enzyme continued to exist even in the absence of substrate (Hanson et al. 2007). Further, the experiment showed that the substrate limited the spatial size of conformational change to increase the shift rate between open and enclose stage for faster catalysis reaction rate.

3.3.6 ADVANTAGES AND DISADVANTAGES

In comparison with X-ray crystallography or Nuclear Magnetic Resonance (NMR), smFRET measurement does not require molecules to take high order structure and applies to dynamical reaction on live cells. Moreover, there are some exclusive advantages as single-molecule sensitivity, nano resolution and real-time detection. However, it is noteworthy that some limitations still exist for application of smFRET technique. Firstly, at least two fluorophores attached on the target molecule is necessary. Secondly, there is not signal when the distance outside the range of 2–8 nm for $R_0 = 0.5$. Thirdly, the time resolution is limited by the frame rate of CCD or CMOS cameras. Lastly, the accurate distance measurement is challenging, as the fluorescence properties and energy transfer depend on the environment and relative orientation of the fluorophores.

3.4 CONCLUSIONS AND PROSPECT

In this chapter, we have reviewed numerous confocal microscopies derived from laser scanning microscope. These microscopes were developed for improving the spatial resolution and sensitivity in biological and medical imaging. Two-photon and multi-photon LSCMs combined with new fluorescent materials pave the way to photodamage-less, polychrome, and deeper imaging in live cells or tissues. Fiber-based and miniaturized LSCMs are flexible for long-term and real-time observation to tissues in freely moving animals. Super-resolution LSCMs aimed at resolving the diffraction limitation of imaging resolution, which shows more detailed structures and events inside cells. As for free-label LSCMs, LSCRSM (including CARSM and SRSM) and SHGM are introduced, and their applications in imaging cells and issues were reviewed and compared. Moreover, several single-molecule detections show the capability of LSCM applied in visualization of biological processes at single-molecule level. In addition, as an important single-molecule analysis technique, smFRET is introduced by describing its principle and applications, which manifests the intermolecular or intramolecular dynamic and configuration. However, a single technique is not adequate for analyzing the complex biological and medical phenomena. Combination of multi-techniques pushes the trend for obtaining more comprehensive characters and analysis of microscopic life.

ACKNOWLEDGMENT

The authors gratefully acknowledge the financial support from Chinese Academy of Sciences (YJKYYQ20190056), Beijing Natural Science Foundation (4192063), and National Nature Science Foundation of China (NSFC) (62275264, 61774167, 62171441), and the Ministry of Science of China (2018YFC2001100, 2020YFC2004502). The authors also greatly thank Prof. Hongyao Liu, and Mr. Xuqing Sun, who attentively proofread the entire chapter and provided constructive suggestions on the book chapter.

REFERENCES

Abdeladim, L., K. S. Matho, S. Clavreul, et al. 2019. "Multicolor multiscale brain imaging with chromatic multiphoton serial microscopy." *Nat Commun* 10(1):1662. doi: 10.1038/s41467-019-09552-9.

Adams, W. R., B. Mehl, E. Lieser, et al. 2020. "Multimodal nonlinear optical and thermal imaging platform for label-free characterization of biological tissue." doi: 10.1101/2020.04.06.023820.

Alhede, M., K. Qvortrup, R. Liebrechts, et al. 2012. "Combination of microscopic techniques reveals a comprehensive visual impression of biofilm structure and composition." *FEMS Immunol Med Microbiol* 65(2):335–342. doi: 10.1111/j.1574-695X.2012.00956.x.

Allgeier, S., A. Bartschat, S. Bohn, et al. 2018. "3D confocal laser-scanning microscopy for large-area imaging of the corneal subbasal nerve plexus." *Sci Rep* 8(1):7468. doi: 10.1038/s41598-018-25915-6.

Barkauskas, D. S., G. Medley, X. Liang, et al. 2020. "Using in vivo multiphoton fluorescence lifetime imaging to unravel disease-specific changes in the liver redox state." *Methods Appl Fluoresc* 8(3):034003. doi: 10.1088/2050-6120/ab93de.

Bautista, G., M.J. Huttunen, J. Makitalo, et al. 2012. "Second-harmonic generation imaging of metal nano-objects with cylindrical vector beams." *Nano Lett* 12(6):3207–3212. doi: 10.1021/nl301190x.

Bayguinov, P. O., D. M. Oakley, C. C. Shih, et al. 2018. "Modern laser scanning confocal microscopy." *Curr Protoc Cytom* 85(1):e39. doi: 10.1002/cpcy.39.

Bethge, P., R. Chereau, E. Avignone, et al. 2013. "Two-photon excitation STED microscopy in two colors in acute brain slices." *Biophys J* 104(4):778–785. doi: 10.1016/j.bpj.2012.12.054.

Bi, Y., C. Yang, Y. Chen, et al. 2018. "Near-resonance enhanced label-free stimulated Raman scattering microscopy with spatial resolution near 130 nm." *Light Sci Appl* 7:81. doi: 10.1038/s41377-018-0082-1.

Blanchard, S. C., H. D. Kim, R. L. Gonzalez, et al. 2004. "tRNA dynamics on the ribosome during translation." *Proc Natl Acad Sci USA* 101(35):12893–12898. doi: 10.1073/pnas.0403884101.

Blom, H., and J. Widengren. 2017. "Stimulated emission depletion microscopy." *Chem Rev* 117(11):7377–7427. doi: 10.1021/acs.chemrev.6b00653.

Born, M., P. C. Clemmow, D. Gabor, et al. 1959. "*Principles of Optics.*" Oxford: Pergamon Press.

Bouevitch, O., A. Lewis, I. Pinevsky, et al. 1993. "Probing membrane potential with nonlinear optics." *Biophys J* 65(2):672–679. doi: 10.1016/s0006-3495(93)81126-3.

Bunt, G., and F.S. Wouters, 2017 "FRET from single to multiplexed signaling events." *Biophys Rev* 9(2):119–129. doi: 10.1007/s12551-017-0252-z.

Camp Jr., C. H., Y. J. Lee, J. M. Heddleston, et al. 2014. "High-speed coherent Raman fingerprint imaging of biological tissues." *Nat Photonics* 8:627–634. doi: 10.1038/nphoton.2014.145.

Campagnola, P. J., M. D. Wei, A. Lewis, et al. 1999. "High-resolution nonlinear optical imaging of live cells by second harmonic generation." *Biophys J* 77(6):3341–3349. doi: 10.1016/s0006-3495(99)77165-1.

Carlson, K., M. Chidley, K. B. Sung, et al. 2005. "In vivo fiber-optic confocal reflectance microscope with an injection-molded plastic miniature objective lens." *Appl Opt* 44(10):1792–1797. doi: 10.1364/ao.44.001792.

Chen, X., O. Nadiarynkh, S. Plotnikov, et al. 2012. "Second harmonic generation microscopy for quantitative analysis of collagen fibrillar structure." *Nat Protoc* 7(4):654–669. doi: 10.1038/nprot.2012.009.

Chen, X., C. Zhang, P. Lin, et al. 2017. "Volumetric chemical imaging by stimulated Raman projection microscopy and tomography." *Nat Commun* 8:15117. doi: 10.1038/ncomms15117.

Cheng, J. X., L. D. Book, and X. S. Xie. 2001. "Polarization coherent anti-Stokes Raman scattering microscopy." *Opt Lett* 26(17):1341–1343. doi: 10.1364/ol.26.001341.

Cheng, J. X., Y. K. Jia, G. Zheng, et al. 2002a. "Laser-scanning coherent anti-Stokes Raman scattering microscopy and applications to cell biology." *Biophys J* 83(1):502–509. doi: 10.1016/S0006-3495(02)75186-2.

Cheng, J. X., A. Volkmer, L. D. Book, et al. 2002b. "Multiplex coherent anti-stokes Raman scattering microspectroscopy and study of lipid vesicles." *J Phys Chem B* 106(34):8493–8498. doi: 10.1021/jp025771z.

Chmyrov, A., J. Keller, T. Grotjohann, et al. 2013. "Nanoscopy with more than 100,000 'doughnuts'." *Nat Methods* 1(8):737–740. doi: 10.1038/nmeth.2556.

Chun, B. S., K. Kim, and D. Gweon. 2009. "Three-dimensional surface profile measurement using a beam scanning chromatic confocal microscope." *Rev Sci Instrum* 80(7):073706. doi: 10.1063/1.3184023.

Dan, F. 2017. "Quantitative chemical imaging with stimulated Raman scattering microscopy." *Curr Opin Chem Biol*, 39, 24–31. doi: 10.1016/j.cbpa.2017.05.002.

Deniz, A. A., M. Dahan, J. R. Grunwell, et al. 1999. "Single-pair fluorescence resonance energy transfer on freely diffusing molecules: Observation of Forster distance dependence and subpopulations." *Proc Natl Acad Sci USA* 96(7):3670–3675. doi: 10.1073/pnas.96.7.3670.

Denk, W., J. H. Strickler, and W. W. Webb. 1990. "Two-photon laser scanning fluorescence microscopy." *Science* 248(4951):73–76. doi: 10.1126/science.2321027.

Dertinger, T., F. Koberling, A. Benda, et al. 2005. "Advanced multifocus confocal laser scanning microscope for single molecule studies." Imaging, Manipulation, and Analysis of Biomolecules and Cells: Fundamentals and Applications III. International Society for Optics and Photonics, 5699, 219–226.

Ding, S. Y., Y. S. Liu, Y. Zeng, et al. 2012. "How does plant cell wall nanoscale architecture correlate with enzymatic digestibility?" *Science* 338(6110):1055–1060. doi: 10.1126/science.1227491.

Doronina-Amitonova, L. V., I. V. Fedotov, A. B. Fedotov, et al. 2015. "Neurophotonics: optical methods to study and control the brain." *Physics-Uspekhi* 58(4):345–364. doi: 10.3367/UFNe.0185.201504c.0371.

Dorosz, A., M. Grosicki, J. Dybas, et al. 2020. "Eosinophils and neutrophils-molecular differences revealed by spontaneous Raman, CARS and fluorescence microscopy." *Cells* 9(9):2041. doi: 10.3390/cells9092041.

Duan, X., H. Li, F. Wang, et al. 2017. "Three-dimensional side-view endomicroscope for tracking individual cells in vivo." *Biomed Opt Express* 8(12):5533–5545. doi: 10.1364/BOE.8.005533.

Dufresne, E. R., E. I. Corwin, N. A. Greenblatt, et al. 2003. "Flow and fracture in drying nanoparticle suspensions." *Phys Rev Lett* 91(22):224501. doi: 10.1103/PhysRevLett.91.224501.

Duncan, M. D., J. Reintjes, and T. J. Manuccia. 1982. "Scanning coherent anti-Stokes Raman microscope." *Opt Lett* 7(8):350–352. doi: 10.1364/ol.7.000350.

Ebersbach, P., F. Stehle, O. Kayser, et al. 2018. "Chemical fingerprinting of single glandular trichomes of Cannabis sativa by Coherent anti-Stokes Raman scattering (CARS) microscopy." *BMC Plant Biol* 18:275. doi:10.1186/s12870-018-1481-4.

Eckhardt, G., R. W. Hellwarth, F. J. McClung, et al. 1962. "Stimulated Raman scattering from organic liquids." *Phys Rev Lett* 9(11):455–457. doi: 10.1103/PhysRevLett.9.455.

Evans, C. L., E. O. Potma, M. Puoris'haag, et al. 2005. "Chemical imaging of tissue in vivo with video-rate coherent anti-Stokes Raman scattering microscopy." *Proc Natl Acad Sci USA* 102(46):16807–16812. doi: 10.1073/pnas.0508282102.

Flusberg, B. A., E. D. Cocker, W. Piyawattanametha, et al. 2005. "Fiber-optic fluorescence imaging." *Nat Methods* 2(12):941–950. doi: 10.1038/nmeth820.

Fölling, J., M. Bossi, H. Bock, et al. 2008. "Fluorescence nanoscopy by ground-state depletion and single-molecule return." *Nat Methods* 5:943–945. doi:10.1038/nmeth.1257.

Förster, T., 1949. "Experimentelle und theoretische Untersuchung des zwischenmolekularen Übergangs von Elektronenanregungsenergie" *Zeitschrift für Naturforschung A*, 4(5):321–327. doi:10.1515/zna-1949-0501.

Förster, T., 1960. "Transfer mechanisms of electronic Eexcitation energy." *Radiat Res Suppl*, 2:326–339. doi: 10.2307/3583604.

Franken, P. A., A. E. Hill, C. W. Peters, et al. 1961. "Generation of optical harmonics." *Phys Rev Lett* 7(4):118–119. doi: 10.1103/PhysRevLett.7.118.

Freudiger, C. W., W. Min, B. G. Saar, et al. 2008. "Label-free biomedical imaging with high sensitivity by stimulated Raman scattering microscopy." *Science* 322(5909):1857–1861. doi: 10.1126/science.1165758.

Freudiger, C. W., R. Pfannl, D. A. Orringer, et al. 2012. "Multicolored stain-free histopathology with coherent Raman imaging." *Lab Invest* 92(10):1492–1502. doi: 10.1038/labinvest.2012.109.

Freund, I., and M. Deutsch. 1986. "Second-harmonic microscopy of biological tissue." *Opt Lett* 11(2):94. doi: 10.1364/ol.11.000094.

Friedrich, M., Q. Gan, V. Ermolayev, and G. S. Harms. 2011. "STED-SPIM: Stimulated emission depletion improves sheet illumination microscopy resolution." *Biophys J* 100(8): L43–L45. doi: 10.1016/j.bpj.2010.12.3748.

Fujita, K., M. Kobayashi, S. Kawano, et al. 2007. "High-resolution confocal microscopy by saturated excitation of fluorescence." *Phys Rev Lett* 99(22):228105. doi: 10.1103/PhysRevLett.99.228105.

Gmitro, A. F., and D. Aziz. 1993. "Confocal microscopy through a fiber-optic imaging bundle." *Opt Lett* 18(8):565. doi: 10.1364/ol.18.000565.

Gong, L., J. Lin, C. Hao, et al. 2018. "Supercritical focusing coherent anti-Stokes Raman scattering microscopy for high-resolution vibrational imaging." *Opt Lett* 43(22):5615–5618. doi: 10.1364/OL.43.005615.

Gong, L., W. Zheng, Y. Ma, et al. 2019a. "Higher-order coherent anti-Stokes Raman scattering microscopy realizes label-free super-resolution vibrational imaging." *Nat Photonics* 14 (2):115–122. doi: 10.1038/s41566-019-0535-y.

Gong, L., W. Zheng, Y. Ma, et al. 2019b. "Saturated stimulated-Raman-scattering microscopy for far-field superresolution vibrational imaging." *Phys Rev Appl* 11(3):034041. doi: 10.1103/PhysRevApplied.11.034041.

Göppert-Mayer, M. 1931. "Über elementarakte mit zwei quantensprüngen". *Annalen der Physik* 401(3):273–294. doi: 10.1002/andp.19314010303.

Gottfert, F., C. A. Wurm, V. Mueller, et al. 2013. "Coaligned dual-channel STED nanoscopy and molecular diffusion analysis at 20 nm resolution." *Biophys J* 105(1): L01–L03. doi: 10.1016/j.bpj.2013.05.029.

Grotjohann, T., I. Testa, M. Leutenegger, et al. 2011. "Diffraction-unlimited all-optical imaging and writing with a photochromic GFP." *Nature* 478(7368):204–208. doi: 10.1038/nature10497.

Grotjohann, T., I. Testa, M. Reuss, et al. 2012. "RsEGFP2 enables fast RESOLFT nanoscopy of living cells." *Elife* 1:e00248. doi: 10.7554/eLife.00248.

Gu, M., H. Bao, and H. Kang. 2014. "Fibre-optical microendoscopy." *J Microsc* 254(1):13–18. doi: 10.1111/jmi.12119.

Ha, T., T. Enderle, D. F. Ogletree, et al. 1996. "Probing the interaction between two single molecules: fluorescence resonance energy transfer between a single donor and a single acceptor." *Proc Natl Acad Sci USA* 93(13):6264–6268. doi: 10.1073/pnas.93.13.6264.

Han, K. Y., K. I. Willig, E. Rittweger, et al. 2009. "Three-dimensional stimulated emission depletion microscopy of nitrogen-vacancy centers in diamond using continuous-wave light." *Nano Lett* 9(9):3323–3329. doi: 10.1021/nl901597v.

Hanson, J. A., K. Duderstadt, L. P. Watkins, et al. 2007. "Illuminating the mechanistic roles of enzyme conformational dynamics." *Proc Natl Acad Sci USA* 104(46):18055–18060. doi: 10.1073/pnas.0708600104.

Hashimoto, M., H. Niioka, K. Ashida, et al. 2015. "High-sensitivity and high-spatial-resolution imaging of self-assembled monolayer on platinum using radially polarized beam excited second-harmonic-generation microscopy." *Appl Phys Express* 8(11):112401. doi: 10.7567/apex.8.112401.

Hell, S. W. 2003. "Toward fluorescence nanoscopy." *Nat Biotechnol* 21(11):1347–1355. doi: 10.1038/nbt895.

Hell, S. W. 2007. "Far-field optical nanoscopy." *Science* 316(5828):1153–1158. doi: 10.1126/science.1137395.

Hell, S.W., and M. Kroug. 1995. "Ground-state-depletion fluorscence microscopy: A concept for breaking the diffraction resolution limit." *Appl. Phys. B* 60, 495–497. doi: 10.1007/BF01081333.

Hell, S. W., and J. Wichmann. 1994. "Breaking the diffraction resolution limit by stimulated-emission-stimulated-emission-depletion fluorescence microscopy." *Opt Lett* 19(11):780–782. doi: 10.1364/Ol.19.000780.

Hellwarth, R. W. 1963. "Theory of stimulated Raman scattering." *Phys Rev* 130(5):1850–1852. doi: 10.1103/PhysRev.130.1850.

Hellwarth, R., and P. Christensen. 1974. "Nonlinear optical microscopic examination of structure in polycrystalline ZnSe." *Opt Commun.* 12(3):318–322. doi: 10.1016/0030-4018(74)90024-8.

Hochreiter, B., A. Pardo-Garcia, and J. A. Schmid. 2015. "Fluorescent proteins as genetically encoded FRET biosensors in life sciences" *Sensors* 15(10):26281–26314. doi: 10.3390/s151026281.

Hohlbein, J., T. D. Craggs, and T. Cordes. 2014. "Alternating-laser excitation: Single-molecule FRET and beyond." *Chem Soc Rev* 43(4):1156–1171. doi: 10.1039/c3cs60233h.

Hu, F., L. Shi, and W. Min. 2019. "Biological imaging of chemical bonds by stimulated Raman scattering microscopy." *Nat Methods* 16(9):830–842. doi: 10.1038/s41592-019-0538-0.

Ishii, H., K. Otomo, J. H. Hung, et al. 2019. "Two-photon STED nanoscopy realizing 100 nm spatial resolution utilizing high-peak-power sub-nanosecond 655 nm pulses." *Biomed Opt Express* 10(7):3104–3113. doi: 10.1364/BOE.10.003104.

Ji, M., M. Arbel, L. Zhang, et al. 2018. "Label-free imaging of amyloid plaques in Alzheimer's disease with stimulated Raman scattering microscopy." *Sci Adv* 4(11): eaat7715. doi: 10.1126/sciadv.aat7715.

Ji, M., D. A. Orringer, C. W. Freudiger, et al. 2013. "Rapid, label-free detection of brain tumors with stimulated Raman scattering microscopy." *Sci Transl Med* 5(201):201ra119. doi: 10.1126/scitranslmed.3005954.

Jing, Y. Y., M. T. Myaing, T. P. Thomas, et al. 2005. "Development of a double-clad photonic-crystal-fiber-based scanning microscope." *Multiphoton Microscopy Biomed. Sci. V.* 5700:23–27.

Jurna, M., J. P. Korterik, C. Otto, et al. 2007. "Shot noise limited heterodyne detection of CARS signals." *Opt Express* 15(23):15207–15213. doi: 10.1364/oe.15.015207.

Kano, H., and H. O. Hamaguchi. 2006. "Dispersion-compensated supercontinuum generation for ultrabroadband multiplex coherent anti-Stokes Raman scattering spectroscopy." *J Raman Spectrosc* 37(1–3):411–415. doi: 10.1002/jrs.1436.

Kapanidis, A. N., N. K. Lee, T. A. Laurence, et al. 2004. "Fluorescence-aided molecule sorting: analysis of structure and interactions by alternating-laser excitation of single molecules." *Proc Natl Acad Sci USA* 101(24):8936–8941. doi: 10.1073/pnas.0401690101.

Karymov, M. A., M. Chinnaraj, A. Bogdanov, et al. 2008. "Structure, dynamics, and branch migration of a DNA Holliday junction: A single-molecule fluorescence and modeling study." *Biophys J* 95(9):4372–4383. doi: 10.1529/biophysj.108.135103.

Kastrup, L., H. Blom, C. Eggeling, et al. 2005. "Fluorescence fluctuation spectroscopy in subdiffraction focal volumes." *Phys Rev Lett* 94(17):178104. doi: 10.1103/PhysRevLett.94.178104.

Keikhosravi, A., J. S. Bredfeldt, A. K. Sagar, et al. 2014. "Second-harmonic generation imaging of cancer." *Methods Cell Biol* 123:531–546. doi: 10.1016/B978-0-12-420138-5.00028-8.

Kim, H., G. W. Bryant, and S. J. Stranick. 2012. "Superresolution four-wave mixing micros-copy." *Opt Express* 20(6):6042–6051. doi: 10.1364/OE.20.006042.

Kinegawa, R., K. Hiramatsu, K. Hashimoto, et al. 2019. "High-speed broadband Fourier-trans-form coherent anti-stokes Raman scattering spectral microscopy." *J Raman Spectrosc* 50(8):1141–1146. doi: 10.1002/jrs.5630.

Klar, T. A., and S. W. Hell. 1999. "Subdiffraction resolution in far-field fluorescence micros-copy." *Opt Lett* 24(14):954–956. doi: 10.1364/ol.24.000954.

Koivistoinen, J., P. Myllyperkio, and M. Pettersson. 2017. "Time-resolved coherent anti-Stokes Raman scattering of graphene: Dephasing dynamics of optical phonon." *J Phys Chem Lett* 8(17):4108–4112. doi: 10.1021/acs.jpclett.7b01711.

Lanin, A. A., A. S. Chebotarev, M. S. Pochechuev, et al. 2020. "Two- and three-photon absorp-tion cross-section characterization for high-brightness, cell-specific multiphoton fluores-cence brain imaging." *J Biophotonics* 13(3):e201900243. doi: 10.1002/jbio.201900243.

Lee, Y. J., Y. Liu, and M. T. Cicerone. 2007. "Characterization of three-color CARS in a two-pulse broadband CARS spectrum." *Opt Lett* 32(22):3370–3372. doi: 10.1364/ol.32.003370.

Lehnhardt, M., T. Riedl, T. Rabe, et al. 2011. "Room temperature lifetime of triplet exci-tons in fluorescent host/guest systems." *Org Electronics* 12(3):486–491. doi: 10.1016/j.orgel.2010.12.017.

Lesoine, M.D., U. Bhattacharjee, Y. Guo, et al. 2013. "Subdiffraction, luminescence-depletion imaging of isolated, giant, CdSe/CdS nanocrystal quantum dots." *J Phys Chem C* 117(7): 3662–3667. doi: 10.1021/jp312231k.

Li, B., C. Wu, M. Wang, et al. 2020. "An adaptive excitation source for high-speed multiphoton microscopy." *Nat Methods* 17(2):163–166. doi: 10.1038/s41592-019-0663-9.

Li, R., L. Wang, X. Mu, et al. 2019. "Nonlinear optical characterization of porous carbon mate-rials by CARS, SHG and TPEF." *Spectrochim Acta A Mol Biomol Spectrosc* 214:58–66 doi: 10.1016/j.saa.2019.02.010.

Li, R., X. X. Wang, Y. Zhou, et al. 2018. "Advances in nonlinear optical microscopy for bio-photonics." *J Nanophotonics* 12(3):033007. doi: Artn 03300710.1117/1.Jnp.12.033007.

Liao, C.-S., K.-C. Huang, W. Hong, et al. 2016. "Stimulated Raman spectroscopic imaging by microsecond delay-line tuning." *Optica* 3(12):1377–1380. doi: 10.1364/optica.3.001377.

Lin, Y., K. Nienhaus, and G. U. Nienhaus. 2018. "Nanoparticle probes for super-resolution flu-orescence microscopy." *Chem Nano Mat* 4(3):253–264. doi: 10.1002/cnma.201700375.

Ling, J., X. Miao, Y. Sun, et al. 2019. "Vibrational imaging and quantification of two-dimensional hexagonal boron nitride with stimulated Raman scattering." *ACS Nano* 13 (12):14033–14040. doi: 10.1021/acsnano.9b06337.

Liu, G., Z. Sun, Z. Fu, et al. 2017a. "Temperature sensing and bio-imaging applications based on polyethylenimine/CaF_2 nanoparticles with upconversion fluorescence." *Talanta* 169:181–188 doi: 10.1016/j.talanta.2017.03.054.

Liu, H., X. Deng, S. Tong, et al. 2019a. "In vivo deep-brain structural and hemodynamic multiphoton microscopy enabled by quantum dots." *Nano Lett* 19(8):5260–5265. doi: 10.1021/acs.nanolett.9b01708.

Liu, L., E. Wang, X. Zhang, et al. 2014. "Mems-based 3d confocal scanning microendoscope using mems scanners for both lateral and axial scan." *Sens Actuators A Phys* 215:89–95 doi: 10.1016/j.sna.2013.09.035.

Liu, T., M. Rajadhyaksha, and D. L. Dickensheets. 2019b. "MEMS-in-the-lens architecture for a miniature high-NA laser scanning microscope." *Light Sci Appl* 8:59. doi: 10.1038/s41377-019-0167-5. Bayguinov

Liu, Y., Y. Lu, X. Yang, et al. 2017b. "Amplified stimulated emission in upconversion nanoparticles for super-resolution nanoscopy." *Nature* 543(7644):229–233. doi: 10.1038/nature21366.

Lu, F. K., S. Basub, V. Igrasc, et al. 2015. "Label-free DNA imaging in vivo with stimulated Raman scattering microscopy." *Proc Natl Acad Sci USA* 112(43):E5902. doi: 10.1073/pnas.1519382112.

Maiman, T. H. 1967. Ruby laser systems: US, US3353115 A[P].

Maruo, M., H. Inagawa, Y. Toratani, et al. 2014. "Three-dimensional laser-scanning confocal reflecting microscope for multicolor single-molecule imaging at 1.5 K." *Cheml Phys Lett* 591:233–236. doi: 10.1016/j.cplett.2013.11.039.

Merchant, K. A., R. B. Best, J. M. Louis, et al. 2007. "Characterizing the unfolded states of proteins using single-molecule FRET spectroscopy and molecular simulations." *P Natl Acad USA* 104(5):1528–1533. doi: 10.1073/pnas.0607097104.

Minsky, M. 1957. US Patent# 3013467, Microscopy Apparatus. The original patent for confocal microscopy by Marvin Minsky.

Minsky, M. 1988. "Memoir on inventing the confocal scanning microscope." *Scanning* 10(4): 128–138. doi: 10.1002/sca.4950100403.

Miyamoto, D., and M. Murayama. 2016. "The fiber-optic imaging and manipulation of neural activity during animal behavior." *Neurosci Res* 103:1–9. doi: 10.1016/j.neures. 2015.09.004.

Moneron, G., and S. W. Hell. 2009. "Two-photon excitation STED microscopy." *Opt Express* 17(17):14567–14573. doi: 10.1364/oe.17.014567.

Müller, M., and J. M. Schins. 2002. "Imaging the thermodynamic state of lipid membranes with multiplex CARS microscopy." *J Phys Chem* B106(14):3715–3723. doi: 10.1021/ jp014012y.

Nadiarnykh, O., R. B. LaComb, M. A. Brewer, et al. 2010. "Alterations of the extracellular matrix in ovarian cancer studied by Second Harmonic Generation imaging microscopy." *BMC Cancer* 10:94. doi: 10.1186/1471-2407-10-94.

Nagel, S. R. 1984. "Review of the depressed cladding single-mode fiber design and performance for the SI undersea system application." *J Lightwave Technol* 2(6):792–801. doi: 10.1109/Jlt.1984.1073723.

Nan, X., W. Y. Yang, and X. S. Xie. 2004. "CARS microscopy lights up lipids in living cells." *Biophotonics Int.* 11:44–47.

Nandakumar, P., A. Kovalev, and A. Volkmer. 2009. "Vibrational imaging based on stimulated Raman scattering microscopy." *New J Phys* 11(3):033026. doi: 10.1088/1367-2630/ 11/3/033026.

Neupane, B., F. S. Ligler, and G. Wang. 2014. "Review of recent developments in stimulated emission depletion microscopy: Applications on cell imaging." *J Biomed Opt* 19(8): 080901. doi: 10.1117/1.JBO.19.8.080901.

Nickerson, A., T. Huang, L. J. Lin, et al. 2014. "Photoactivated localization microscopy with bimolecular fluorescence complementation (BiFC-PALM) for nanoscale imaging of protein-protein interactions in cells." *PLoS One* 9(6):e100589. doi: 10.1371/journal. pone.0100589.

Nuriya, M., H. Yoneyama, K. Takahashi, et al. 2019. "Characterization of intra/extracellular water states probed by ultrabroadband multiplex coherent anti-Stokes Raman scattering (CARS) spectroscopic imaging." *J Phys Chem A* 123(17):3928–3934. doi: 10.1021/acs. jpca.9b03018.

Okamoto, K. 2013. "Introduction of FRET application to biological single-molecule experiments". *Int J Biophys.* 3(1A):9–17. doi:10.5923/s.biophysics.201311.02.

Olivier, N., M. A. Luengo-Oroz, L. Duloquin, et al. 2010. "Cell lineage reconstruction of early zebrafish embryos using label-free nonlinear microscopy." *Science* 329(5994):967–971. doi: 10.1126/science.1189428.

Oron, D., N. Dudovich, and Y. Silberberg. 2003. "Femtosecond phase-and-polarization control for background-free coherent anti-Stokes Raman spectroscopy." *Phys Rev Lett* 90(21): 213902. doi: 10.1103/PhysRevLett.90.213902.

Ozeki, Y., F. Dake, S. Kajiyama, K. Fukui, et al. 2009. "Analysis and experimental assessment of the sensitivity of stimulated Raman scattering microscopy." *Opt Express* 17(5):3651– 3658. doi: 10.1364/oe.17.003651.

Ozeki, Y., W. Umemura, Y. Otsuka, et al. 2012. "High-speed molecular spectral imaging of tissue with stimulated Raman scattering." *Nat Photonics* 6(12):845–851. doi: 10.1038/nphoton.2012.263.

Paddock, S. W. 1999. "Confocal laser scanning microscopy." *Biotechniques* 27(5):992–996, 998–1002, 1004. doi: 10.2144/99275ov01.

Paddock, S. W. 2000. "Principles and practices of laser scanning confocal microscopy." *Mol Biotechnol* 16(2):127–149. doi: 10.1385/MB:16: 2: 127.

Pantazis, P., J. Maloney, D. Wu, et al. 2010. "Second harmonic generating (SHG) nanoprobes for in vivo imaging." *Proc Natl Acad Sci USA* 107(33):14535–14540. doi: 10.1073/pnas.1004748107.

Pawlicki, M., H. A. Collins, R. G. Denning, et al. 2009. "Two-photon absorption and the design of two-photon dyes." *Angew Chem Int Ed Engl* 48(18):3244–3266. doi: 10.1002/anie.200805257.

Peron, S., T. W. Chen, and K. Svoboda. 2015. "Comprehensive imaging of cortical networks." *Curr Opin Neurobiol* 32:115–123. doi:10.1016/j.conb.2015.03.016.

Pestov, D., R. K. Murawski, G. O. Ariunbold, et al. 2007. "Optimizing the laser-pulse configuration for coherent Raman spectroscopy." *Science* 316(5822):265–268. doi: 10.1126/science.1139055.

Peters, K. 2011. "Polymer optical fiber sensors-a review." *Smart Mater Struct* 20(1). doi: 10.1088/0964-1726/20/1/013002.

Petrov, G. I., R. Arora, V. V. Yakovlev, et al. 2007. "Comparison of coherent and spontaneous Raman microspectroscopies for noninvasive detection of single bacterial endospores." *Proc Natl Acad Sci USA* 104(19):7776–7779. doi: 10.1073/pnas.0702107104.

Piyawattanametha, W., and T. D. Wang. 2010. "MEMS-based dual axes confocal microendoscopy." *IEEE J Sel Top Quantum Electron* 16(4):804–814. doi: 10.1109/JSTQE.2009.2032785.

Pletnev, S., D. Shcherbo, D. M. Chudakov, et al. 2008. "A crystallographic study of bright far-red fluorescent protein mKate reveals pH-induced cis-trans isomerization of the chromophore." *J Biol Chem* 283(43):28980–28987. doi: 10.1074/jbc.M800599200.

Ploetz, E., S. Laimgruber, S. Berner, et al. 2007. "Femtosecond stimulated Raman microscopy." *Appl Phys B* 87(3):389–393. doi: 10.1007/s00340-007-2630-x.

Polzer, C., S. Ness, M. Mohseni, et al. 2019. "Correlative two-color two-photon (2C2P) excitation STED microscopy." *Biomed Opt Express* 10(9):4516–4530. doi: 10.1364/BOE.10.004516.

Potcoava, M. C., G. L. Futia, J. Aughenbaugh, et al. 2014. "Raman and coherent anti-Stokes Raman scattering microscopy studies of changes in lipid content and composition in hormone-treated breast and prostate cancer cells." *J Biomed Opt* 19(11):111605. doi: 10.1117/1.JBO.19.11.111605.

Potma, E. O., and X. S. Xie. 2005. "Direct visualization of lipid phase segregation in single lipid bilayers with coherent anti-Stokes Raman scattering microscopy." *Chemphyschem* 6(1):77–79. doi: 10.1002/cphc.200400390.

Rittweger, E., K. Y. Han, S. E. Irvine, et al. 2009. "STED microscopy reveals crystal colour centres with nanometric resolution." *Nat Photonics* 3(3):144–147. doi: 10.1038/nphoton.2009.2.

Roy, S., J. R. Gord, and A. K. Patnaik. 2010. "Recent advances in coherent anti-Stokes Raman scattering spectroscopy: Fundamental developments and applications in reacting flows." *Prog Energ Combust* 36(2):280–306. doi: 10.1016/j.pecs.2009.11.001.

Saar, B. G., Y. Zeng, C. W. Freudiger, et al. 2010. "Label-free, real-time monitoring of biomass processing with stimulated Raman scattering microscopy." *Angew Chem Int Ed Engl* 49(32):5476–5479. doi: 10.1002/anie.201000900.

Sarkar, S. K., N. M. Andoy, J. J. Benitez, et al. 2007. "Engineered Holliday junctions as single-molecule reporters for protein-DNA interactions with application to a MerR-family regulator." *J Am Chem Soc* 129(41):12461–12467. doi: 10.1021/ja072485y.

Sarri, B., R. Canonge, X. Audier, et al. 2019. "Fast stimulated Raman and second harmonic generation imaging for intraoperative gastro-intestinal cancer detection." *Sci Rep* 9(1):10052. doi: 10.1038/s41598-019-46489-x.

Shaw E., P. St-Pierre, K. McCluskey, et al. 2014. "Using sm-FRET and denaturants to reveal folding landscapes." *Methods Enzymol.* 549:313–341. doi: 10.1016/B978-0-12-801122-5.00014-3.

Shcherbakova, D. M., and V. V. Verkhusha. 2014. "Chromophore chemistry of fluorescent proteins controlled by light." *Curr Opin Chem Biol* 20:60–68. doi: 10.1016/j.cbpa.2014.04.010.

Sheppard, C. J. R., M. Castello, G. Tortarolo, et al. 2020. "Pixel reassignment in image scanning microscopy: a re-evaluation." *J Opt Soc Am A Opt Image Sci Vis* 37(1):154–162. doi: 10.1364/JOSAA.37.000154.

Silva, W. R., C. T. Graefe, and R. R. Frontiera. 2015. "Toward label-free super-resolution microscopy." *ACS Photonics* 3(1):79–86. doi: 10.1021/acsphotonics.5b00467.

Sivaguru, M., S. Durgam, R. Ambekar, et al. 2010. "Quantitative analysis of collagen fiber organization in injured tendons using Fourier transform-second harmonic generation imaging." *Opt Express* 18(24):24983–24993. doi: 10.1364/OE.18.024983.

Solgaard, O., A. A. Godil, R. T. Howe, et al. 2014. "Optical MEMS: From micromirrors to complex systems." *J Microelectromech S* 23(3):517–538. doi: 10.1109/jmems.2014.2319266.

Stryer, L., and R. P. Haugland, 1967. "Energy transfer: a spectroscopic ruler." *Proc Natl Acad Sci USA*, 58(2), 719–726. doi: 10.1073/pnas.58.2.719.

Sung, K. B., C. Liang, M. Descour, et al. 2002. "Fiber-optic confocal reflectance microscope with miniature objective for in vivo imaging of human tissues." *IEEE Trans Biomed Eng* 49(10):1168–1172. doi: 10.1109/TBME.2002.803524.

Tai, S. P., M. C. Chan, T. H. Tsai, et al. 2004. "Two-photon fluorescence microscope with a hollow-core photonic crystal fiber." *Opt Express* 12(25):6122–6128. doi: 10.1364/opex.12.006122.

Terhune, R. W., P. D. Maker, and C. M. Savage. 1965. "Measurements of nonlinear light scattering." *Phys Rev Lett* 14(17):681–684. doi: 10.1103/PhysRevLett.14.681.

Thiele, J. C., D. A. Helmerich, N. Oleksiievets, et al. 2020. "Confocal fluorescence-lifetime single-molecule localization microscopy." *ACS Nano* 14(10):14190–14200. doi: 10.1021/acsnano.0c07322.

Tian, N., L. Fu, and M. Gu. 2015. "Resolution and contrast enhancement of subtractive second harmonic generation microscopy with a circularly polarized vortex beam." *Sci Rep* 5:13580. doi: 10.1038/srep13580.

Tokarz, D., R. Cisek, A. Joseph, et al. 2019. "Characterization of pancreatic cancer tissue using multiphoton excitation fluorescence and polarization-sensitive harmonic generation microscopy." *Front Oncol* 9:272. doi: 10.3389/fonc.2019.00272.

Tomishige, M., N. Stuurman, and R. D. Vale. 2006. "Single-molecule observations of neck linker conformational changes in the kinesin motor protein." *Nat Struct Mol Biol* 13(10): 887–894. doi: 10.1038/nsmb1151.

Tsafas, V., E. Gavgiotaki, M. Tzardi, et al. 2020. "Polarization-dependent second-harmonic generation for collagen-based differentiation of breast cancer samples." *J Biophotonics* e202000180. doi: 10.1002/jbio.202000180.

Valdmanis, J. A., and G. Mourou. 1986. "Subpicosecond electrooptic sampling-principles and applications." *IEEE J Quantum Elect* 22(1):69–78. doi: 10.1109/Jqe.1986.1072867.

Vicidomini, G., P. Bianchini, and A. Diaspro. 2018. "STED super-resolved microscopy." *Nat Methods* 15(3):173–182. doi: 10.1038/nmeth.4593.

Vicidomini, G., G. Moneron, K. Y. Han, et al. 2011. "Sharper low-power STED nanoscopy by time gating." *Nat Methods* 8(7):571–573. doi: 10.1038/nmeth.1624.

Vicidomini, G., A. Schonle, H. Ta, et al. 2013. "STED nanoscopy with time-gated detection: theoretical and experimental aspects." *PLoS One* 8(1): e54421. doi: 10.1371/journal.pone.0054421.

Virga, A., C. Ferrante, G. Batignani, et al. 2019. "Coherent anti-Stokes Raman spectroscopy of single and multi-layer graphene." *Nat Commun* 10(1):3658. doi: 10.1038/s41467-019-11165-1.

Vukojevic, V., M. Heidkamp, Y. Ming, et al. 2008. "Quantitative single-molecule imaging by confocal laser scanning microscopy." *Proc Natl Acad Sci USA* 105(47):18176–18181. doi: 10.1073/pnas.0809250105.

Waharte, F., K. Steenkeste, R. Briandet, et al. 2010. "Diffusion measurements inside biofilms by image-based fluorescence recovery after photobleaching (FRAP) analysis with a commercial confocal laser scanning microscope." *Appl Environ Microbiol* 76(17):5860–5869. doi: 10.1128/AEM.00754-10.

Wang, D., W. Fu, Y. Lei, et al. 2019a. "A theoretical study of diffraction limit breaking via coherent control of the relative phase in coherent anti-Stokes Raman scattering micros-copy." *Opt Express* 27(4):5005–5013. doi: 10.1364/OE.27.005005.

Wang, F., D. Banerjee, Y. Liu, et al. 2010. "Upconversion nanoparticles in biological labeling, imaging, and therapy." *Analyst* 135(8):1839–1854. doi: 10.1039/c0an00144a.

Wang, G., Y. Sun, Y. Chen, et al. 2020. "Rapid identification of human ovarian cancer in second harmonic generation images using radiomics feature analyses and tree-based pipeline optimization tool." *J Biophotonics* 13(9):e202000050. doi: 10.1002/jbio.202000050.

Wang, J. F., M. Yang, Y. Li, et al. 2015. "A confocal endoscope for cellular imaging." *Engineering* 1(3):351–360. doi:10.15302/J-ENG-2015081.

Wang, W., B. Wu, P. Liu, et al. 2019b. "Calculations of second harmonic generation with radi-ally polarized excitations by elliptical mirror focusing." *J Microsc* 273(1):36–45. doi: 10.1111/jmi.12758.

Wang, W., B. Wu, B. Zhang, et al. 2021. "Second harmonic generation microscopy using pixel reassignment." *J Microsc* 281(1):97–105. doi: 10.1111/jmi.12956.

Wei, M., L. Shi, Y. Shen, et al. 2019. "Volumetric chemical imaging by clearing-enhanced stimulated Raman scattering microscopy." *Proc Natl Acad Sci USA* 116(14):6608–6617. doi: 10.1073/pnas.1813044116.

Wen, B., K. R. Campbell, K. Tilbury, et al. 2016. "3D texture analysis for classification of second harmonic generation images of human ovarian cancer." *Sci Rep* 6:35734. doi: 10.1038/srep35734.

Wilson, T., and C. Sheppard 1984. *"Theory and Practice of Scanning Optical Microscopy"*. London: Academic Press.

Wu, S., Y. Huang, Q. Tang, et al. 2018. "Quantitative evaluation of redox ratio and collagen characteristics during breast cancer chemotherapy using two-photon intrinsic imaging." *Biomed Opt Express* 9(3):1375–1388. doi: 10.1364/BOE.9.001375.

Xiong, H., L. Shi, L. Wei, et al. 2019. "Stimulated Raman excited fluorescence spectroscopy and imaging." *Nat Photonics* 13(6):412–417. doi: 10.1038/s41566-019-0396-4.

Xu, C., W. Zipfel, J. B. Shear, et al. 1996. "Multiphoton fluorescence excitation: New spectral windows for biological nonlinear microscopy." *Proc Natl Acad Sci USA* 93(20):10763–10768. doi: 10.1073/pnas.93.20.10763.

Yamato, N., H. Niioka, J. Miyake, et al. 2020. "Improvement of nerve imaging speed with coherent anti-Stokes Raman scattering rigid endoscope using deep-learning noise reduc-tion." *Sci Rep* 10(1):15212. doi: 10.1038/s41598-020-72241-x.

Yang, W., and R. Yuste. 2017. "In vivo imaging of neural activity." *Nat Methods* 14(4):349–359. doi: 10.1038/nmeth.4230.

Yariv, A. 1976. "Three-dimensional pictorial transmission in optical fibers." *Appl Phys Lett* 28(2):88–89. doi: 10.1063/1.88650.

Ye, S., W. Yan, M. Zhao, et al. 2018. "Low-saturation-intensity, high-photostability, and high-resolution STED nanoscopy assisted by CsPbBr3 quantum dots." *Adv Mater* 30(23):e1800167. doi: 10.1002/adma.201800167.

Yeh, C. H., C. Z. Tan, C. A. Cheng, et al. 2018. "Improving resolution of second harmonic generation microscopy via scanning structured illumination." *Biomed Opt Express* 9(12):6081–6090. doi: 10.1364/BOE.9.006081.

Yu, M., F. Li, Z. Chen, et al. 2009. "Laser scanning up-conversion luminescence microscopy for imaging cells labeled with rare-earth nanophosphors." *Anal Chem* 81(3):930–935. doi: 10.1021/ac802072d.

Yue, S., J. Li, S. Y. Lee, et al. 2014. "Cholesteryl ester accumulation induced by PTEN loss and PI3K/AKT activation underlies human prostate cancer aggressiveness." *Cell Metab* 19(3):393–406. doi: 10.1016/j.cmet.2014.01.019.

Zeitoune, A. A., J. S. Luna, K. S. Salas, et al. 2017. "Epithelial ovarian cancer diagnosis of second-harmonic generation images: A semiautomatic collagen fibers quantification protocol." *Cancer Inform* 16. doi: 10.1177/1176935117690162.

Zhen, Q., and P. Wibool. 2019. "MEMS actuators for optical microendoscopy." *Micromachines (Basel)* 10(2). doi: 10.3390/mi10020085.

Zhen, Q., and P. Wibool. 2015. "MEMS based fiber optical microendoscopes." *Displays* 37:41–53. doi: 10.1016/j.displa.2014.12.001.

Zong, W., R. Wu, M. Li, et al. 2017. "Fast high-resolution miniature two-photon microscopy for brain imaging in freely behaving mice." *Nat Methods* 14(7):713–719. doi: 10.1038/nmeth.4305.

4 Flow Cytometry-Based Detection

James F. Leary
Purdue University

CONTENTS

DOI: 10.1201/9781003409472-4

4.1 INTRODUCTION

A number of rapidly evolving technologies intersect in terms of flow cytometry instrumentation. Biological and clinical applications of these technologies continue to grow as the importance of single-cell analyses is appreciated. Data analysis, after many years of being limited to the original conventional analyses, now embraces bioinformatics as the amount of data has extended beyond manual analyses of individual data files into the world of "big data."

4.1.1 FLOW CYTOMETRY AND PRINCIPLES OF "SINGLE-CELL ANALYSIS"

Flow cytometry involves measurements of various properties of a single cell in a liquid suspension. Single cells are rapidly moved through a light excitation source (typically one or more lasers) at typical speeds of 1–10 m/seconds at rates of 1000–10,000 cells/seconds. The resolution of measurements is typically 1%–5%, so small biological differences can be rapidly and reproducibly measured.

If a secondary cell sorting stage is added, cells can be isolated at a single cell level at the same speeds and resolutions. Typically, these "cell sorters" use a variation of inkjet technology whereby cells are ejected from a nozzle into droplets (typically a single cell per droplet), which can be subsequently analyzed or grown in the case of live cells. It has been one of the great biomedical technologies of the twentieth century and has revolutionized both basic biomedical research and clinical diagnostics and therapeutics. The size of these flow cytometers and cell sorters started with room-sized instruments and gradually evolved into benchtop instruments as many of the components were miniaturized. An example of a multicolor cell sorter is shown in Figure 4.1.

A number of useful books and reviews explain flow cytometry and cell sorting for the beginner (Shapiro, 2001, Vembadi et al., 2019, Givan, 2001, Omerod, 1994).

4.1.2 EARLY HISTORY AND EVOLUTION OF CYTOMETRY

Flow cytometry started with electrical impedance measurements of cell volume and rapid measurements of this parameter on single cells, the so-called "Coulter Counter"

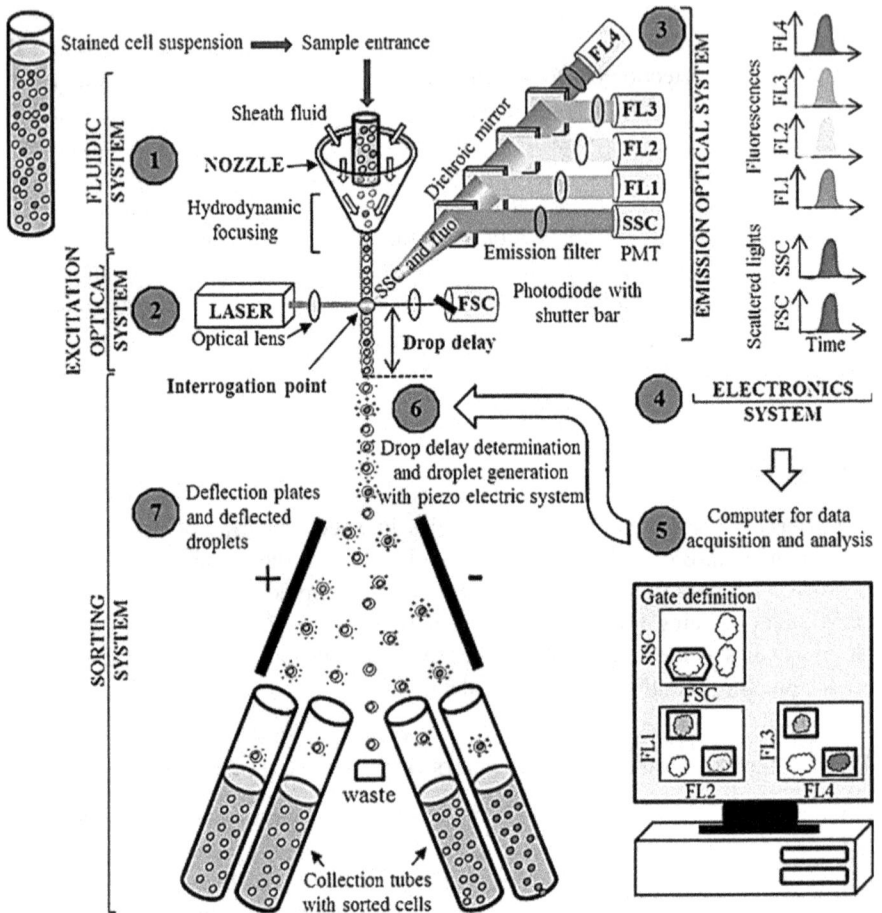

FIGURE 4.1 The front end is a "flow cytometer" which runs single cells past a light source and then collects light scatter and multiple colors of fluorescence data. The back end is the subsequent cell sorter that electrostatically deflects droplets containing cells of interest, based on these measurements, into containers or even single cells into microwells for cell cloning. (Reprinted with permission from Picot et al., 2012.)

invented by Wallace Coulter (Graham, 2013, 2016, 2021). The Coulter principle was originally discovered by W. Coulter in 1940, though US Patent 2,656,508 was not awarded until October 20, 1953. But then other scientists took light scatter and fluorescence measurements, based on staining cells with fluorescent dye molecules or fluorescent antibodies. A useful and readable book on Flow Cytometry was written by one of its pioneers and has since gone through many different editions (Shapiro, 2001). The invention of monoclonal antibodies and their use in AIDs detection and treatment greatly expanded the use of the technologies in both biomedical research departments and institutes as well as in medical clinics.

4.1.3   HIGH-SPEED, MULTIPARAMETER FLOW CYTOMETRY

Initially, the technology was fast, but not fast enough, to sample enough cells to find very rare cells. This changed in the 1980s and 1990s as James Leary's group developed extremely high-speed (>100,000 cells/seconds) instruments capable of detecting and isolating cells with frequencies below one cell in a million (Leary et al., 1993a,b, 1994, 1998a,b,d, 1999b, 2001a,b, 2002a,b, 2005, Leary, 1994b, 2000, 2005, Corio and Leary, 1993, 1996). An important application of this capability was minimal residual disease monitoring (Cooper and Leary, 2015, Leary et al., 1996, 1999a, 2001a,b, 2002b, Leary and McLaughlin, 1995, Hokanson et al., 1999) whereby flow cytometry rare cell sampling statistics was elucidated (Rosenblatt et al., 1997, Hokanson et al., 1999).

4.1.4   TIME-RESOLVED CYTOMETRY

Time-resolved cytometry (Steinkamp, 2001, Bitton et al., 2021) uses special "delayed fluorescence" probes, discussed later in this chapter. Since the fluorescence is separated by hundreds of nanoseconds or longer, the background autofluorescence, which is not delayed, has already occurred and gone before the delayed fluorescence is measured. This gives essentially a fluorescence signal over a zero background.

4.1.5   MICROFLUIDIC CYTOMETRY

Microfluidic cytometry had an almost completely different developmental timeline with different researchers who did not communicate much with the macro instrument people. This is a shame because both groups could have benefitted from the expertise and perspective of each group. Instead, they developed separately and only started to converge as a small number of researchers did both types of development.

Typically, microfluidic flow cytometers were very slow both in terms of cell velocities and numbers of cells/s that could be processed and were not capable of making high-resolution fluorescence measurements on single cells. Since microfluidic cytometers were very small in size with small channel cross sections, the volume of cell suspensions analyzed in reasonable periods of time was very small and, frequently, impractical, particularly for the detection of rare cell subpopulations. But now there are microfluidic chips being used on macro flow cytometers and cell sorters (Lapsley et al., 2013).

Another form of microfluidic cytometry involves using fluids of different viscosities that do not mix with each other. Droplets containing cells are flowing inside droplets (Shang et al., 2017). One advantage of such droplet systems is that the contents of a single cell can be examined because the contents fill only the original droplet.

High-speed microfluidic cytometry has also been developed. The use of multilevel branched microfluidic channels showed the capability to perform high-speed, including rare-cell applications, on a microfluidic cell sorter (Leary and Frederickson, 2008). Forms of branched channel microfluidic cytometry processing, including real-time cell sorting with on-chip magnetic cell sorting, of whole human blood through a microfluidic cytometer/cell sorter (Leary and Frederickson, 2008, Leary, 2018, 2019, 2021), as shown in Figure 4.2.

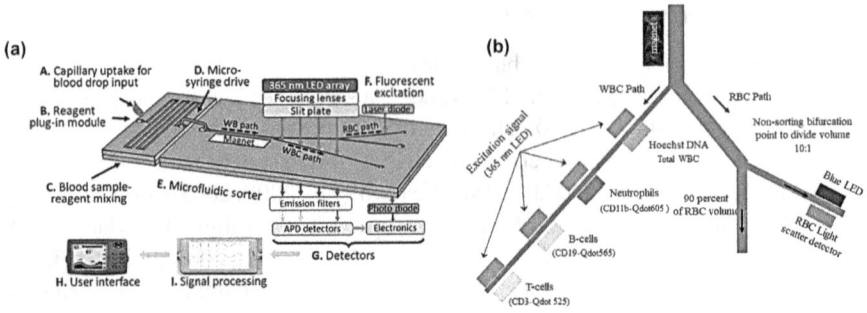

FIGURE 4.2 (a) A microfluidic cytometer is very similar to the flow cytometer described in Figure 4.1, but within enclosed microchannels. But it is much smaller and potentially portable if excitation light sources and detector can be powered by batteries. The microfluidic cytometer in this figure is also a cell sorter, in this case by bifurcating microchannels to perform on-chip real-time sorting of leukocytes from whole blood. (b) Branched microchannels with separate excitation and detection stations allow higher throughput rates of multicolor fluorescence measurements of cells.

4.1.6 CYTOF

Recently, there has been the development and commercialization of a new type of flow cytometer, the so-called "CyTOF" (Cytometry by Time-of-Flight) (Bandura et al., 2009). Mass cytometry, or CyTOF (by Fluidigm Corporation), is a variation of flow cytometry in which antibodies are labeled with heavy metal ion tags rather than fluorochromes. Analysis occurs in the Time-of-Flight (TOF) chamber, allowing the detection of over 40 different parameters simultaneously.

This figure provides a summary of how mass cytometry works. A liquid sample containing cells labeled with antibodies tagged with heavy metal isotope conjugated probes (i) are introduced into a nebulizer creating an aerosol (ii) which is directed towards a plasma torch (iii) where the cells are vaporized. Low-mass ions are removed (iv), and the ion cloud enters the TOF chamber where probes are separated based on their mass-to-charge ratio as they accelerate towards the detector (v). The time-resolved detector measures a mass spectrum that represents the identity and quantity of each isotopic probe on a per-cell basis (vi). Data is generated in FCS format (vii) and can be analyzed using third-party software (viii).

The number of CyTOF parameters, corresponding to multicolor fluorescence, is limited only by the number of isotope labels that can be generated. Samples must be processed through a multi-step procedure, as shown in Figure 4.3.

A typical workflow of a mass cytometer is shown in Figure 4.4.

4.2 FLOW CYTOMETRY AND SINGLE-CELL ANALYSIS TECHNOLOGIES

A variety of technologies including photonics, lasers, and digital signal processing has led to the current state-of-the-art. A critical aspect of flow cytometry is that is

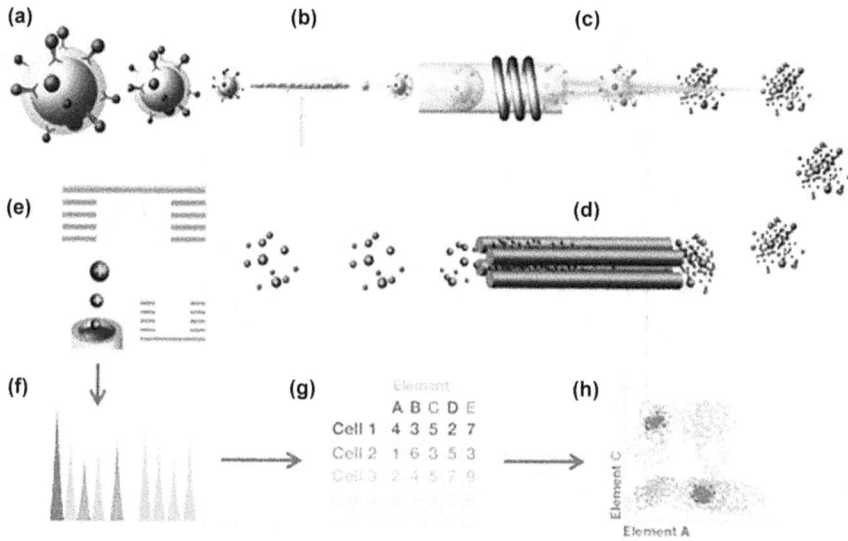

FIGURE 4.3 The multi-step process of a CyTOF mass cytometer. (Source: UCSF Cytometry Facility (https://flow.ucsf.edu/mass-cytometry).)

a "tightly designed" system, meaning that, when one aspect of the flow cytometer changes, the consequences ripple through the rest of the system design. This chapter also tries to provide some perspective on how flow cytometry evolved to its present state of sophistication. As in all areas of science, we stand on the shoulders of early giants in the field who deserve to be honored for their ground-breaking work. But these descriptions also show how flow cytometry has harnessed the many technologies invented for other fields as well as simply reusing consumer-level technologies developed for completely different uses.

The rapid development and growth in sophistication of flow cytometry happened because it was able to "piggyback" on the rapid technological developments of other fields including computers, spectroscopy, and digital signal processing.

4.2.1 OPTICS

The optics of "macro" flow cytometers are used both the focus the exciting light in single cells and to collect the light scatter or fluorescence measurements on detectors, usually after passing through optical filters in the case of fluorescence measurements. The excitation light is typically focused on an elliptical spot on the order of the diameters of the single cells and with a longer axis designed to minimize the excitation intensity variation spatially as the cell moves in the channel. Most macro flow cytometers use "hydrodynamic focusing" (Golden et al., 2012) whereby the spatial location of cells is limited to a small subregion of the overall channel by use of the concentric laminar flow of a sheath fluid around a smaller diameter channel of sample fluid containing cells (Crosland-Taylor, 1953). An important more recent

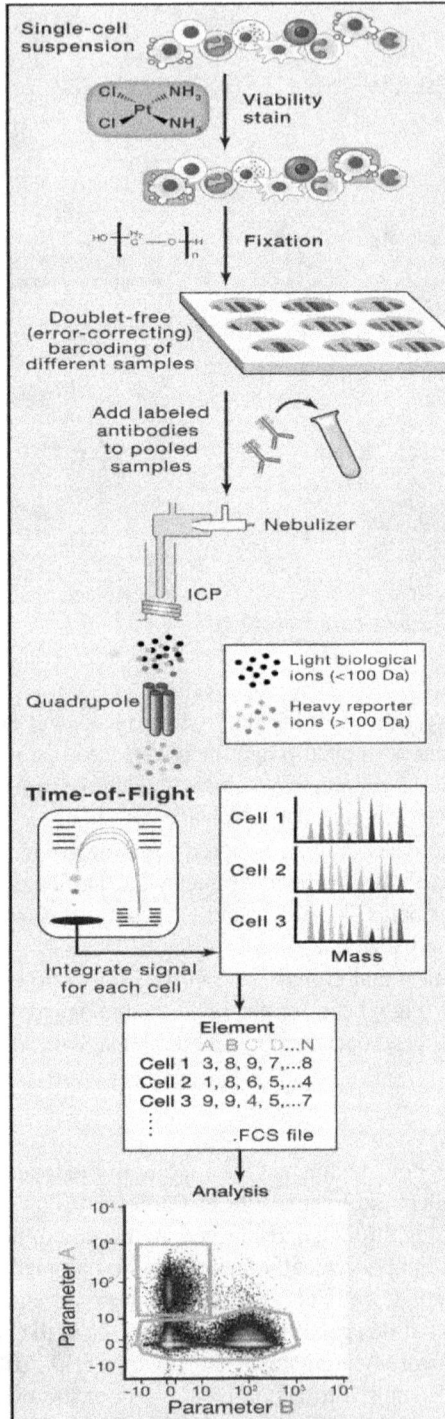

alternative is the use of "acoustic focusing" (Kaduchak, Ward and Kaduchak, 2018), described in Figure 4.5, whereby sound waves are used to focus the cells in a narrow line.

Since little or no sheath is needed to position the cells in the channel, this technology has important implications for the production of truly portable microfluidic cytometers to be discussed later in this chapter. Acoustic waves are sent through the sample channel. There are two-fold results. First, the cells are driven to the center of the stream in a single file line. Second, in the process, the inter-cell spacings are made far more uniform.

The results are dramatic and important for portable microfluidic cytometers. Concentric laminar flow is not needed, which is good for portable cytometers in the field. It is a system of "sheath-less" focusing of cells, essentially allowing the system to run with little or no sheath (Goddard et al., 2006). Second, the regularizing of spacing allows higher throughputs due to much lower cell-cell coincidence.

When lasers are used in macro flow cytometers and cell sorters, there is a need to confine the cells passing through that laser by either hydrodynamic focusing or acoustic focusing. This requirement is due to the approximately Gaussian laser beam profile whereby small deviations in cell position can lead to different cells receiving different excitation intensities. Since the desire is to have an ultimate fluorescent signal that depends on the amount of fluorochrome per cell, that relationship only holds true if all cells receive the same exciting intensities.

In microfluidic cytometry systems using light excitation sources other than lasers (e.g., LEDs), those LED light sources can have so-called "top hat" beams profiles described earlier that make the fluorescence per cell less dependent on position in the microfluidic channels. This means that we can, under these circumstances, reduce or even eliminate the need for hydrodynamic focusing with a sheath fluid. That is particularly important since microfluidic cytometers are usually designed for portability into the field in which case the overall weight of the system is important (Leary, 2018, 2019, 2021).

FIGURE 4.4   Workflow of a Typical Mass Cytometry Experiment. Single cells are acquired, and a viability stain is applied to mark dead cells for exclusion from analyses. Fixation can optionally be applied at this point to preserve the cell state. Multiple samples can be barcoded with unique combinations of heavy metal tags, enabling them to be pooled together prior to staining to minimize technical variability at this step. After pooling samples into one tube, cells are then incubated with antibodies targeted against proteins of interest. Cell permeabilization can be performed if intracellular targets are to be measured. Cells are nebulized into droplets as they are introduced into the mass cytometer. They then travel into an inductively coupled argon plasma (ICP), in which covalent bonds are broken and ions are liberated. The ion cloud is filtered by a quadrupole to remove common biological elements and enrich the heavy metal reporter ions to be quantified by time-of-flight mass spectrometry. Ion signals are integrated on a per-cell basis, resulting in single-cell measurements for downstream analysis. Data are compiled in an FCS file that can then be parsed and plotted in a variety of ways. (Reprinted with permission from Spitzer and Nolan, 2016.)

hydrodynamic focus only   hydrodynamic + acoustic focus

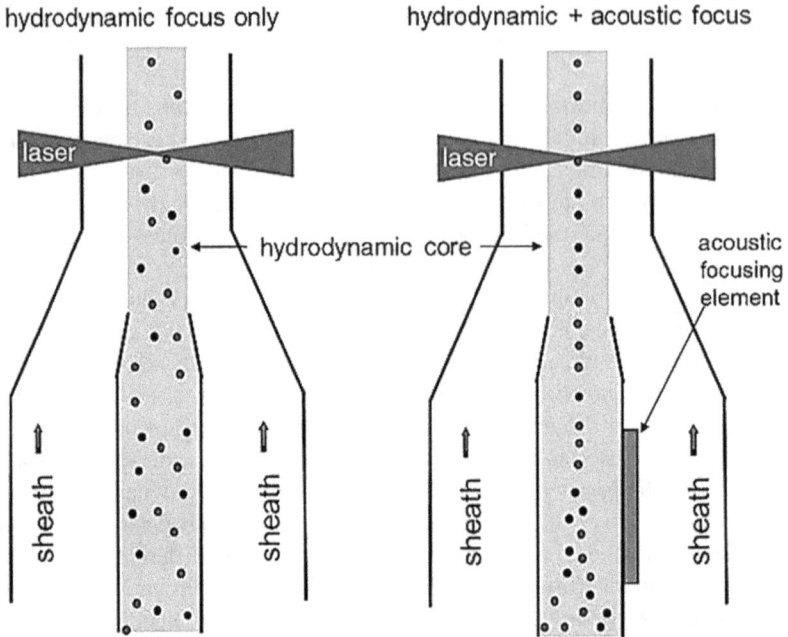

FIGURE 4.5 The acoustic focusing principle. (Reprinted with permission from Ward and Kaduchak, 2018.)

4.2.2 Optical Measurements on Single Cells

The essence of flow cytometry is optical measurements of single cells. To do this involves focusing a light source down to the dimensions of a single cell and the optical detection sensitivity structures light to detect "processed excitation light" in terms of single-cell light scatter and fluorescence.

4.2.3 "Label-Free" Detection of Cells

"Label-free" analysis means that properties of a single cell are deduced on the processed light without the need for extrinsic probes – particularly important if cells are to be sorted for subsequent applications such as transplantations where there is concern about the possible negative effects of these probes on the patient.

Early label-free systems in flow cytometry were based on either electrical impedance or light scatter analyses of single cells. Later intrinsic "autofluorescence" was used to distinguish the type of cell subpopulation and its characteristics. Extrinsic fluorescence measurements quickly superseded these label-free methods due to the much greater sophistication that fluorescence measurements can provide.

4.2.3.1 Conventional Light Scatter

One of the earliest uses of light scatter, still used today, is to distinguish lymphocytes, monocytes and granulocytes in human peripheral blood. Forward angle ($2°$–$8°$) and right angles (90°) light scatter, when combined in correlated 2-D scattergrams was sufficient to distinguish these three basic cell types. Other blood cell types were more difficult, or impossible, to distinguish based on simple light scatter measurements.

4.2.3.2 Multi-Angle Light Scatter

Multi-angle light scatter (32 different angles simultaneously) (Salzman et al., 1975, Mullaney, 1974) measurement done on a very special instrument at Los Alamos National Laboratory, one of the birthplaces of flow cytometry, was an early example of "big data" that was beyond the capability of researchers to do simple pattern recognition of the data. Analysis of this data was one of the first uses of cluster analysis techniques to recognize cell subpopulations in high-dimensional data space. While multi-angle light scatter was initially promising, it took a lot of work to analyze the complex data. But ultimately those analyses led to the application of cluster analysis of complex multicolor fluorescence data which has revolutionized the use of flow cytometry.

4.2.3.3 Impedance Measurements of Electronic Cell Volume

As mentioned earlier, the first single cell volume measurements were done by Wallace Coulter in his development of the first commercial "Coulter Counter" US Patent 2,656,508– the world's first "bench top" flow cytometer. These early manifestations of flow cytometers showed the value and importance of single-cell analysis and were quickly introduced to hospitals around the world and paved the way for the introduction of flow cytometers providing light scatter and fluorescence measurement to further distinguish different types of lymphocytes into B-cell and T-cell subsets important for diagnosing a wide variety of human diseases as well as a way of monitoring the effects of various treatments.

Further development of impedance measurements has led to the development of multi-frequency impedance cytometers that can provide some of the advantages of multicolor fluorescence measurements without the need for extrinsic labels.

4.2.3.4 Fluorescence

Early flow cytometry used one-color fluorescence measurements to perform cell cycle analysis. Before flow cytometry, cell cycle analysis was very laborious (hours to count a few hundred cells) and involved the counting of grains on photographic film caused by the release of radioactivity from tritiated thymidine DNA precursors taken up during DNA synthesis. No wonder the world was astounded at the title "Cell Cycle Analysis in 20 minutes" (Crissman and Tobey, 1974), one of the first papers using fluorescence and flow cytometry emanating from Los Alamos National Laboratory. Thousands of cells per second could be analyzed for cell cycle information. It was also an example of the need for more sophisticated analyses using mathematical deconvolution techniques to extract the basic cell cycle stages (Dean and Jett, 1974). Later developments using fluorescent monoclonal antibodies against BudR (Gratzner et al., 1975) were able to duplicate the initial tritiated thymidine cell cycle analysis measurements at thousands of cells/seconds.

Once a second color of fluorescence was introduced, flow cytometrists had to tackle the problem of fluorescence emission overlaps (Loken et al., 1977). In the early

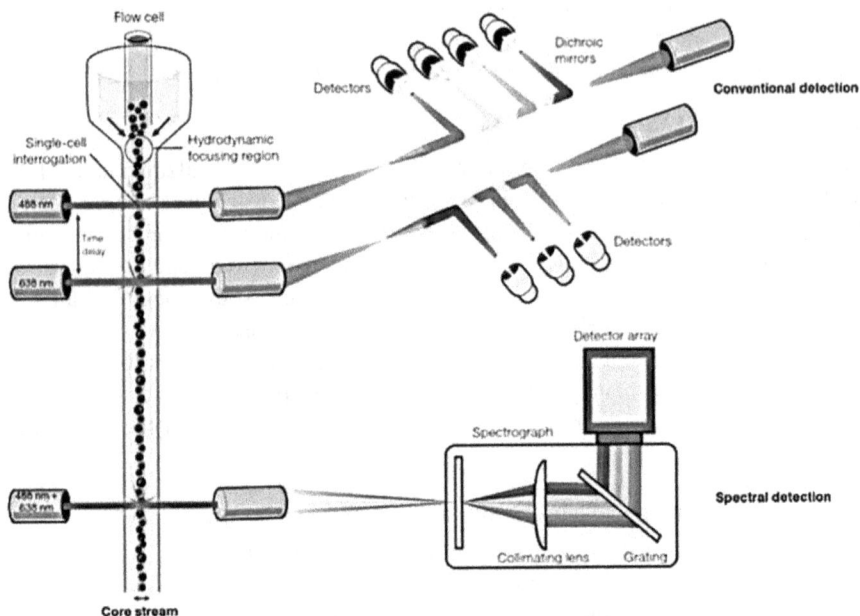

FIGURE 4.6 Multicolor flow cytometry can either use (Top) optical filters to direct light to separate detectors, or (Bottom) a multichannel (typically 32 channels) detector collecting light after a spectral diffraction grating.

days of flow cytometry, there were only a relatively small number of probes mainly used for fluorescence microscopy. Initially, spectral deconvolution was done by a simple but ingenious analog circuit. But the most important advance was the realization that many colors could be used. Initially, these multicolor fluorescence measurements were accomplished using a complex, multi-step separation of the colors using fluorescence filters similar to what was already being used in fluorescence microscopes. Dye chemists (e.g., Richard Haugland, Lubert Stryer, and Alan Waggoner) created thousands of new dyes, and filter manufacturers made increasingly sophisticated narrow bandpass filters to separate the various fluorescence colors. Removing remaining fluorescence overlaps necessitated the development of mathematical multicolor compensation ("spectral unmixing") algorithms as used today (Vo-Dinh et al., 2005).

More recently engineers have used multispectral analyses from spectroscopy to eliminate the need for complex optical filters leading to the current state-of-the-art "multispectral flow cytometry" (Novo et al., 2013) found in the more advanced flow cytometers. A comparison of conventional fluorescence cytometry and multispectral cytometry is shown in Figure 4.6.

4.3   FLUORESCENT PROBE TECHNOLOGIES

While early use of fluorescent probes in flow cytometry was simply reusing probes for fluorescence microscopy, dye chemists started developing new probes that were easily excited by the lasers used in flow cytometers. Blue light (488nm) was the major laser used in early flow cytometers and allowed the use of a number of fluorescent dyes. But as soon as flow cytometrists learned how to spatially separate two or more lasers in a form of "time-resolved flow cytometry," an explosion of new fluorescent dyes developed (Cunningham, 2010, Johnsom, 1998).

4.3.1 Fluorophores

Due to the fundamental design of flow cytometers, cells are excited for less than a millisecond. To get an amplification factor due to multiple excitations and fluorescent emissions of dyes from a single cell, most fluorophores have fluorescent lifetimes of 1–10 ns leading to an amplification factor of 1000–10,000 times of fluorescent signals from a single cell as it passes through the flow cytometer excitation light source. This is the fundamental basis of the exquisite sensitivity of flow cytometry to detect relatively small numbers of fluorescent molecules per cell.

4.3.1.1 Conventional Fluorophores

The first fluorescent probe, fluorescein isothiocyanate (FITC), attached to an antibody was developed by Albert Coons at Harvard University. I had the honor of giving Dr. Coons, then an old man and in his final years before his death, a tour of the flow cytometry at Los Alamos National Laboratory where I was a postdoctoral fellow at the time. He was astounded by how far flow cytometry had progressed. I asked him why FITC? With a twinkle in his eye, he said it was the only fluorescent dye he could get to work. As a desperate graduate student, he wanted to graduate! Never underestimate the ingenuity of desperate graduate students!

Rhodamine, another fluorescent dye used for fluorescence microscopy, was really the second major dye used for many flow cytometry applications. But the spectral overlap of rhodamine and fluorescein caused a lot of measurement problems and required the invention of spectral "compensation" (Loken et al., 1977).

4.3.1.2 New "Designer" Fluorophores

With the advent of cyanine dyes (Mujumdar et al., 1989, Ernst et al., 1989, Rastede et al., 2015, Constantin and Fague, 2008), chemists were able to fine-tune the emission properties of dyes by altering the length of the hydrocarbon chains within the dye molecule to get essentially any desired fluorescent emission spectrum. This led to the rapid proliferation of new dyes that could be excited with different color lasers.

4.3.1.3 Energy Transfer Probes

The overlap of fluorescein and rhodamine is not the only important characteristic of these tow dyes. When they are in close proximity (governed by a sixth-order inverse square distance) these two dyes transfer energy to each other. Since the emission of fluorescein (donor) overlaps the excitation of rhodamine (acceptor), one can measure and elegantly determine the relative distance of cell molecules labeled with these two probes (Szollosi et al., 1998, Tron et al., 1984).

This led to the development of so-called "tandem energy transfer probes," now commercially available. The early available tandem conjugate dyes had problems since the energy transfer depends not only on the relative distance between the donor and acceptor molecules but also on their relative orientation. If their electric dipole moments are spatially aligned (in three dimensions) there is good energy transfer, but if they are oriented differently, the energy transfer drops sharply. Chemists solved this problem by chemically introducing spatial orientation constraints.

4.3.1.4 SERS

Surface-enhanced Raman spectroscopy (SERS) is a powerful vibrational spectroscopy technique that allows for highly sensitive structural detection of low-concentration analytes through the amplification of electromagnetic fields generated by the excitation of localized surface plasmons. Raman scattering is an optical process where incoming excitation light interacting with a sample produces scattered light that is lessened in energy by the vibrational modes of the chemical bonds of the specimen.

The so-called "SERS effect" involves the interaction of the metal surface plasmons with the electronic and vibrational states of the adsorbed molecules, under the influence of oscillating electromagnetic radiation. When Raman active molecules are near gold or silver nanoparticles, the Raman intensity is significantly amplified. This phenomenon is referred to as surface-enhanced Raman spectroscopy (SERS).

4.3.1.4.1 Principles of SERS for Flow Cytometry

SERS for flow cytometry (Krutzik et al., 2004, Krutzik and Nolan, 2003, Nolan and Sebba, 2011). These surface plasmon resonance properties of gold and silver nanoparticles can be used as Surface Enhanced Raman Spectroscopy (SERS) probes. The SERS effect can increase the Raman signal intensities by up to 10^{14}–10^{15} times, resulting in a detection sensitivity comparable to fluorescence (Nie and Emory, 1997).

4.3.1.4.2 SERS Probes

SERS tags can also be made for flow cytometry probes (Krutzik et al., 2004, Krutzik and Nolan, 2003, Nolan and Sebba, 2011) as shown in Figure 4.7.

Instead of using different color fluorochromes, plasmonic nanoparticles with desired resonant wavelengths ("SERS tags") can be used. These SERS tags are Raman active compounds. Since their spectral signatures are quite narrow, more parameters can be measured.

4.3.2 FLUORESCENT PROBES

There are currently thousands of fluorescent probes or phosphorescence probes commercially available. The phosphorescence probes are primarily used for fluorescence microscopy because they have very long fluorescent lifetimes (typically milliseconds) but often low rates of photobleaching. But flow cytometry requires much lower fluorescent lifetimes (typically 1–10 ns) to be excited during the short transit time through a flow cytometer's excitation beam. For this reason, the fluorescent probes described in this chapter will be mostly organic dyes with nanosecond fluorescent lifetimes.

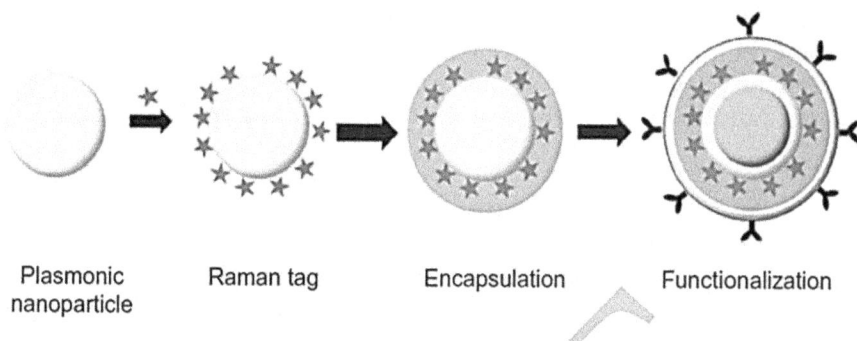

Plasmonic nanoparticle   Raman tag   Encapsulation   Functionalization

FIGURE 4.7 Anatomy of a SERS tag. A SERS tag generally consists of plasmonic nanoparticles with the desired resonant wavelengths, a Raman active compound that confers a particular spectral signature, a coating to stabilize and protect the SERS tag, and a targeting molecule such as an antibody or nucleic acid that confers molecular specificity to the SERS tag. (Reprinted with permission from Nolan and Sebba, 2011.)

An exception will be made to discuss so-called "delayed fluorescent" probes (Jin et al., 2009, Wang et al., 2020), primarily lanthanide heterocyclic probes which can still be used in a flow cytometer with fluorescent lifetimes on the order of 100 ns. Their importance is that very small numbers of molecules can be detected since the time delay for fluorescence allows the better separation of fluorescence from background light scatter as well as cellular autofluorescence which occurs on the timescale of nanoseconds, well separable from the delayed fluorescence signal which arrives later in time.

4.3.2.1 DNA Dyes

Initial so-called "DNA dyes" intercalated either reversibly (important for live "vital" cell analysis by flow cytometry and isolation by cell sorting) in the case of Hoechst 33342 (Arndt-Jovin and Jovin, 1977) or irreversibly, as in the case of ethidium bromide or propidium iodide. There are now a number of commercially available vital DNA dyes (e.g., Vybrant Dye Cycle dyes) that can be excited by different color lasers (Bradford et al., 2006).

Later fluorescent dyes were developed that bound specifically to either AT- or GC base pairs in DNA. These dyes were critical to the early chromosome analysis by flow cytometry in so-called "flow karyotyping" which was subsequently used in the early stages of the Human Genome Project at Los Alamos National Laboratory and Lawrence Livermore National Laboratory (Van Dilla et al., 1986). These techniques were then used to analyze polymorphisms in these human chromosomes (Trask et al., 1989, Pinkel et al., 1998).

4.3.2.2 Other Fluorescent Stains

A wide variety of other fluorescent dyes, mainly binding as fluorescent ligands, have been developed. One of the most useful was the development of fluorescent Annexin

V ligands which could bind to phosphatidyl serine to help dissect the process of programmed cell death (apoptosis) in cells (for reviews see Darzynkiewicz et al., 1997, 2008). Later fluorescently labeled probes for cyclins were found to be even more sensitive to subtle changes in cell cycle dynamics (Gong et al., 1995, Darzynkiewicz et al., 1996).

4.3.2.3 Antibodies

Initially, antibodies were produced by inoculating animals (typically mouse, goat or sheep) with antigens and harvesting the serum of these animals which hopefully contained antibodies generated from those animals. The major problem with this approach was that the serum antibodies were "polyclonal" meaning they were all recognizing a variety of different epitopes which could be in common to many antigens.

A revolution occurred with the introduction of so-called "monoclonal antibodies" (Kohler and Milstein, 1975), whereby hybridoma cells (typically myelogenous leukemia cells) were created in the laboratory and either grown in-vitro or in-vivo within the peritoneal cavity of mice (now banned in many Western countries due to perceived cruelty to these animals).

4.3.2.4 "Nanobodies"

"Nanobodies," a relatively recent term, are a novel class of proprietary therapeutic proteins based on single-domain antibody fragments that contain the unique structural and functional properties of naturally-occurring heavy-chain-only antibodies. They may have major therapeutic potential (Jovcevska and Muyldermans, 2020). The simplest forms are fragments of antibodies (Fab fragments, F(ab)2 as well as other shape-recognizing molecules such as peptides (Figure 4.8).

4.3.2.5 Peptides, Aptamers, etc.

Antibodies are not the only molecule that can recognize other molecules. Peptides fold themselves in ways that can allow them to recognize other molecules. They can also be constructed and manufactured entirely in-vitro. Peptides as small as seven amino acids can recognize specific biomarkers (Haglund et al., 2008). The size of the recognition agent is important because antibodies, particularly the complete immunoglobulin, are more than 180 kDa (depending on Ig subtype). For multicolor immunofluorescence labeling strategies, it allows more different antigens on the same cell to be recognized without interfering with each other. Such interference is called "steric hindrance." Cell surface real estate is limited! While few researchers are labeling more than four or five specific biomarkers on the same cells, at that level one must be careful to determine whether the labeling of some antigens, governed by the dynamics of avidity do not overwhelm other antigens that are fewer in number and/or are of lower avidity. One way to test this is to vary the order of labeling in steps rather than trying to label all antigens simultaneously.

Aptamers are small molecules made from DNA or RNA (or in some cases DNA: RNA hybrids. They are comparatively easy to chemically construct and can be constructed against virtually any antigen, some of which it is not possibly to raise antibodies against. Vast libraries (e.g., 100 million or more) of aptamers can be

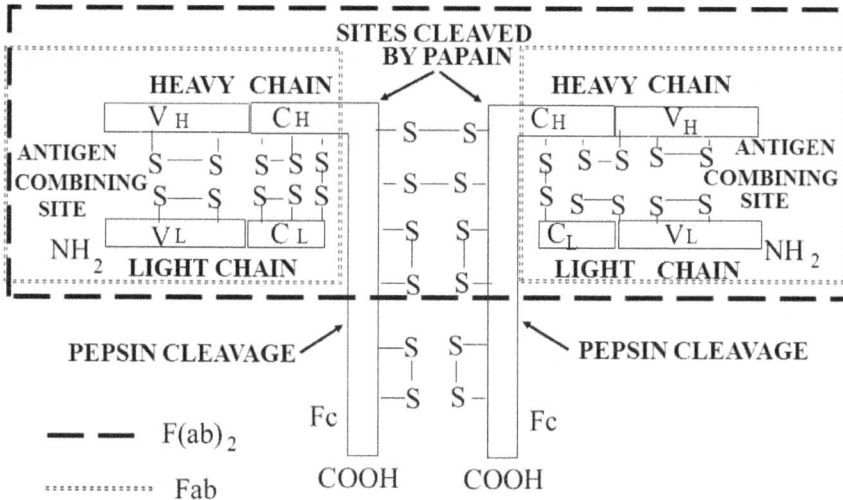

FIGURE 4.8 Whole immunoglobulins are very large (180–220 kDa). Using different enzymatic cleavages, this large protein complex can be broken down into much smaller subunits (e.g., Fab, F(ab)2, Fc) that can still label cells with lower instances of steric hindrance.

chemically constructed and then screened by either panning techniques or high-speed flow cytometry and cell sorting (Yang et al., 2002, 2003, 2011, Leary et al., 2005, Gorenstein et al., 2011) to choose the "winners" in terms of recognizing specific antigens (Figure 4.9).

A database by the Ellington group at UT-Austin (Lee et al., 2004) continues to gather aptamer sequences from around the world and organize them into a searchable library.

4.3.2.6 Nanoparticles

Nanoparticles with recognition molecules on board, particularly peptides or aptamers since the real estate on nanoparticles is very limited! These nanoparticles can be fluorescent, magnetic, or both. Why will the subject of nanoparticle labeling of biomarkers, and their screening by flow cytometry, become increasingly important? Because the future of drug delivery is the use of nanoparticles containing recognition molecules to guide drugs to specific diseased cells – a new field of clinical science known as nanomedicine (Leary, 2022) which involves massive parallel processing single cell therapies to eliminate diseased cells at the single cell level – a form of "nanosurgery." It is a natural wedding of single cell flow cytometry and single cell medicine. Drugs can be guided to specific diseased cells and at the proper dose. This allows the drug exposure of patients to be reduced by more than 90%, eliminating most "side-effects" which cause patients to stop taking those drugs that could deal with their diseases. Nanomedicine is single-cell medicine for which single-cell flow cytometry is particularly appropriate.

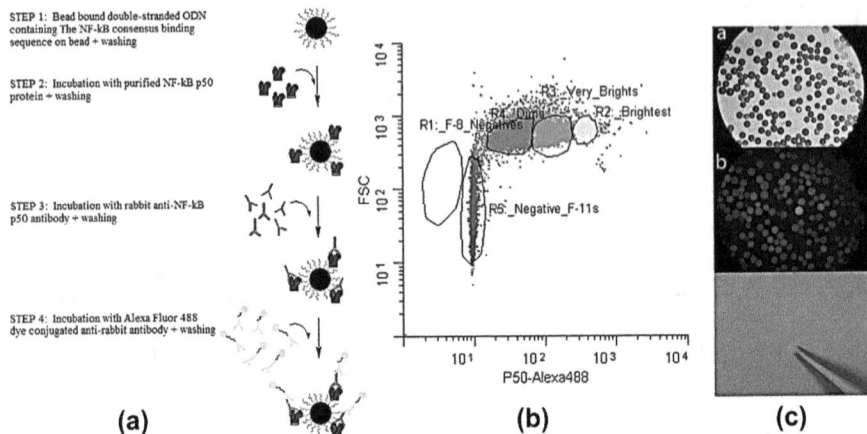

STEP 1: Bead bound double-stranded ODN containing The NF-kB consensus binding sequence on bead + washing

STEP 2: Incubation with purified NF-kB p50 protein + washing

STEP 3: Incubation with rabbit anti-NF-kB p50 antibody + washing

STEP 4: Incubation with Alexa Fluor 488 dye conjugated anti-rabbit antibody + washing

(a)   (b)   (c)

FIGURE 4.9 Combinatorial chemistry can be used to generate millions of probes attached to bead libraries which can then be rapidly screened by high-speed flow cytometry and bead sorting. Beads which bind with highest avidities can be selected by this process and then sequenced to determine the peptide (or in this case, thioaptamer sequences so they can then be synthesized in large quantities. (Reprinted with permission from Yang et al., 2003.)

4.4 FLUIDICS

The essence of flow cytometry is the "flow" which is a way of delivering single cells for analysis. In macro flow cytometers there is a lot of fluid going through a flow cell for high-speed single-cell analyses. To ensure that all cells receive the same light, the cells in the single-cell sample suspension are hydrodynamically focused with sheath fluid to be in the same location of the excitation beam – usually the center. They must be dispersed to single-cell level and also run at a fluid flow rate that ensures that the cells are spaced apart enough to allow only one cell at a time in the excitation light. They also need, in general, to allow enough time between cells to finish the signal processing of that cell, although systems have been built with buffer circuitry to deal with cells that are within the same time period to perform the signal processing. Since the purpose of flow cytometry is single, and not multiple, cell analysis cells that are too close together are usually eliminated from analysis by anti-coincidence circuitry. All of this can be predicted by queuing theory models (Leary, 1994b, 2005, Hokanson et al., 1999, Gross and Harris, 1985). A similar situation applies to cell sorting except that the queue lengths are much longer than the analysis queue lengths. A sorting anti-coincidence circuitry tries to detect of two or more cells are residing in the same sorting unit (usually one, two, or three droplets) and not sort those cells, although more sophisticated sorting strategies have been developed to allow for so-called "enrichment" strategies which try to balance yield versus purity. Since the identity of these cells is known during the analysis stage of the process, it is possible to design much more sophisticated strategies to identify sort unit coincident cells as "friend" or "foe" (Corio and Leary, 1993, 1996, Leary et al., 1998a). If the neighboring cell or cells in the same sorting unit are similar then a decision to sort, rather than reject, is warranted. But even if the sort coincidence cell is "foe," meaning dissimilar, then a decision to allow this contaminate may be wise to improve yield at a slight loss of purity. The decision to sort, or not sort, one or more cells in a given sort unit also depends on the negative consequences of sorting that cell foe. In some cases, it does not matter. In other cases, it may matter a great deal! These situations are well described in another series of published papers and patents from the Leary lab (Corio and Leary, 1993, 1996, Leary, 1994a,b, Leary et al., 1994, 1996, 1997, 1998a,b,c, 1999a, 2001b, 2002a,b, 2005, Leary and McLaughlin, 1995, Rosenblatt et al., 1997, Hokanson et al., 1999, Leary and Reece, 2000, Leary, 2000). This is particularly important for the high-speed sorting of rare cells pioneered in the Leary lab. The inter-cell spacing in flow is typically governed by Poisson statistics, although if the cells are sticky they will come through in clumps and bursts – very bad for flow cytometry! Throughput rates in macro flow cytometry can vary between a few thousand cells/seconds up to 100,000 cells/seconds or more.

4.4.1 Conventional Hydrodynamic Focusing

Conventional hydrodynamic focusing uses the concentric laminar flow of an outside "sheath" fluid to pitch or otherwise confine an inner column of cell sample solution. By adjusting the relative sheath to sample pressure one can confine the cell sample to as close to a narrow single file as desired, but at the cost of throughput if the sample stream is very narrow. This is particularly important to detect small differences in DNA/cell or to analyze and sort chromosomes where the instrumental variations need to be less than the biological variations, typically 2%–3%. In cases of analyzing and/or sorting cells labeled with antibodies for immunophenotyping, it is still important, but less so, since both the true biological variation and staining variation are typically on the order of 10% or more. The instrumental parameter regarding precision and reproducibility is known as the "coefficient of variation" (CV). CVs are also a function of fluid flow rates and the level of hydrodynamic focusing. As the sample stream is expanded to allow for higher throughput rates, the CV becomes larger. This along, with most other flow cytometry variables, is an engineering trade-off of yield versus purity which has been much described elsewhere (Leary, 1994a, 2005).

4.4.2 Orienting Fluid Flow

In most cases, the orientation of the cells in the flow stream as it intersects the excitation light source is unimportant because most cells in suspension are either spherical or prolate ellipsoid. But of the cells are very non-spherical, as in the case of sperm, orientation becomes very important. This was first observed when supposedly haploid sperm showed a flow cytometry distribution looking like cycling diploid cells (Pinkel et al., 1982). These unusual flow cytometry distributions were discovered to be due to cell orientation differences. Once the sperm where not only hydrodynamically focused but also re-oriented using a beveled sample injection needle, the haploid distributions looked haploid. The first flow cytometry nozzle with orientation capability became known as the Rens nozzle (Rens et al., 1998) which greatly improved the bull sperm sorting business (Johnson et al., 1999), perhaps the first "industrial sorting application, and has been the subject of many lawsuits in the bull sperm sorting industry. Today, sperm cell sorting as a form of sex selection is a very big industry. To distinguish between male (Y) and female (X) sperm, the reorientation of sperm is necessary to detect the approximately 4% differences in total DNA between male and female sperm. Mis-oriented Y-sperm can be confused and misclassified as X-sperm. The consequences of mistaken sex selection are very costly to the industry. While 100% pure sex sperm cannot be accomplished the goal is usually to obtain at least 90% and preferably 95% purity of the desired sex sperm. Such purity comes at a cost of yield.

4.4.3   ACOUSTIC FOCUSING

Hydrodynamic focusing works but at a cost, particularly in the design of portable instruments where the extra weight and bulk of sheath fluid compromise the portability of the instrument. "Acoustic focusing" largely solves this problem by creating acoustic waves in the sample fluid which causes these cells to move to the center of the stream. Acoustic focusing has been done in macro flow cytometers (Goddard et al., 2006, Kaduchak et al., 2006, Ward and Kaduchak, 2018, Fong et al., 2014) but is perhaps even more important for the field of microfluidics.

4.4.4   HIGH-SPEED FLOW CYTOMETRY

The early history of flow cytometry in the 1960s and 1970s involved the development of instruments capable of a few thousand cells/seconds, a limitation both of technology and the lack of biological and clinical applications requiring much higher speeds. During the 1980s and 1990s a few applications, including minimal residual disease monitoring and sorting chromosomes for the precursor project to the Human Genome Project drove the development of high-speed cell sorters (Van Dilla et al., 1986). The rare-event analysis required very high-speed sampling of many millions of cells to find subpopulations at frequencies of 10^{-5} to 10^{-6} (Cupp et al., 1984, Leary, 1994b, 2000, Leary et al., 1994, 1996, 1997, 1998b,c, 1999a, 2001a,b, 2002b, 2005, Leary and McLaughlin, 1995, Leary and Reece, 2000). The sorting of human cells and chromosomes required large enough amounts to produce chromosome libraries and enough DNA to sequence the human genome (Van Dilla et al., 1986). These academic and national labs paved the way for the development of high-speed flow cytometers and cell sorters available today.

4.4.5   IN-VIVO FLOW CYTOMETRY

It was perhaps inevitable that people would use the blood vessels in living animals (Georgakoudi et al., 2004). Their application in humans is at a very early stage. The most important thing about in-vivo flow cytometry is that it has the potential for sampling a significant fraction of the entire human blood supply (5–7 L), rather than a few tubes of blood, for minimal residual disease monitoring in cancer and for monitoring patients with septicemia.

4.4.6 MICROFLUIDICS

Microfluidics, for reviews, see (Bene, 2017, Shrirao et al., 2018, Honrado et al., 2021, Yang et al., 2018) for much more detail. Much of the work, until recently, has been performed using PDMS in academia in "mom-and-pop" laboratories. More recently many commercial entities started manufacturing and providing microfluidic chips (e.g., Agilent, uFluidics, ChipShop).

While a different form of microfluidic cytometry, new "organ-on-a-chip" technologies use microfluidic structures to model human organs. Recently new "organ-on-a-chip" approaches (van der Meer and van den Berg, 2012, Bischel et al., 2015, Huh et al., 2010, Grafton et al., 2011, Vidi et al., 2014, Wang et al., 2016, Lelievre et al., 2018) whereby in-vitro 3D engineered tissues approximate in-vivo organs and allow for higher throughput testing techniques. In this example, we created a "breast on a chip" microfluidic organ-on-a-chip to study how nanomedicine might be delivered through nanodrug delivery systems to treat ductal breast cancer (Figure 4.10).

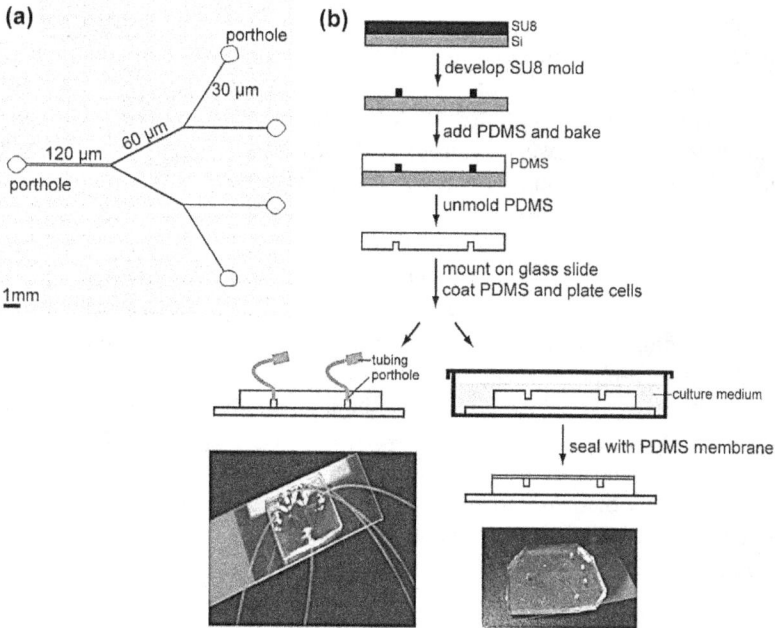

FIGURE 4.10 Engineering of PDMS channels on a chip. A microchannel system was molded in PDMS, coated with laminin 111, and used as a substrate for the culture of HMT-3522 S1 cells. (a) Schematic of the branched channel system. (b) Two independent approaches were developed: In the first approach (left, drawing and picture of the system), PDMS microchannels were sealed onto a glass coverslip, coated with dried or dripped laminin 111, and used for the culture of S1 cells in a closed environment. Cells were injected through tubing connected to the portholes using a syringe pump and the medium was changed by immersion. In the second approach (right), cells were cultured in an open 'hemichannel' system (top side of microchannel left open). The channels can be completed using a PDMS membrane on the day of the experiment. (Reprinted with permission from Grafton et al., 2011a,b.)

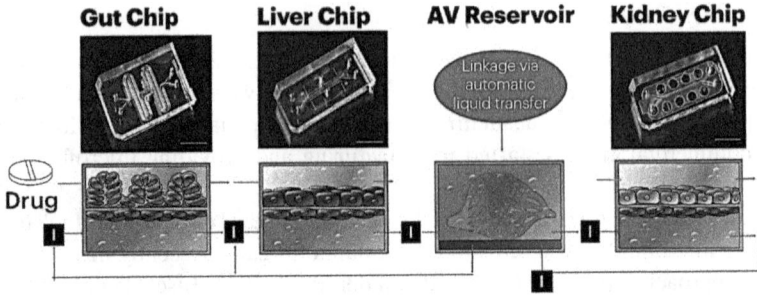

FIGURE 4.11 Artificial human organs-on-a-chip can be constructed to provide 3D in-vitro models of real human organs, rather than animal models. Drug treatments can then be performed safely and more rapidly in these organs-on-a-chip. (Source: Wyss Institute.)

Human "organs-on-a-chip" may provide a better prediction of toxicity in humans than conventional animal testing. People opposed to the use of animals for drug testing should be delighted by the development which should greatly reduce the number of animals used for this purpose. This advance is partially changing the paradigm of first testing in animals, then in humans. Since most animal model systems are not very good predictors of human responses to drugs, there was a push to develop "artificial 3D human organs in-vitro whereby drugs could be tested on human cells and "pseudo-organs" without endangering humans. It also provides the advantage of being able to test on human cells of different races and ethnicities, even making use of human genome repositories. Artificial 3D tissue-engineered "organs-on-a-chip" can provide a partial bridge between in-vitro and in-vivo testing platforms. These systems have many of the advantages of both in-vitro and in-vivo systems and can be a "closed system" platform allowing easier accounting of all experimental inputs and outputs. Some of these "organs-on-a-chip" are quite sophisticated and multi-organ systems have been developed that even allow for trafficking of both molecules and cells between several different connected "organs-on-a-chip" whereby trafficking cells can be studied by on-chip flow cytometry within the actual organ-on-a-chip. A good source of the latest information on the subject can be obtained by visiting the website of the Wyss Institute at Harvard University (http://wyss.harvard.edu/) which is currently the premier research institute in the world in this important new area (Figure 4.11).

4.5 LIGHT EXCITATION SOURCES

Flow cytometry made early use of lasers, first gas ion ones and then semi-conductor lasers as they became available. A few flow cytometers made use of arc lamps, particularly mercury ones. Recently superluminescent LEDs have started to become excitation light sources for microfluidic cytometers.

4.5.1 ARC-LAMP SYSTEMS

An alternative to laser excitation sources for flow cytometry, all that is needed is an epi-fluorescence microscope using arc lamps (Steen, 1983). Cells can flow in an angled stream across a microscope slide and fluorescence data can be measured. Arc lamp sources are gradually being phased out of fluorescence microscopes in favor of semi-conductor diodes which are very stable and have very long lifetimes, unlike arc lamp bulbs.

4.5.2 Lasers

Interestingly, the first laser wavelengths chosen for flow cytometry were laser equivalents to the mercury arc lamp spectra so they could excite the same fluorescent dyes that had been used for fluorescence microscopy of cells and tissues.

4.5.2.1 Water-Cooled Gas Systems

The first lasers used in flow cytometry were argon-ion and Helium ion gas lasers. The argon ion lasers were only efficient at below 1%, so they required huge amounts of current. That meant that they needed a circulating water jacket to keep from melting down. Some of these lasers consumed more power than many houses!

4.5.2.2 Dye Lasers

Since the colors of gas lasers were pretty limited, dye lasers provided a way to produce laser outputs in almost any color. But the use of these systems never became really widespread due to the huge maintenance overhead and, more importantly, the hazard due to the flowing of large amounts of potentially mutagenic fluorescent dyes used for dye laser excitations. But these dye lasers could provide tremendous output powers because they could be pumped with all of the many different modes of the upstream laser, rather than just the TEMoo mode.

4.5.2.3 Semiconductor Lasers

Semi-conductor lasers have revolutionized flow cytometry and have replaced all of the gas ion lasers. They are much less expensive to purchase and maintain, as well as have lifetimes in the many tens of thousands of hours. They are also much more compact laser light sources. The advent of frequency-doubled and tripled laser allowed near UV lasers from frequency-tripled near-infrared lasers.

4.5.2.4 Polarized or Unpolarized?

For most of the first 25 years flow cytometry used gas, water-cooled lasers that brought the laser light out through Brewster windows that in the process produced polarized versions of the laser output. Most of the time vertically polarized light was the norm. This also allowed for polarized fluorescence to provide measures of cell membrane microviscosity as well as another way to measure energy transfer between probes. The only problem is that vertically polarized light only maximally excites fluorescent probes oriented in the same direction. As the angle between the polarization plane of the laser and the polarization direction of the fluorescent probes become misaligned the probability of exciting a fluorescent photon decreases until the probability becomes zero when the fluorescent probe and the laser polarization plane are at 90° with respect to each other. Interestingly, LED light sources are usually not polarized. They also have a wider spectral bandwidth than lasers so that they may excite the fluorescent probe more efficiently.

4.5.2.5   Constant Power Lasers or Pulsed Lasers?

Flow cytometry started with constant lasers. But with ultra-fast pulsed semiconductor lasers, it is possible to achieve much higher powers and on a time scale where the fluorophores are being excited and re-excited by these lasers giving results much like constant power lasers (Shapiro and Telford, 2018).

4.5.2.6   Beam Shaping

It is important to tailor the laser energy to the application, so-called "beam-shaping optics." "Structured light" can come in a variety of forms (e.g., light sheets, elliptical and spherical). Laser light beam profiles usually have a gaussian profile. That can be problematical for flow cytometry because the fundamental premise if that all cells receive the same exciting light. Otherwise, if the cells flow through different parts of the Gaussian beam profile they will receive different excitations and the single cell data will be meaningless.

4.5.2.6.1   Elliptical

Elliptically-shaped laser beam profiles are particularly important for DNA cell cycle and chromosome sorting applications because the coefficient of variation of this data must be low enough (typically 4% or less, the lower the better!) to allow mathematical analyses to distinguish between cell cycle phases or chromosome analysis and sorting purity. Since labeling of cells or chromosomes with fluorescent DNA-specific dyes tends to produce very brightly labeled cells or chromosomes, it is possible to get by with only a tiny fraction of the total energy of the light source.

4.5.2.6.2   Spherical

Spherically focused laser beams were particularly important for immunology applications. In this type of application, the immunofluorescently labeled cells are relatively dim, so you want the cells to use the largest possible fraction of the exciting source. For immunological applications, the biological variation can vary from 10% to 15% of several orders of magnitude.

4.5.3   LEDs

Until relatively recently, it was not possible to use LED (Light Emitting Diode) technology because they were simply not bright enough. LEDs also have a more unique beam shape that widely disperses the light in a cone rather than a gaussian profile like lasers. Top-hat shapes, similar to a step function, can have a relatively flat beam profile (Figure 4.12).

This also allows elliptical and spherical shapes.

No longer is it necessary to hydrodynamically focus the cells to the center of the stream because the cells will receive approximately the same amount of exciting light no matter where they are in the stream. This is particularly important for the

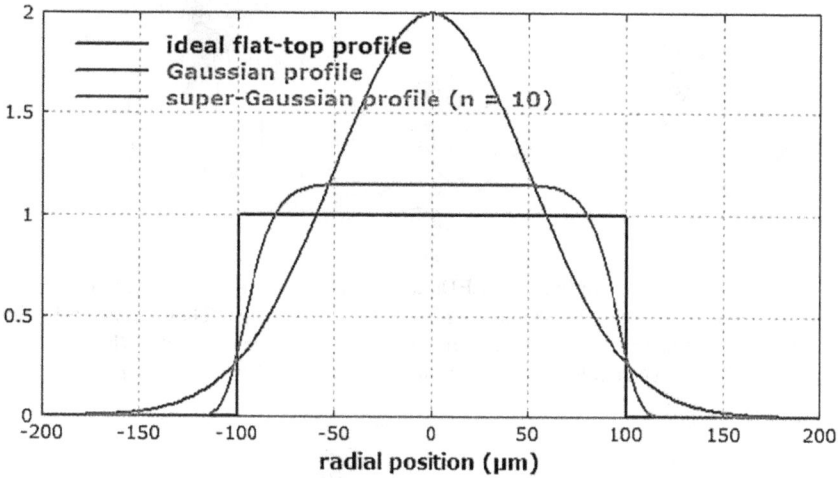

FIGURE 4.12 Lasers naturally have an approximately Gaussian beam profile, making cell position in eh beam critical in order to receive the same intensity and duration of exciting light. But with appropriate beam shaping optics one can obtain an approximately "top hat" beam profile that is very forgiving of cell position in the beam and ideally suited to microfluidic cytometry. (https://www.rp-photonics.com/flat_top_beams.htm.)

development of portable medical diagnostic cytometers as well as portable environmental monitoring devices (Leary, 2018, 2019, 2021).

4.5.3.1 Superluminescent LEDs

"Superluminescent LEDs" are now commercially available in a wide variety of colors and give light power outputs similar to lasers but at a tiny fraction of the price or size. Superluminescent LEDs are being used to design portable microfluidic cytometers (Leary, 2018 2019, 2021).

4.5.3.2 Multispectral Superluminescent LEDs

It is now possible to buy commercial superluminescent LEDs with multiple colors. Sometimes these multispectral LEDs have red, green and blue (RGB) outputs using three LEDs on the same chip, but more complicated chips can have so-called "white light" which offers a collection of colors from violet to red. There are even Superluminescent LEDs that provide near-UV outputs sufficient to excite Hoechst 33342 and so-called Violet dyes (Bradford et al., 2006). Examples of these LEDs are shown in Figure 4.13.

4.6 DETECTORS

In the early days of flow cytometry, photodetectors were usually of two types: photovoltaic and photomultiplier tubes (PMTs). The photovoltaic detectors were mainly used for forward and other low-angle light scatter detection because they were not

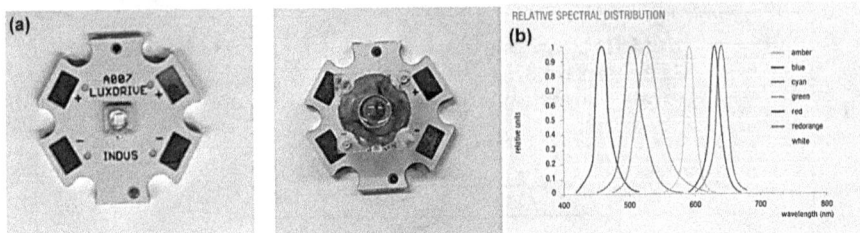

FIGURE 4.13 (a) Superluminescent LEDs as a flow cytometry light source can provide cheap alternatives to lasers, particularly for microfluidic cytometry. They can be easily fitted for clip-on plastic beam-shaping optics suitable for microfluidic cytometry. (b) these LEDs come in a wide variety of colors, mimicking many of the wavelengths of lasers used for flow cytometry.

very sensitive. Fluorescence, which is on the order of 10,000 times lower intensity was detected using a multi-electrode PMT that could amplify millions of times the incoming fluorescent signal int output electrical pulses. These PMTs typically needed 1000–2000 V from a serious voltage source. They were gradually made much smaller for bench-top flow cytometers but still required major voltage sources making them not practical for truly portable flow cytometry.

4.6.1 PHOTODIODES

Photovoltaic sensors are typically just slabs of photosensitive material that are crude in their ability to control angles which are very important to measure different properties of cells by light scatter. As a graduate student, I used to travel to Los Alamos National Laboratory to use their multiangle (32-angle) light scatter cytometers (Salzman et al., 1975). Then I would analyze this very high dimensionality data to choose the most important angle. Then I would simply use electrical tape to create specific angular regions on my much more primitive flow cytometer!

4.6.2 PMTs

The essential design of a conventional PMT measuring fluorescence in the conventional way is one of the cascading electrodes (dynodes) that amplify the initial light input millions of times, creating more and more photoelectrons. If run properly, the photoelectron output for each PMT gathering a particular color of fluorescence should be proportional to the color of that fluorescence of the cell. Usually, flow cytometry does not worry about absolute fluorescence measurements, but by running calibration standards one can determine the absolute numbers of receptors per cell. PMTs are highly sensitive but require 1000–2000 V. While they have been made smaller and more energy efficient in recent years they are still in the realm of benchtop cytometers. Portable cytometers require much smaller, lightweight types of PMTs that can be powered by batteries, as described in the next section.

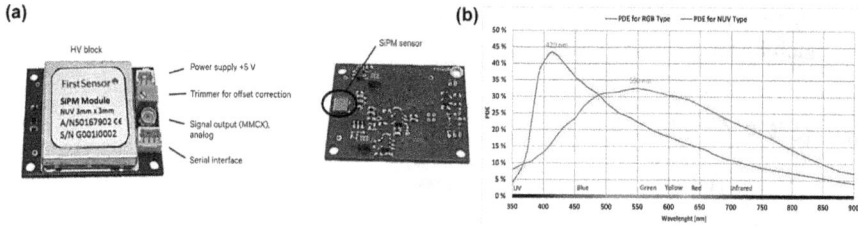

FIGURE 4.14 Silicon photodetectors (SiPM) are solid-state analogue to conventional photomultiplier tubes (PMTs), including sensitivity in the visible spectrum, but they are very small and can be powered by batteries making them ideal detector systems for microfluidic cytometry. (Courtesy First Sensor, Gmbh (now part of TE Connectivity).)

4.6.3 SiPMTs

To reduce the voltage requirements and miniature the size and weight, silicon PMTs were invented. They can be powered by batteries after boosting the voltage to about 25–30 V. An example of such SiPMTs is described in a series of papers (Grafton et al., 2008, 2009, 2010, 2011a,b, Maleki et al., 2012, Leary, 2018, 2019, 2021). Our current SiPMT is from First Sensor (Figure 4.14).

4.6.4 Multispectral Detectors

Until recently, flow cytometers still used complicated combinations of optical filters to choose different parts of the fluorescent emission patterns from fluorescently labeled cells. Flow cytometers were just a new form of epi-fluorescent microscopes. This worked well enough until people starting collecting very large numbers of colors (>10 colors). The spectral unmixing algorithms did a pretty good job of separating the colors to appropriate detectors, but the cost also skyrocketed. For this reason, flow cytometrists borrowed from multispectral spectroscopy, another example of piggy-backing on technologies for other purposes. Essentially one gets a multispectral readout of all wavelengths of fluorescence. Multispectral flow cytometers (Feher et al., 2016) are now commercially available.

4.6.5 "Imaging-In-Flow"

In the early days of flow cytometry, the Wheeless lab at the University of Rochester Medical Center developed "slit-scan" flow cytometry (Wheeless et al., 1990) in order to get at least partial information of location of fluorescent dyes within individual cells, in particular for analyzing cervical cytology suspensions of cells for cervical cancer screenings. It captured these "flow images" in real time. Due to limitations in the sensitivity of detectors at the time, only very brightly acridine orange labeled cells, which labeled the nucleus a different color from the cytoplasm. Commercialization of this type of imaging in flow cytometer took many years, eventually being developed and sold by the Amnis Corporation. Using more modern technology Amnis was able to image-in-flow immunofluorescently labeled cells. Amnis also took advantage of the much faster electronics to allow for cell sorting (Luminex, 2021, Basiji, 2007).

4.7 SIGNAL PROCESSING

4.7.1 OLD ANALOG-TO-DIGITAL SYSTEMS

For more than 25 years, flow cytometry mostly produced analog signals that needed to be processed through an analog-to-digital converter (ADC). It took a revolution in digital signal processing technology to reduce the size of flow cytometers to bench-top and even portable cytometers for use in the field.

4.7.2 DIGITAL SIGNAL PROCESSING CHIPS

Again, piggy-backing onto technologies developed for completely different purposes, allowed flow cytometry to make use of the tremendous advantages in integrated circuit chips and now complete digital signal processing boards.

4.8 DATA ANALYSIS

Data analysis in flow cytometry began with simple analysis of 1-parameter histograms or by 2-D scattergrams of two correlated parameters. But it quickly became evident that this form of human pattern recognition was not going to be useful beyond 3D data. Initially, people attempted to reduce the dimensionality of the data using principal component analysis (PCA) or discriminant function analyses (DFA). Using a new research-level sorter based on a neural network architecture one group was able to generate the first three PCA and DFA multivariate statistical classifiers in real-time, permitting their use as cell sorting parameters (Leary et al., 1997, 1998b, 1999b, 2002a, Hokanson et al., 1999). Indeed, these methods could be thought of as "real-time" data analysis. Analysis of listmode data in a data stream is analogous to real-time classifications for cell sorting. But most reductions of dimensionality used to allow human pattern recognition are ultimately doomed to failure for high-dimensional data.

What we really want to do is to compare two or more data files to see where there are differences in cell subpopulations in high-dimensionality data space. One such attempt was the use of "subtractive clustering," a novel data mining technique (Smith et al., 2005, Leary, 2006) which could be either guided or else let the algorithm determine whether two clusters in different files were different or "similar" whereby measures of similarity could be determined by different parameters including those generated from neural networks or artificial intelligence.

4.8.1 CONVENTIONAL DATA ANALYSIS

About the only remaining histogram analysis is cell cycle analysis based on distributions of single cells by their DNA content per cell. The norm for most of the rest of the flow cytometry data is a series of two-dimensional scattergrams. By gating through chosen parameters one can see the populations identified by other parameters. It was certainly an advance that served the field well for many years as illustrated by this example in the literature (Bonilla et al., 2020). But as people started doing >15 parameter experiments it became evident that much of the insight into the multidimensional data was lost in these 2D projections of multidimensional clouds of data.

4.8.2 BIOINFORMATIC ANALYSES

Investigators then began using bioinformatic algorithms that were well tuned toward the analysis of "big data" with high-dimensionality (e.g., 15–20 parameter flow cytometry data). This software known as "R" or "R-Bioconductor" software is readily available on the Internet (Montante and Brinkman, 2019). It includes a number of multidimensional cluster analysis algorithms as well as many other routines. An excellent overview showing the trends of flow cytometry data analysis is described here (Kimball et al., 2018, Cheung et al., 2021). A paper showing how the different R routines can be combined is shown in Figure 4.15.

4.9 CELL SORTING

4.9.1 MECHANICAL CELL SORTING

In the early days of cell sorter development, mechanical valves were envisioned as the way to sort cells. But the technology of that era (the early 1960s) was not adequate

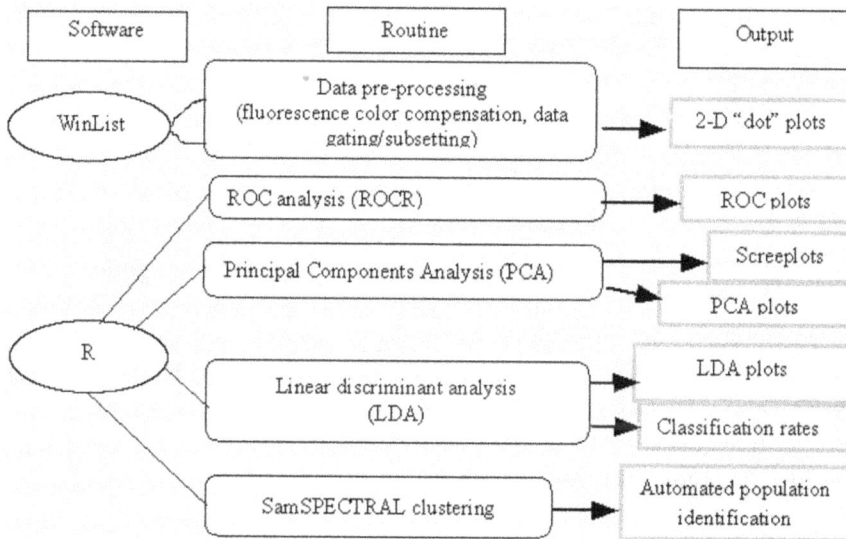

FIGURE 4.15 R Libraries of a wide variety of statistical classification routines to find subpopulations of cells within high-dimensionality flow cytometric data are readily available. (http://cran.r-project.org/web/packages.)

for sorting cells faster than about 100 cells per second. It was obvious that sorting speeds had to be at least in the thousands of cells per second to be practical. Developers quickly abandoned this approach.

4.9.2   CONVENTIONAL "INK-JET" DROPLET SORTING OF CELLSUBSETS

The use of droplet-based sorting for flow cytometry/cell sorting began as an outgrowth of ink-jet printing technologies (Sweet, 1965). Indeed, an ink-jet home computer printer can be thought of as a sorter, except sorting ink droplets that do not contain cells.

4.9.3   SINGLE-CELL SORTING (STOVEL AND SWEET, 1979) FOR CELL SEPARATION OF COMPLEX MIXTURES OF CELLS

A logical extension of the preceding cell sorting methods is to sort individual cells into microwells. By doing this individual cell separation into separate microwells, one achieves single-cell cloning. This has been highly developed and is now commercially available in a variety of instruments, usually as add-ons to regular cell sorters.

4.9.4   FLUIDIC SWITCHING

The microfluidic cytometry field is re-imagining the early attempts to use mechanical valves to sort cells within microfluidic chip architectures. Elastomeric valves and other approaches within microfluidic chips are being re-invented. But it is still hard to sort at rates much greater than a few hundred cells per second. That suffices for a limited number of biomedical and clinical applications.

4.9.5   ON-CHIP MAGNETIC SORTING

Another way to approach the problem of sorting within microfluidic chips is to label cells with magnetic nanoparticles and divert them down through branched channels (Leary and Frederickson, 2008, Grafton et al., 2009, 2010, 2011, Maleki et al., 2012), as described earlier in this chapter. The beauty of this approach is its simplicity and very high throughputs. It is most useful for analyzing and sorting leukocyte subsets from small volume samples of human peripheral blood. Because it needs minimal, if any, hydrodynamic focusing, it is promising for the development of small battery-powered flow cytometer/cell sorters that are truly portable and can be taken to a point-of-care (Leary, 2018, 2019, 2021).

4.10   BIOLOGICAL AND CLINICAL APPLICATIONS

The first obvious application of flow cytometry was human blood cell analysis. Initially, this was done in simple terms, namely dividing human blood cell subsets into lymphocytes, monocytes, and granulocytes. Later, considerably more sophisticated cell subset analysis was performed using antibodies which distinguished between B and T lymphocytes. The advent of monoclonal antibodies continues to provide increasingly more sophisticated delineation of cell subpopulations within human blood.

4.10.1   DNA

One of the most important measures of a cell is its DNA. Researchers started using DNA-specific fluorescent dyes, for example acridine orange, which stains both DNA and RNA in different colors (Darzynkiewicz, 1990), propidium iodide (Deitch et al., 1982), mithramycin and chromomycin (Crissman and Tobey, 1990) on fixed cells, and Hoechst 33342 on living cells (Arndt-Jovin and Jovin, 1977). Several of these dyes were also used to identify different human chromosomes based on their relative AT and GC base pair compositions which are slightly different for human chromosomes and were the basis of the National Chromosome assorting Project at Los Alamos and Lawrence Livermore National Labs. It was many years until the development of more "vital" visible spectrum excited DNA dyes (Bradford et al., 2006) that could be used for the analysis of living cells without the use of near-UV lasers which were initially very expensive, before the invention of modern frequency-tripled lasers.

4.10.1.1   Cell Cycle Analysis

Near the beginning of flow cytometry was the use of DNA-specific dyes which rapidly replaced tritiated thymidine autoradiography as a way of determining the cell cycle phase. This application also represents one of the earliest instances of using mathematical deconvolution methods to extract information not readily available by conventional gating analysis techniques (Dean and Jett, 1974).

4.10.1.2   Ploidy

As soon as people started analyzing human cancer cells, it became obvious that many of these cancer cells were not normal diploids. Many tumors are hyperdiploid, created by the initial failure to fully divide and then tossing out of extra chromosomes. Occasionally tumors are hypodiploid, meaning they lost chromosomes from the original diploid cell. Since DNA dyes will also show increased, or decreased, chromosome number, so-called "ploidy analysis," whereby these different aneuploid cancer cells could be measured by flow cytometry.

4.10.1.3   Chromosome Analyses

The Human Chromosome Project was devoted to the analysis and isolation of different human chromosomes by DNA analyses of metaphase chromosomes. It was discovered at the use of both AT- and GC-base pair-specific, fluorescent DNA dyes could distinguish all of the different human chromosomes (Van Dilla et al., 1986, Van Dilla and Deaven, 1990) in an application known as flow karyotyping (Van Dilla et al., 1986, Van Dilla and Deaven, 1990). Human gene regions could then be mapped onto these isolated metaphase chromosomes using fluorescence in-situ hybridization and chromosome-painting techniques (Trask et al., 1989).

4.10.2 DETECTION OF MOLECULES ON THE SURFACE OF CELLS

One of the earliest applications of flow cytometry was begin in the Herzenberg laboratory at Stanford University – one of several birthplaces of flow cytometry and where
the term "fluorescence-activated cell sorting (FACS)" originated. Once Albert Coons
had managed to connect fluorescein isothiocyanate to antibodies, these fluorescently
labeled antibodies could be used to identify different immune cell types. Rhodamine
isothiocyanate (RITC), another dye used early on in fluorescence microscopy, could
also be easily linked to antibodies. This created the problem of dealing with the
fluorescence-spillover problem initially for two-color immunofluorescence of FITC
and RITC labeled cells (Loken et al., 1977). The invention of monoclonal antibodies (Kohler and Milstein, 1975) and the medical urgency of the AIDS epidemic led
to twin explosions of the use of antibodies to detect many cell subpopulations in the
human immune system.

4.10.2.1 Antibody Detection of "biomarkers"

Cell subset-specific surface molecules became known as "biomarkers." With the
explosion of both monoclonal antibodies and new fluorescent dyes, biomarker analysis at the level of 17-colors or more (Perfetto et al., 2004), also spurred the development of spectral unmixing deconvolution software that could fix the optical spillover
problems of such a complicated fluorescent dye mixture of cells. Once the problem
had grown to this many colors, it drive the development of multi-spectral flow cytometry allowing the replacement of these complicated combinations of detectors and
optical filters (Robinson et al., 2005, Vo-Dinh et al., 2005, Basiji, 2007, Feher et al.,
2016, RoZanc et al., 2021, Orporation, 2021).

4.10.2.2 Use of Peptides and Aptamers

Initially, people used whole antibodies to label biomarkers on cell subsets. But as
soon as some cells were co-labeled with multiple different antibodies, steric hindrance (whereby the linking of one antibody to the cell surface partially inhibited
the binding of subsequent different antibodies, the search for ways to reduce the size
of antibodies began. Initially, researchers cleaved the whole immunoglobulin molecules into sub-region fragments (e.g., F(ab), F(ab)2, and Fc) as described previously.
This process has continued to the point where the term "nanobodies" was applied to
smaller and smaller pieces of antibodies.

Researchers soon discovered that peptides could be used for the shape-recognition
of antigen epitopes. Sequences as small as seven amino acids can recognize a human
breast cancer biomarker (Haglund et al., 2008). There are libraries of thousands of
peptides which can be used to identify cell subsets that require multiple different
biomarkers on a given cell. Another advantage of this approach is that it is far faster
and less expensive than generating monoclonal antibodies since they can be created
on standard peptide synthesis instruments once the proper sequence is known.

It turns out that DNA and RNA sequences, so-called "aptamers," can also be created that have shape recognition properties. Combinatorial chemistry, as described
earlier in this chapter, can be used to generate many millions of these molecules (Yang
et al., 2002) which can then be screened by high-speed flow cytometry and cell sorting

of microsphere labeled with each of these aptamers (Yang et al., 2003, Leary et al., 2005). Another advantage of this approach is that these molecules can be created and found even in situations where it is impossible to generate a monoclonal antibody.

4.10.2.3 Multicolor, High-Dimensionality Detection of Phenotypes

As was mentioned earlier the number of different biomarkers per cell led to the development of multi-color flow cytometry which has expanded to more than 17 colors (Perfetto et al., 2004). This complexity of colors and optical spillovers led to the development of spectral unmixing algorithms (Novo et al., 2013). It also led to the development of high-dimensionality data analysis techniques to be discussed a little later in this chapter.

4.10.2.4 Immunophenotyping of Cancers in the Clinic

The development of biomarker probes and multicolor fluorescence cytometry led quickly to their application in the clinic for diagnosing lymphomas and leukemias. Immunophenotyping of cancers has become the norm in most areas of the world.

4.10.2.5 Minimal Residual Disease Monitoring

Practical minimum residual disease, which is commonly used to monitor the tumor load in cancer patients to see if therapies are working, could not begin until the invention of high-speed flow cytometry (Corio and Leary, 1993, 1996, Leary et al., 1993a, 1998a, Leary, 2005). Then it was possible to not only enumerate rare circulating tumor cells (Leary, 1994a,b, 2000, , Leary and McLaughlin, 1995, Leary et al., 1996, 1999a, Leary and Reece, 2000) but to even isolated these cells and sequence them for mutations (Leary et al., 1999a), which was an extremely arduous process at the time. But the advent of single cell sequencing technologies has made the process of molecularly characterizing rare tumor cells at the level of single cells practical.

4.10.3 DETECTION OF INTRACELLULAR MOLECULES

Detection of intracellular molecules proved to be a much harder task. First the dyes had to be permeable (most were not!), so chemical fixation of the cells was required. More difficult was the fact that prior chemical fixation of the cells sufficiently changed the antigenic structures of cell surface and intracellular biomarkers so that most conventional antibodies could no long be used. One must choose only those antibodies for use on fixed cells.

4.10.4 DETECTION OF ACTIVE OR INACTIVE GENES WITH PHOSPHO-SPECIFIC ANTIBODIES

One of the most important applications of this is the study of the phosphorylation of genes. Active or inactive gene products (proteins) can be identified by so-called phosphor-antibodies which are able to distinguish between phosphorylated and un-phosphorylated proteins (Krutzik and Nolan, 2003, Krutzik, 2004). This allows gene activation or de-activation in response to drugs to be easily studied by flow cytometry.

4.10.5   Detection of Drug Delivery

Studying the effects of drug treatments on individual diseased cells, and the side effects on normal cells, is becoming increasing important. Human disease ultimately occurs at the single cell level, and nanomedicine is the new field of massively parallel and highly targeted single cell drug delivery. All of the features of flow cytometry make it the ideal way to study targeting of drugs and multiparameter measurements of efficacy including not only the death of diseased cells but the more sophisticated inducement of programmed cell death (apoptosis).

4.10.5.1   Conventional Drugs

Conventional drugs are delivered orally, intradermally, or intravenously. There is usually very little active targeting to diseased cells so side effects can be substantial, in many cases causing the patient to stop taking those drugs. The reasons for side effects are two-fold. First, the patient must be exposed to 10–100 times as much drug as they would if the drug were targeted.

4.10.5.2   Nanodrug Delivery Using Flow Cytometry

There is a better way – highly targeted, through a multi-step targeting process, delivery of drugs encapsulated within nano-sized drug delivery systems which can be quite sophisticated. This author has spent the past 20 years of his career developing sophisticated nanomedical drug delivery systems and has recently written a major textbook on nanomedicine (Leary, 2022). Flow cytometry enjoyed an important role in evaluating the targeting of these systems to diseased cells (Cooper and Leary, 2014, 2015, Haglund et al., 2008) as well as studying the efficacy of these delivered drugs at the single cell level.

The new CRISPR gene editing technology (Pennisi, 2013, Cong et al., 2013) has exciting potential for treating many human genetic problems at the somatic cell level. Part of this use of CRISPR to provide gene editing to diseased single cells to fix, rather than kill, these cells. CRISPR technology has already been used to treat disease in patients (Chadwick and Musunuru, 2018, Frangoul et al., 2020).

4.11   CONCLUSIONS

Flow cytometry has been one of the great medical technologies of the 20th century, and is becoming even more important in the 21st century. It is used in more than 65 countries and is being used in hospitals and clinics throughout most of the first and second world countries. It has become a multi-billion-dollar business, including both instruments and reagents. Small portable cytometers are now being used in third-world countries. Publications number in the tens of thousands. There is a growing number of patents involving flow cytometry and cells sorting.

From humble origins, flow cytometry has become one of the major technologies in generating "big data" for biomedical and clinical applications. In the 1980s days, my lab was generating "big data" but had no place to store the data. We had to do "real-time data analysis" by incorporating it into the flow cytometry cell sorting hardware through a neural network design. This allowed us to do not only real-time data analysis but also real-time sorting of those same cells at high speeds (Leary et al., 1998b, 1999b, 2001a, 2002a, 2005). Now with disk storage space very inexpensive, very large data files of listmode flow cytometry data can be readily stored. Most analyses for these "big data" flow cytometry data are now done by automated supervised or unsupervised clustering algorithms that can identify subpopulations in high-dimensionality data space.

Initially, flow cytometers were large, dumb instruments, initially built by the early researchers themselves. This meant that very smart operators were needed! This limited the market to those few institutions that could afford to train and retain these very smart operators. To overcome this bottleneck, the flow cytometry industry had their engineers re-tool their instruments to make their operation easier and easier. This has reached the level of "walk-away" instruments that can automatically deliver large numbers of different cell samples, use pre-set flow cytometry protocols and analyze and store the data. The trend toward miniaturization has also continued leading to flow cytometers with smaller and smaller footprints.

4.12 FUTURE SCOPE

The future for flow cytometry remains bright. It now has an important role in the field of nanomedicine, a form of single-cell medicine well linked to single cell analysis by flow cytometry, for advanced drug delivery by targeted nanoparticles (Leary, 2022)

4.12.1 SMALL, PORTABLE CYTOMETERS FOR POINT-OF-CARE MEDICAL APPLICATIONS

This has reached the point where it is now feasible to develop portable, even handheld, flow cytometers that can be designed to run in field, for example, in emergency medical situations (Maleki et al., 2012, Leary, 2018, 2019, 2021).

4.12.2 SMALL PORTABLE CYTOMETERS FOR ENVIRONMENTAL MONITORING

Not only will these portable cytometers be used for medical applications. They will also be used for environmental monitoring as the world has become aware that we must study the rapidly changing ecology of our surroundings due to climate change. An early application of a cytometer in the field was a combination of microfluidics and real-time surface plasmonic detection of pathogens in the field (Zordan et al., 2009).

4.12.3 CYTOMETERS EVERYWHERE!

As with computers which were initially mainframes then personal computers, there will be a market for flow cytometers that are sufficiently smart and portable that they will be known only for the application, not as flow cytometers per se. This stage of development will cause the market to explode outside the research and clinical labs. Personal flow cytometers for specific applications will be possible for individuals.

REFERENCES

Arndt-Jovin, D.J., & Jovin, T.M. 1977. Analysis and sorting-of living cells according to deoxy-ribonucleic acid content. *J. Histochem. Cytochem.*, 25, 585–589.

Bandura, D.M., Baranov, V.I., Ornatsky, O.I., Antonov, A., Kinach, R., Lou, X., Pavlov, S., Vorobiev, S., Dick, J.E., & Tanner, S.D. 2009. Mass cytometry: Technique for real time single cell multitarget immunoassay based on inductively coupled plasma time-of-flight mass spectrometry. *Anal. Chem.*, 81, 6813–6822.

Basiji, D. 2007. Multispectral imaging flow cytometry," *2007 4th IEEE International symposium on biomedical imaging: From nano to macro*, Arlington, VA, USA, pp. 1100–1103, doi: 10.1109/ISBI.2007.357048.

Bene, M.C. 2017. Microfluidics in flow cytometry and related techniques. *Int. J. Lab. Hematol.*, 39, 93–97.

Bischel, L.L., Beebe, D.J., & Sung, K.E. 2015. Microfluidic model of ductal carcinoma in situ with 3D, organotypic structure. *BMC Cancer*, 15, 1–10.

Bitton, A., Sambrano, J., Valentino, S. & Houston, J.P. 2021. A review of new high-throughput methods designed for fluorescence lifetime sensing from cells and tissues. *Front. Phys.*, 9, 648553.

Bonilla, D.L., Reinin, G. & Chua, E. 2020. Full spectrum flow cytometry as a powerful technology for cancer immunotherapy research. *Front. Mol. Biosci.*, 7, 612801.

Bradford, J.A., Whitney, P., Huang, T., Pinson, P., Cheung, C.-Y., Yue, S., & Godfrey, W.L. 2006. Novel Vybrant® DyeCycle ™ Stains provide cell cycle analysis in live cells using flow cytometry with violet, blue, and green excitation. *Blood*, 108, 4234–4234.

Chadwick, A.C., & Musunuru, K. 2018. CRISPR-Cas9 genome editing for treatment of atherogenic dyslipidemia. *Arterioscler Thromb Vasc Biol*, 38, 12–18.

Cheung, M., Campbell, J.J., Whitby, L., Thomas, R.J., Braybrook, J. & Petzing, J. 2021. Current trends in flow cytometry automated data analysis software. *Cytometry A*, 99, 1007–1021.

Cong, L., Ran, F.A., Cox, D., Lin, S., Barretto, R., Habib, N., Hsu, P.D., Wu, X., Jiang, W., Marraffini, L.A., & Zhang, F. 2013. Multiplex genome engineering using CRISPR/Cas systems. *Science*, 339, 819–823.

Constantin, T.P., Silva, G.L., Robertson, K.L., Hamilton, T.P., Fague, K., Waggoner, A.S., & Armitage, B.A. 2008. Synthesis of new fluorogenic cyanine dyes and incorporation into RNA fluoromodules. *Org. Lett.*, 10, 1561–1564.

Cooper, C.L., & Leary, J.F. 2014. High-speed flow cytometric analysis of nanoparticle targeting to rare leukemic stem cells in peripheral human blood: Preliminary in-vitro studies. *SPIE Imaging, Manipulation, and Analysis of Biomolecules, Cells, and Tissues XII.* SPIE.

Cooper, C.L., & Leary, J.F. 2015. Advanced flow cytometric analysis of nanoparticle targeting to rare leukemic stem cells in peripheral human blood in a defined model system. *SPIE Imaging, Manipulation, and Analysis of Biomolecules, Cells, and Tissues XIII.* SPIE.

Corio, M.A.., & Leary, J.F. *"System for Flexibly Sorting Particles"*. U.S. Patents 5,199,576(1993) and 5,550,058 (1996) and 5,998,212 (1999).

Corio, M.A., & Leary, J.F. 1996. *System for Flexibly Sorting Particles*. USA patent application.

Crissman, H.A., & Tobey, R.A. 1974. Cell-cycle analysis in 20 minutes. *Science*, 184, 1297–1298.

Crissman, H.A., & Tobey, R.A. 1990. Specific staining of DNA with the fluorescent antibiotics, mithramycin, chromomycin, and olivomycin. *Methods Cell Biol.*, 33, 97–103.

Crosland-Taylor, P. 1953. A device for counting small particles suspended in a fluid through a tube. *Nature*, 171, 37–38.

Cunningham, R.E. 2010. Overview of flow cytometry and fluorescent probes for flow cytometry. *Methods Mol. Biol.*, 588, 319–26.

Cupp, J.E., Leary, J.F., Cernichiari, E., Wood, J.C., & Doherty, R.A. 1984. Rare-event analysis methods for detection of fetal Red blood cells in maternal blood. *Cytometry*, 5, 138–44.

Darzynkiewicz, Z. 1990. Differential staining of DNA and RNA in intact cells and isolated cell nuclei with acridine Orange. *Methods Cell Biol.*, 33, 285–298.

Darzynkiewicz, Z., Galkowski, D., & Zhao, H. 2008. Analysis of apoptosis by cytometry using Tunel assay. *Methods*, 44, 250–254.

Darzynkiewicz, Z., Gong, J., Juan, G., Ardelt, B., & Traganos, F. 1996. Cytometry of cyclin proteins. *Cytometry*, 25, 1–13.

Darzynkiewicz, Z., Juan, G., Li, X., Gorczyca, W., Murakami, T., & Traganos, F. 1997. Cytometry in cell necrobiology: Analysis of apoptosis and accidental cell death (Necrosis). *Cytometry*, 27, 1–20.

Dean, P.N., & Jett, J.H. 1974. Mathematical analysis of DNA distributions derived from flow microfluorometry. *Cell Biol.*, 60, 523–527.

Deitch, A.D., Law, H., & White, R.D. 1982. <Deitch-1982-A-stable-propidium-iodide-staining-.pdf>. *J. Histochem. Cytochem.*, 30, 967–972.

Ernst, L.A., Gupta, R.K., Mujumdar, R.B., & Waggoner, A.S. 1989. Cyanine dye labeling reagents for sulfhydryl groups. *Cytometry*, 10, 3–10.

Feher, K., Von Volkmann, K., Kirsch, J., Radbruch, A., Popien, J., & Kaiser, T. 2016. Multispectral flow cytometry: The consequences of increased light collection. *Cytometry A*, 89, 681–9.

Fong, E.J., Johnston, A.C., Notton, T., Jung, S.Y., Rose, K.A., Weinberger, L.S., & Shusteff, M. 2014. Acoustic focusing with engineered node locations for high-performance microfluidic particle separation. *Analyst*, 139, 1192–200.

Frangoul, H., Altshuler, D., Cappellini, M.D., Chen, Y.-S., Domm, J., Eustace, B.K., Foell, J., De La Fuente, J., Grupp, S., Handgretinger, R., Ho, T.W., Kattamis, A., Kernytsky, A., Lekstrom-Himes, J., Li, A.M., Locatelli, F., Mapara, M.Y., De Montalembert, M., Rondelli, D., Sharma, A., Sheth, S., Soni, S., Steinberg, M.H., Wall, D., Yen, A., & Corbacioglu, S. 2020. Crispr-Cas9 gene editing for sickle cell disease and β-Thalassemia. *NEJM*, 384, 1–9.

Georgakoudi, I., Solban, N., Novak, J., Rice, W.L., Wei, X., Hasan, T., & Lin, C.P. 2004. In vivo flow cytometry: A new method for enumerating circulating cancer cells. *Cancer Res.*, 64, 5044–5047.

Givan, A.L. 2001. *Flow Cytometry: First Principles*. Hoboken, NJ: Wiley-Liss, Inc.

Goddard, G., Martin, J.C., Graves, S.W., & Kaduchak, G. 2006. Ultrasonic particle-concentration for sheathless focusing of particles for analysis in a flow cytometer. *Cytometry A*, 69, 66–74.

Golden, J.P., Justin, G.A., Nasir, M., & Ligler, F.S. 2012. Hydrodynamic focusing–a versatile tool. *Anal. Bioanal. Chem.*, 402, 325–35.

Gong, J., Traganos, F., & Darzynkiewicz, Z. 1995. Growth imbalance and altered expression of cyclins B1, A, E, and D3 in MOLT-4 cells synchronized in the cell cycle by inhibitors of DNA replication. *Cell Growth Differ.*, 6(11), 1485–1493.

Gorenstein, D.G., Luxon, B.A., Herzog, N.K., Yang, X.-B., Aronson, J.F., Shope, R., Beasley, D., Barrett, A.D., & Leary, J.F. 2011. *Structure-based and combinatorially selected oligonucleoside phosphoro-thioate and phosphorodithioate aptamers targeting AP-1 transcription factors*. United States patent application U.S. Patent 7,910,523.

Grafton, M., Reece, L.M., Irazoqui, P.P., Jung, B., Summers, H.D., Bashir, R., & Leary, J.F. 2008. Design of a multi-stage microfluidics system for high-speed flow cytometry and closed system cell sorting for cytomics. *Imaging, Manipulation, and Analysis of Biomolecules, Cells, and Tissues VI*.

Grafton, M.G., Maleki, T., Zordan, M.D., Reece, L.M., Byrnes, R., Jones, A., Todd, P., & Leary, J.F. 2011a. Microfluidic MEMS hand-held flow cytometer. *Microfluidics, BioMEMS, and Medical Microsystems IX*.

Grafton, M.M., Geheb, B., Jang, J.H., Chuang, H.-S., Rajdev, P., Reece, L.M., Irazoqui, P.P., Wereley, S.T., Jung, B., & Leary, J.F. 2009. Microfabrication of a two-stage Biomems microfluidic cell sorter. *Microfluidics, BioMEMS, and Medical Microsystems VII.*

Grafton, M.M., Wang, L., Vidi, P.A., Leary, J., & Lelievre, S.A. 2011b. Breast on-a-chip: Mimicry of the channeling system of the breast for development of theranostics. *Integr. Biol. (Camb)*, 3, 451–459.

Grafton, M.M., Zordan, M.D., Chuang, H.-S., Rajdev, P., Reece, L.M., Irazoqui, P.P., Wereley, S.T., Byrnes, R., Todd, P., & Leary, J.F. 2010. Portable microfluidic cytometer for whole blood cell analysis. *Microfluidics, BioMEMS, and Medical Microsystems VIII.*

Graham, M.D. 2013. The Coulter Principle: Imaginary origins. *Cytometry A*, 83, 1057–61.

Graham, M.D. 2016. The Coulter Principle: Foundation of an industry. *JALA: J. Assoc. Lab. Autom.*, 8, 72–81.

Graham, M.D. 2021. The Coulter principle: A history. *Cytometry A*, 101, 8–11.

Gratzner, H.G., Leif, R.C., Ingraham, D.J., & Castro, A. 1975. The use of antibody specific for bromodeoxyuridine for the immunofluorescent determination of DNA replication in single cells and chromosomes. *Exp. Cell Res.*, 95, 88–94.

Gross, D., & Harris, C.M. 1985. *Fundamentals of Queuing Theory.* New York: John Wiley and Sons.

Haglund, E.M., Seale-Goldsmith, M-M., Dhawan, D., Stewart, J., Ramos-Vara, J., Cooper, C., Reece, L.M., Husk, T., Bergstrom, D., Knapp, D., & Leary, J.F. 2008. Peptide targeting of quantum dots to human breast cancer cells. *SPIE.* SPIE.

Hokanson, J.A., Rosenblatt, J.I., & Leary, J.F. 1999. Some theoretical and practical considerations for multivariate statistical cell classification useful in autologous stem cell transplantation and tumor cell purging. *Cytometry*, 36, 60–70.

Honrado, C., Bisegna, P., Swami, N.S., & Caselli, F. 2021. Single-cell microfluidic impedance cytometry: From raw signals to cell phenotypes using data analytics. *Lab Chip*, 21, 22–54.

Huh, D., Matthews, B.D., Mammoto, A., Montoya-Zavala, M., Hsin, H.Y., & Ingber, D.E. 2010. Reconstituting organ-level lung functions on a chip. *Science*, 328, 1662–1668.

Jin, D., Piper, J.A., Leif, R.C., Yang, S., Ferrari, B.C., Yuan, J., Wang, G., Vallarino, L.M., & Williams, J.W. 2009. Time-gated flow cytometry: An ultra-high selectivity method to recover ultra-rare-event mu-targets in high-background biosamples. *J. Biomed. Opt.*, 14, 024023.

Johnsom, I. 1998. Fluorescent probes for living cells. *Histochem. J.*, 30, 123–140.

Johnson, L.A., Welch, G.R., & Rens, W. 1999. The Beltsville sperm sexing technology: High-speed sperm sorting gives improved sperm output for in vitro fertilization and AI. *J. Anim. Sci.*, 77, 213–220.

Jovcevska, I., & Muyldermans, S. 2020. The therapeutic potential of nanobodies. *BioDrugs*, 34, 11–26.

Kaduchak, G., Goddard, G., Salzman, G., Sinha, D.N., Martin, J.C., Kwiatkowski, C., & Graves, S. *Ultrasonic analyte concentration and application in flow cytometry.* US Patent 9,074,979.

Kaduchak, G., Goddard, G., Salzman, G., Sinha, D.P., Martin, J.C., Kwiatkowski, C., & Graves, S. 2006. *Ultrasonic analyte concentration and application in flow cytometry.* USA Patent application 8,783,109.

Kimball, A.K., Oko, L.M., Bullock, B.L., Nemenoff, R.A., Van Dyk, L.F., & Clambey, E.T. 2018. A beginner's guide to analyzing and visualizing mass cytometry data. *J. Immunol.*, 200, 3–22.

Kohler, G., & Milstein, C. 1975. Continuous cultures of fused cells secreting antibody of pre-defined specificity. *Nature*, 256, 495–497.

Krutzik, P.O., Irish, J.M., Nolan, G.P., & Perez, O.D. 2004. Analysis of protein phosphorylation and cellular signaling events by flow cytometry: Techniques and clinical applications. *Clin. Immunol.*, 110, 206–221.

Krutzik, P.O., & Nolan, G.P. 2003. Intracellular phospho-protein staining techniques for flow cytometry: Monitoring single cell signaling events. *Cytometry A*, 55, 61–70.

Lapsley, M.I., Wang, L., & Huang, T.J. 2013. On-chip flow cytometry: Where is it now and where is it going? *Biomarkers Med.*, 7, 75–78.

Leary, J.F. 1994a. Strategies for rare cell detection and isolation. In: Z. Darzynkiewicz, J.P. Robinson, H.A. Crissman (eds.) *Methods in Cell Biology: Flow Cytometry*. San Diego, CA: Academic Press.

Leary, J.F. 1994b. Strategies for rare cell detection and isolation. *Methods Cell Biol.*, 42, 331–358.

Leary, J.F. 2000. Rare event detection and sorting of rare cells. In Durack, G., Robinson, J.P. (ed.) *Emerging Tools for Single Cell Analysis: Advances in Optical Measurement Technologies*. New York: Wiley-Liss.

Leary, J.F. 2005. Ultra high-speed sorting. *Cytometry. Part A*, 67, 76–85.

Leary, J.F. 2006. *Subtractive Clustering for Use in Analysis of Data" (advanced data mining algorithms for comparing two or more large multidimensional data sets)*. U.S. Patent 7,043,500.

Leary, J.F. 2018. Design of point-of-care (POC) microfluidic medical diagnostic devices. *Microfluidics, BioMEMS, and Medical Microsystems XVI*.

Leary, J.F. 2019. Design of portable microfluidic cytometry devices for rapid medical diagnostics in the field. *Imaging, Manipulation, and Analysis of Biomolecules, Cells, and Tissues XVII*.

Leary, J.F. 2021. Systems-level designs of portable microfluidic cytometers for operation in the field. *Imaging, Manipulation, and Analysis of Biomolecules, Cells, and Tissues XIX*.

Leary, J.F. 2022. *Fundamentals of Nanomedicine*. Cambridge: Cambridge University Press.

Leary, J.F., & Frederickson, C.J. 2008. *Flow Sorting System and Methods Regarding Same (high-speed microfluidics cell sorter)*. USA patent application U.S. Patent Application 7,452,725.

Leary, J.F., & McLaughlin, S.R. 1995. New technology for ultrasensitive detection and isolation of rare cells for clinical diagnostics and therapeutics. *J. Opt. Soc. Am. (SPIE)*, 150–163.

Leary, J.F. & Reece, L.M. 2000. Application of advanced cytometric and molecular technologies to minimal residual disease monitoring. *SPIE*, 36–44.

Leary, J.F., Corio, M.A., & Mclaughlin, S.R. 1993b. *System for High-Speed Measurement and Sorting of Particles*. USA patent application US Patent 5,204,884.

Leary, J.F., Corio, M.A., & McLaughlin, S.R. "System for High-Speed Measurement and Sorting of Particles" U.S. Patents 5,204,884 (1993) and 5,804,143 (1998).

Leary, J.F., Corio, M.A., & Mclaughlin, S.R. 1998d. *System for High-Speed Measurement and Sorting of Particles*. US Patent 5,804,143.

Leary, J.F., He, R., & Reece, L.N. 1999a. Detection and isolation of single tumor cells containing mutated DNA sequences. *SPIE Proceedings Systems and Technologies for Clinical Diagnostics and Drug Discovery II*, 3603, 93–101.

Leary, J.F., Hokanson, J.A., & Mclaughlin, S.R. 1997. High speed cell classification systems for real-time data classification and cell sorting. *J. Opt. Soc. Am. (SPIE)*, 2982, 342–352.

Leary, J.F., Hokanson, J.A., Rosenblatt, J.I., & Reece, L.N. 2001a. Real-time decision-making for high-throughput screening applications. *SPIE*, Vol. 4260, 219–225.

Leary, J.F., Mclaughlin, S.R., Hokanson, J.A., & Rosenblatt, J.I. 1998b. High speed real-time data classification and cell sorting using discriminant functions and probabilities of misclassification for stem cell enrichment and tumor purging. *J. Opt. Soc. Am. (SPIE)*, 3260, 274–281.

Leary, J.F., Mclaughlin, S.R., Hokanson, J.A., & Rosenblatt, J.I. 1998c. New high speed cell sorting methods for stem cell sorting and breast cancer cell purging. *J. Opt. Soc. Am. (SPIE)*, 3259, 114–121.

Leary, J.F., Mclaughlin, S.R., & Kavanau, K. 1996. New methods for detection, analysis and isolation of rare cell populations. *J. Opt. Soc. Am.(SPIE)*, 2678, 240–253.

Leary, J.F., Mclaughlin, S.R., Reece, L.N., Rosenblatt, J.I., & Hokanson, J.A. 1999b. Real-time multivariate statistical classification of cells for flow cytometry and cell sorting: A data mining application for stem cell isolation and tumor purging. *SPIE*, 3604, 158–169.

Leary, J.F., Reece, L.M., Hokanson, J.A., & Rosenblatt, J.I. (ed.) 2002a. *Advanced Real-Time Classification Methods for Flow Cytometry Data Analysis and Cell Sorting*. doi: 10.1117/12.468345.

Leary, J.F., Reece, L.M., Yang, X., & Gorenstein, D.G. 2005. High-throughput flow cytometric screening of combinatorial chemistry bead libraries for proteomics and drug discovery. *Proc. SPIE*, 216–223.

Leary, J.F., Reece, L.N., Szaniszlo, P., Prow, T., & Wang, N. 2001b. High-throughput cell analysis and sorting technologies for clinical diagnostics and therapeutics. *SPIE*, 16–27.

Leary, J.F., Schmidt, D., Gram, J.G., Mclaughlin, S.R., Dallatorre, C., & Burde, S. 1994. High-speed flow cytometric analysis and sorting of human fetal cells from maternal blood for molecular characterization. In: Elias, J.L.S.A.S. (ed.) *Fetal Cells in Maternal Blood: Prospects for Non-invasive Prenatal Diagnosis*. New York: Annals of the New York Academy of Sciences.

Leary, J.F., Szaniszlo, P., Prow, T., Reece, L.M., Wang, N., & Asmuth, D.M. 2002b. The Importance of High-Throughput Cell Separation Technologies for Genomics/Proteomics-Based Clinical Therapeutics. *SPIE*, 1–8.

Lee, J.F., Hesselberth, J.R., Meyers, L.A., & Ellington, A.D. 2004. Aptamer database. *Nucleic Acids Res.*, 32, D95–D100.

Lelievre, S.A., Vidi, P.-A., Leary, J.F., & Maleki, T. 2018. *Disease on a Chip*. US Patent 9,9969,964

Loken, M.R., Parks, D.R., & Herzenberg, L.A. 1977. Two-color immunofluorescence using a fluorescence-activated cell sorter. *Histochem. Cytochem.*, 25, 899–907.

Luminex 2021. Amnis imaging fl ow cytometers.

Maleki, T., Fricke, T., Quesenberry, J.T., Todd, P. & Leary, J.F. 2012. Point-of-care, portable microfluidic blood analyzer system. *Microfluidics, BioMEMS, and Medical Microsystems X*. doi: 10.1117/12.909051.

Montante, S., & Brinkman, R.R. 2019. Flow cytometry data analysis: Recent tools and algorithms. *Int. J. Lab. Hematol.*, 41(Suppl 1), 56–62.

Mujumdar, R.B., Ernst, L.A., Mujumdar, S.R., & Waggoner, A.S. 1989. Cyanine dye labeling reagents containing isothiocyanate groups. *Cytometry*, 10, 11–19.

Mullaney, B.A. 1974. Differential light scattering from spherical mammalian cells. *Biophys. J.*, 14, 439–453.

Nie, S., & Emory, S.R. 1997. Probing single molecules and single nanoparticles by surface-enhanced raman scattering. *Science*, 275, 1102–1106.

Nolan, J.P., & Sebba, D.S. 2011. Surface-enhanced Raman scattering (SERS) cytometry. *Methods Cell Biol.*, 102, 515–532.

Novo, D., Gregori, G., & Rajwa, B. 2013. Generalized unmixing model for multispectral flow cytometry utilizing nonsquare compensation matrices. *Cytometry A*, 83, 508–20.

Omerod, M.G. 1994. *Flow Cytometry: A Practical Approach*. Oxford: Oxford University Press.

Orporation, A.-B. 2021. True Multispectral Flow-Cytometry & Pulse Shape Analysis.

Pennisi, E. 2013. The CRISPR craze. *Science*, 341, 833–836.

Perfetto, S.P., Chattopadhyay, P.K., & Roederer, M. 2004. Seventeen-colour flow cytometry: Unravelling the immune system. *Nature Rev. Immunol.*, 4, 648–655.

Picot, J., Guerin, C.L., Le Van Kim, C., & Boulanger, C.M. 2012. Flow cytometry: Retrospective, fundamentals and recent instrumentation. *Cytotechnology*, 64, 109–30.

Pinkel, D., Lake, S., Gledhill, B.L., Van Dilla, M.A., Stephenson, D., & Watchmaker, G. 1982. High resolution DNA content measurements of mammalian sperm. *Cytometry*, 3, 1–9.

Pinkel, D., Landegent, J., Collins, C., Fusco, J., Segraves, R., Lucas, J., & Gray, J. 1998. Fluorescence-in-situ-hybridization. *Proc. Nati. Acad. Sci. USA*, 85, 9138–9142.

Rastede, E.E., Tanha, M., Yaron, D., Watkins, S.C., Waggoner, A.S., & Armitage, B.A. 2015. Spectral fine tuning of cyanine dyes: Electron donor-acceptor substituted analogues of thiazole orange. *Photochem. Photobiol. Sci.*, 14, 1703–1712.

Rens, W., Welch, G.R., & Johnson, L.A. 1998. A novel nozzle for more efficient sperm orientation to improve sorting efficiency of X and Y chromosome-bearing sperm. *Cytometry*, 33, 476–481.

Robinson, J.P., Patsekin, V., Gregori, G., Rajwa, B., & Jones, J. 2005. Multispectral flow cytometry: Next generation tools for automated classification. *Microsc. Microanal.*, 11, 2–3.

Rosenblatt, J.I., Hokanson, J.A., Mclaughlin, S.R., & Leary, J.F. 1997. Theoretical basis for sampling statistics useful for detecting and isolating rare cells using flow cytometry and cell sorting. *Cytometry*, 27, 233–238.

Rozanc, J., Finsgar, M., & Maver, U. 2021. Progressive use of multispectral imaging flow cytometry in various research areas. *Analyst*, 146, 4985–5007.

Salzman, G.C., Crowell, J.M., Goad, C.A., Hansen, K.M., Hiebert, R.D., Labauve, P.M., Martin, J.C., Ingram, M.L., & Mullaney, P.F. 1975. A flow-system multiangle light-scattering instrument for cell characterization. *Clin. Chem.*, 21, 1297–1304.

Shang, L., Cheng, Y., & Zhao, Y. 2017. Emerging droplet microfluidics. *Chem. Rev.*, 117, 7964–8040.

Shapiro, H.M. 2001. *Practical Flow Cytometry*. Hoboken, NJ: John Wiley & Sons.

Shapiro, H.M., & Telford, W.G. 2018. Lasers for flow cytometry: Current and future trends. *Curr. Protoc. Cytom.*, 83, 1.9.1–1.9.21.

Shrirao, A.B., Fritz, Z., Novik, E.M., Yarmush, G.M., Schloss, R.S., Zahn, J.D., & Yarmush, M.L. 2018. Microfluidic flow cytometry: The role of microfabrication methodologies, performance and functional specification. *Technology (Singap World Sci)*, 6, 1–23.

Smith, J.N., Reece, L., Szaniszlo, P., Leary, R.C., & Leary, J.F. 2005. Subtractive clustering analysis: A novel data mining method for finding cell subpopulations. *Imaging, Manipulation, and Analysis of Biomolecules and Cells: Fundamentals and Applications III*.

Spitzer, M.H. & Nolan, G.P. 2016. Mass cytometry: Single cells, many features. *Cell* 165, 780–791.

Steen, H.B. 1983. A microscope-based flow cytophotometer. *Histochem. J.*, 15, 147–160.

Steinkamp, J.A. 2001. Time-resolved fluorescence measurements. *Current Protocols in Cytometry*. Chapter 1:Unit 1.15.

Stovel, R.T., & Sweet, R.G. 1979. Individual cell sorting. *J. Histochem. Cytochem.*, 27, 284–288.

Sweet, R.G. 1965. High frequency recording with electrostatically deflected ink jets. *Rev. Sci. Instrum.*, 36, 131–136.

Szollosi, J., Damjanovich, S., & Laszlo, M. 1998. Application of fluorescence resonance energy transfer in the clinical laboratory: Routine and research. *Cytometry (Commun. Clin. Cytom.)*, 34, 159–179.

Trask, B., Van Den Engh, G., Mayall, B., & Gray, J.W. 1989. Chromosome heteromorphism quantified by high-resolution bivariate flow karyotyping. *Am. J. Hum. Genet.*, 45, 739–752.

Tron, L., Szollosi, J., & Damjanovich, S. 1984. Flow cytometric measurement of fluorescence resonance energy transfer. *Biophys. J.*, 45, 939–946.

Van Der Meer, A.D., & Van Den Berg, A. 2012. Organs-on-chips: Breaking the in vitro impasse. *Integr. Biol. (Camb)*, 4, 461–70.

Van Dilla, M.A., & Deaven, L.L. 1990. Construction of gene libraries for each human chromosome. *Cytometry*, 11, 208–218.

Van Dilla, M.A., Deaven, L.L., Albright, K.L., Allen, N.A., Aubuchon, M.R., Bartholdi, M.F., Brown, N.C., Campbell, E.W., Carrano, A.V., Clark, L.M., Cram, L.S., Crawford, B.D., Fuscoe, J.C., Gray, J.W., Hildebrand, C.E., Jackson, P.J., Jett, J.H., Longmiret, J.L., Lozes, C.R., Luedemann, M.L., Martin, J.C., Mcninch, J.S., Meincke, L.J., Mendelsohn, M.L., Meyne, J., Moyzis, R.K., Munk, A.C., Perlman, J., Peters, D.C., Silva, A.J., & Trask, B.J. 1986. Human chromosome-specific DNA libraries: Construction and availability. *Nature Biotechnol.*, 4, 537–552.

Vembadi, A., Menachery, A., & Qasaimeh, M.A. 2019. Cell cytometry: Review and perspective on biotechnological advances. *Front. Bioeng. Biotechnol.*, 7, 147.

Vidi, P.A., Maleki, T., Ochoa, M., Wang, L., Clark, S.M., Leary, J.F., & Lelievre, S.A. 2014. Disease-on-a-chip: Mimicry of tumor growth in mammary ducts. *Lab Chip*, 14, 172–177.

Vo-Dinh, T., Robinson, J.P., Grundfest, W.S., Rajwa, B., Gregori, G., Benaron, D.A., Cohn, G.E., Jones, J., & Patsekin, V. 2005. Multispectral cytometry of single bio-particles using a 32-channel detector. *Advanced Biomedical and Clinical Diagnostic Systems III*.

Wang, Y., Sayyadi, N., Zheng, X., Woods, T.A., Leif, R.C., Shi, B., Graves, S.W., Piper, J.A., & Lu, Y. 2020. Time-resolved microfluidic flow cytometer for decoding luminescence lifetimes in the microsecond region. *Lab Chip*, 20, 655–664.

Wang, Z., Samanipour, R., & Kim, K. 2016. Organ-on-a-chip platforms for drug screening and tissue engineering. In Jo, H., Jun, H.W., Shin, J., Lee, S. (eds.) *Biomedical Engineering: Frontier Research and Converging Technologies, Biosystems & Biorobotics*. Switzerland: Springer International Publishing.

Ward, M.D., & Kaduchak, G. 2018. Fundamentals of acoustic cytometry. *Curr. Protoc. Cytom.*, 84, e36.

Wheeless, L., Reeder, J.E., & O'connell, M.J. 1990. Slit-scan flow analysis of cytologic specimens from the female genital tract. *Methods Biol.*, 33, 501–507.

Yang, R.J., Fu, L.-M., & Hou, H.-H. 2018. Review and perspectives on microfluidic flow cytometers. *Sens. Actuators B Chem.*, 266, 26–45.

Yang, X., Bassett, S.E., Li, X., Luxon, B.A., Herzog, N.K., Shope, R.E., Aronson, J., Prow, T.W., Leary, J.F., Kirby, R., Ellington, A.D., & Gorenstein, D.G. 2002. Construction and selection of bead-bound combinatorial oligonucleoside phosphorothioate and phosphorodithioate aptamer libraries designed for rapid PCR-based sequencing. *Nucleic Acids Res.*, 30, 1–8.

Yang, X., Li, N., & Gorenstein, D.G. 2011. Strategies for the discovery of therapeutic aptamers. *Expert Opin. Drug Discov.*, 6, 75–87.

Yang, X., Li, X., Prow, T.W., Reece, L.M., Bassett, S.E., Luxon, B.A., Herzog, N.K., Aronson, J., Shope, R.E., Leary, J.F., & Gorenstein, D.G. 2003. Immunofluorescence assay and flow-cytometry selection of bead-bound aptamers. *Nucleic Acids Res.*, 31, e54.

Zordan, M.D., Grafton, M.M., Acharya, G., Reece, L.M., Cooper, C.L., Aronson, A.I., Park, K., & Leary, J.F. 2009. Detection of pathogenic E. coli O157:H7 by a hybrid microfluidic SPR and molecular imaging cytometry device. *Cytometry A*, 75, 155–62.

5 Single-Molecule Separation

Ching-Te Kuo and Yu-Chia Lin
National Sun Yat-sen University

CONTENTS

5.1 INTRODUCTION

"Great oaks from hhghgjlittle acorns grow" can be expended to explain the construction of nature biological system. While a living system is typically made up of single cells, the sole and complex unit of life. Altered by gene mutations and environmental stresses, the genomic, transcriptomic, and proteomic levels will be modulated with heterogeneity in individual cells (Saadatpour, Lai, Guo, & Yuan, 2015). Such cellular heterogeneity may attenuate the treatment efficiency against diseases if the therapy only targets one cell population (Leung et al., 2019). Similarly, single molecules (e.g., DNA, RNA, proteins, biomolecules) are the key building components for cells. For industry, discovering a new single-molecule therapeutic target, termed biologic, would cost averagely $2.6 billion. At least of which, one half may be attributed to separating and purifying the active compounds (Bishop & Landes, 2018). It suggests that the single-molecule separation from a pool of analytes or cells may play a pioneer for the study of new drug development or the exploration of single molecules' kinematics (Agrawal, Zhang, Byassee, Tripp, & Nie, 2006; Chiesl, Forster, Root, Larkin, & Barron, 2007).

Technology involved in the separation of single molecules includes traditional and advanced approaches. For the traditional separation, aqueous two-phase system

(ATPS) was first used to identify small molecules from both starch and gelatin (Bockelmann, Essevaz-Roulet, & Heslot, 1998). Following such technology, the separation, purification, and analysis of protein, DNA, and other important biomolecules from a mixture of analytes were also achieved (Grilo, Aires-Barros, & Azevedo, 2014; Streitner, VoB, & Flaschel, 2007). Other traditional approaches such as using capillary electrophoresis (CE) (Castro & Shera, 1995; Swinney & Bornhop, 2000), liquid/gas chromatography (Pasti, Cavazzini, Felinger, Martin, & Dondi, 2005; Senavirathne Jr., Messer, Fishel, & Yoder, 2018), and protein crystallization (Navarro, Wu, & Wang, 2009; Suzuki et al., 2010) were performed to separate DNA bands with an efficiency ranging from 70 to 15,000 bp.

The advanced separation approach was mainly contributed by the incorporation of hydrodynamics (Friedrich, Liu, & Wang, 2016; Liu, Rane, Zhang, & Wang, 2011), electrohydrodynamics (Green & Morgan, 1998; Yamamoto & Fujii, 2010), and/or microfluidics/nanofluidics technology (Haab & Mathies, 1999; Shimizu et al., 2021; Tabuchi, Kuramitsu, Nakamura, & Baba, 2003) with the traditional approach described above. Several advantages of adopting such a technique were investigated, including extremely low analyte volume needed, rapid separation, low signal-to-noise ratio, and superior reliability during operating.

Detection with the single molecules separated from the above technology is also eye-catching since the separation efficiency is needed to be verified. The emerging approaches for single molecule sensing can be contributed through the measurements using rapid near-field detection with engineered nanopore (Shimizu et al., 2021), near-field massively parallel detection (Gooding & Gaus, 2016), wide field sampling-near field detection (Gooding & Gaus, 2016), wide field detection through a fluorescence microscopy (Calabrase et al., 2020; Kamagata, 2021), super-resolution fluorescence microscopy (Zhang, Shao, & Sun, 2022), and single-molecule mass photometry (Li, Struwe, & Kukura, 2020).

In this chapter, recent technology and its advances in single-molecule separation will be discussed. The summary of these separation approaches is listed in Table 5.1.

5.2 TECHNOLOGY INVOLVED IN SINGLE-MOLECULE SEPARATION

5.2.1 AQUEOUS TWO-PHASE SYSTEMS

The aqueous two-phase system (ATPS) has been attracted with a huge interest owing to its potential applications, in terms of extraction, fractionation, purification, and enrichment of proteins, nucleic acids, viruses, enzymes, and other critical biomolecules in medicine (Bockelmann et al., 1998; Grilo et al., 2014; Hutti-Kaul, 2001; Iqbal et al., 2016; Laskowski et al., 2019; Mukherjee, Knop, Mobitz, & Winter, 2020). Back in 1896, Martinus Willem Beijerinck accidentally discovered the ATPS from a mixed aqueous solution with both starch and gelatin. Since water plays the main bridge between both phases in ATPS, the ATPS provides a gentle environment for the separation of biomolecules and the stabilization of polymers' structures and biological activities. Furthermore, an additional and complex equipment is not necessarily required for the ATPS.

TABLE 5.1

Summary of Technology Involved in Single-Molecule Separation

Technology	Target	Time	Volume	Resolution	References
ATPS	DNA, protein	> 10 hours	Nil	Nil	Hutti-Kaul (2001) Streitner et al. (2007) Laskowski et al. (2019) Mukherjee et al. (2020)
CE	DNA, protein, fluorophore	12~810 seconds	1~5 µL	4.6~7.25 kbp DNA 1fM-rhodamine-6G	Castro and Shera (1995) Haab and Mathies (1996) Swinney and Bornhop (2000) Fang et al. (2007)
Chromatography	DNA, protein	0.1~1.45 minutes	0.5~2 mL	440~670 kDa protein	Pasti et al. (2005) Senavirathne et al. (2018) Lu and Bianco (2021)
Protein crystallization	Protein	Nil	0.5 µL	7.1~28 kDa protein	Navarro et al. (2009) Suzuki et al. (2010) McPherson and Gavira (2014)
Hydrodynamics	DNA, protein, fluorophore	20 seconds ~ 100 minutes	45 pL ~ 0.5 µL	2~48.5 kbp DNA 10~50 aM DNA	Friedrich et al. (2016) Li et al. (2018) Lundgren et al. (2018) Friedrich et al. (2019)
Electrohydrodynamics	DNA, protein, fluorophore	5 seconds ~ 15 minutes	Nil	7~50 kbp DNA 93 and 216 nm beads	Green and Morgan (1998) Ranchon et al. (2016) Chami et al. (2018) Ghosh et al. (2020)
Micro/nanofluidics	DNA, protein, fluorophore	14~300 seconds	100 pL ~ 10 µL	100~1000 bp DNA 14~200 kDa protein	Haab and Mathies (1999) Tabuchi et al. (2003) Kim et al. (2007) Zrehen et al. (2020)

Types of ATPS are typically made by two polymers with polyethylene glycol (PEG) and dextran or formed by a polymer and a salt (such as phosphate, sulfate, and citrate). Ionic liquids and short-chain alcohols, in addition to ionic and/or non-ionic surfactants are also adopted to make the micellar and reverse micellar ATPSs. For example, the PEG-dextran-based ATPS was successfully used for the partition of chitinase towards the bottom phase (Iqbal et al., 2016). The purification of plasmid DNA and thus the removal of RNA was performed by the incorporation of isooctane/ ethylhexanol/methyltrioctyl ammoniumchloride-based ATPS (Streitner et al., 2007). Another study for the conformational dynamics of α–Syn interacting with DNA harpins (DNA-HPs) was performed by a single-molecule Forster resonance energy transfer (smFRET) system, in which the formation of protein-bound intermediate conformations is stimulated in the presence of ATPS (Mukherjee et al., 2020).

5.2.2 CAPILLARY ELECTROPHORESIS

Capillary electrophoresis (CE) is a family of electrokinetic separation methods, which is conventionally adopted in submillimeter diameter capillaries. Upon an electric field applied, in general, the analytes in electrolyte solutions would migrate along the capillaries (Figure 5.1). Meanwhile, the analytes can be separated due to their ionic mobility and may be concentrated and immobilized in specific zones following the gradients in conductivity and/or pH. According to the working principles of CE, it can be categorized with capillary zone electrophoresis (CZE), capillary gel electrophoresis (CGE), capillary isotachophoresis, capillary isoelectric focusing (CIEF), and micellar electrokinetic chromatography (MEKC). So far, the CE technology is widely applied for the separation and detection of protein and DNA-associated analysis such as oligonucleotides analysis, DNA sequencing, and dsDNA fragments analysis (Fang, Zhang, Li, Li, & Yeung, 2007; Swinney & Bornhop, 2000).

For the CE application of single-molecule separation and detection, Castro and Shera presented an integrated CE method with a two-laser beams system (Castro & Shera, 1995). The separation efficiency was demonstrated to lie in the range of 2%– 3%, satisfying most laboratory separation. Such the free-solution method may be further applied for the fluorescence immunoassay, DNA/RNA hybridization, and DNA fingerprinting, in which no extensive DNA amplification using polymerase chain

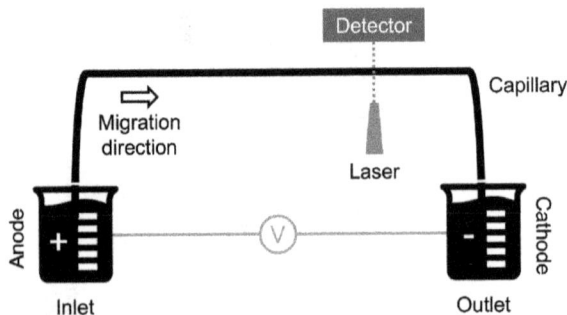

FIGURE 5.1 Capillary electrophoresis setup.

reaction (PCR) was needed. Another application of the CE separation was revealed for the detection of separated DNA fragments by single-molecule fluorescence burst (Haab & Mathies, 1995, 1996). Through the incorporation of CE and single molecule photon burst counting, it can be applied to separate M13 DNA and to fractionalize pBR 322 DNA from pRL 277 DNA with an on-column sensitivity of 50–100 molecules of DNA per band.

5.2.3 CHROMATOGRAPHY

Bio/chemical molecule separation by the chromatography is usually accomplished by the differential rate of migration induced by fluidic solvent flowing through solid substrates. The fluidic solvent called the mobile phase is typically gas or liquid phase. The solid substrate called the stationary phase is typically a column, a capillary, a plate, or a sheet (Figure 5.2). The separation mechanism is mainly based on the different affinities of the components in the mobile phase against the stationary phase. The individual components are eventually retained in the solid phase for different lengths of time and be separated, according to their interactions with solid inner surfaces and different traveling velocities in the mobile fluid. Chromatography can further provide critical knowledge of the underlying thermodynamic information by the sorption isotherm characteristic (Fang et al., 2007). Based on the separation mechanism, chromatography can be categorized with ion exchange, size exclusion, and expanded bed adsorption separations. In addition, such chromatography techniques can be further contributed with reversed-phase, hydrophobic interaction,

FIGURE 5.2 Two typical chromatography techniques including thin-layer chromatography and column chromatography.

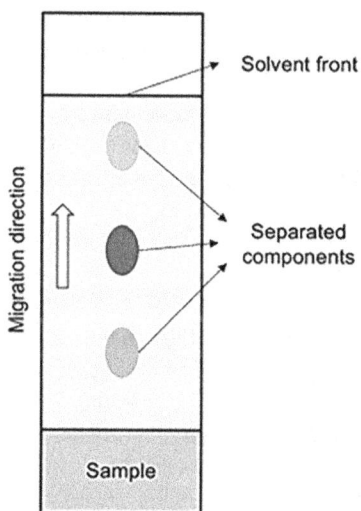

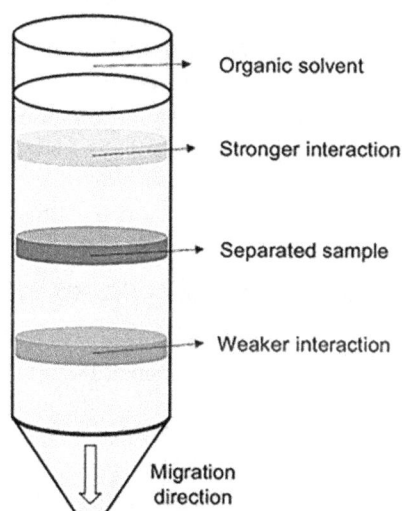

hydrodynamic, two-dimensional, simulated moving-bed, pyrolysis gas, fast protein liquid, countercurrent, period counter-current, chiral, and aqueous normal-phase separations.

For applications of single-molecule separation, Pasti et al. (2005) presented a renewed stochastic algorithm of chromatography for conceptually bridging between the single-molecule dynamics in an established chromatography system and its resulted data. The improved model interpreted from the Levy approach also opened a new way for explaining the fundamental steps in separation and thus sensors design behind it. Senavirathne et al. (2018) presented the single-molecule fluorescence imaging with nuclease-free protocatechuate 3,4-dioxygenase (PCD) by the use of two chromatographic steps. The size exclusion column chromatography and the following gel filtration of the recombinant PCD from *Pseudomonas putida* in *E. coli* were adopted. Significant enrichment of >99% purity for the oxygen-scavenging enzyme PCD was achieved to study the nucleic acid-protein interactions. No damage and contamination of the extracted nucleic acids was also demonstrated. Lu and Bianco (2021) presented a fully optimized method available to perform high yields of a single-molecule hairpin DNA substrate. Through the incorporation of the separated substrates and the presented magnetic tweezer-associated experiments, the single-molecule studies of DNA were demonstrated ranging in size from 70 to 15,000 bp.

5.2.4 PROTEIN CRYSTALLIZATION

More than 150 years ago, the first protein crystals of hemoglobin resulted from wild worms and fishes were discovered (McPherson & Gavira, 2014). Crystallization of proteins is the process of crystal contact in which the individual protein molecules stabilize in the presence of a regular array. During protein crystallization, proteins are initially dissolved in an aqueous solution and then turned into an array of solid crystals following the supersaturated state is reached. Several approaches have been applied to achieve such crystallization, including vapor diffusion, microbatch, microdialysis, and free-interface diffusion (McPherson & Gavira, 2014). Key factors affecting the protein crystallization include pH of solution, environmental temperature, and chemical additives (Navarro et al., 2009). One example of using the single-molecule protein crystallization was demonstrated for assisting the crystallization of small proteins fused by green fluorescent protein (GFP) (Suzuki et al., 2010).

5.2.5 HYDRODYNAMICS-BASED SEPARATION

Separation of single molecules and its interactions between DNA/RNA and other biomolecules is of fundamental importance in life science, diagnostics of disease, and drug development for increasing human health (Friedrich, Bang, Li, & Wang, 2019). The utilization of CE is widely applied to determine the affinity and stoichiometry for interactions among DNA–DNA, DNA–protein, and DNA–small molecule. However, the intrinsic drawbacks of using such method gain with time and funding waste, labor-intensive, and relatively large amount of samples required (Friedrich et al., 2016). To overcome the above issue, the electrophoretic mobility shift assays (EMSAs) have been developed to be of not only the qualitative conformational analysis of

DNA-biomolecule bindings but also the large-scale conformational changes. Despite the above merits of CE-based approach, challenges remain. For example, the EMSAs only consider with a semi-quantitative sensitivity. The use of gel to separate and detect biomolecules that contain differential mobilities in analytes with similar charge or mass ratios may attenuate the stability of inter-molecular interactions. Furthermore, the behavior of molecules may differ between the artificial gel and the native solution.

To address the above issue, the hydrodynamics-based separation of single molecules has become an alternative approach to analyze biomolecules in free solution. The principle of separation mainly depends on the molecule size sampling, which is caused by the flow-induced collision of particles in a capillary tube or a bead-packed column. The input work is contributed by the hydraulic pressure difference within the tube or column. It has been demonstrated that the column format packed with nanoporous beads presented a particular performance in the characterization of particles and polymers (Li, Friedrich, & Wang, 2018). In contrast, the hydrodynamic open tube-based separation possessed effectivity, especially for the separation of bio-macromolecules, such as DNA (Friedrich et al., 2016).

For the application of single-molecule separation, Liu et al. (2011) presented a single-molecule free solution hydrodynamic separation (SML-FSHS) to be able to separate and cohesively detect the *E. coli* 16s rRNA targets. The usage of SML-FSHS can achieve a wide dynamic range and a high sampling resolution for both large DNA (23 *versus* 27 kbp) and small DNA (100 *versus* 200 bp). The efficiency of the separation can be improved up to 240 yoctomoles (approximately 150 molecules) from a critical sample volume of 5 pL. Based on the SML-FSHS technique, Friedrich et al. furthermore demonstrated that it can be applied for the in-line preconcentration of biomolecules, such as DNA, with a relatively high performance as compared with other preconcentrated approaches (e.g. isotachophoresis, field-amplified sample stacking, and electrokinetic methods) (2017). Only pressure-driven flow was loaded into a micro-sized capillary tube for the single-molecule separation by the SML-FSHS. It thus provided a completely electrode-free platform to be capable of DNA preconcentration with a 10,000-fold increment with *HindIII*–digested γ DNA beginning from an extremely low concentration of 150 aM.

Another separation example of using a hydrodynamic approach was presented for the single-molecule analysis of integral membrane proteins (Lundgren et al., 2018). Through the incorporation of nanoparticles, conjugated against the ectodomain of β–secretase 1 (BACE1), and a weak hydraulic flow (approximately in the range of fN force), it thus enabled a significant thousand-fold enrichment of the target membrane proteins.

5.2.6   ELECTROHYDRODYNAMICS-BASED SEPARATION

In the past decades, electrohydrodynamics has been widely applied in microelectromechanical systems (MEMS) for controlling fluidic flow in a microscale channel (Kuo & Liu, 2008a, b). Such applications are called as micropumps. The developed micropumps can be categorized into two types, including membrane displacement-derived micropump and field-induced-flow micropump. The electrohydrodynamic micropump is typically derived from field-based pumping, which is typically powered

Electrohydrodynamic

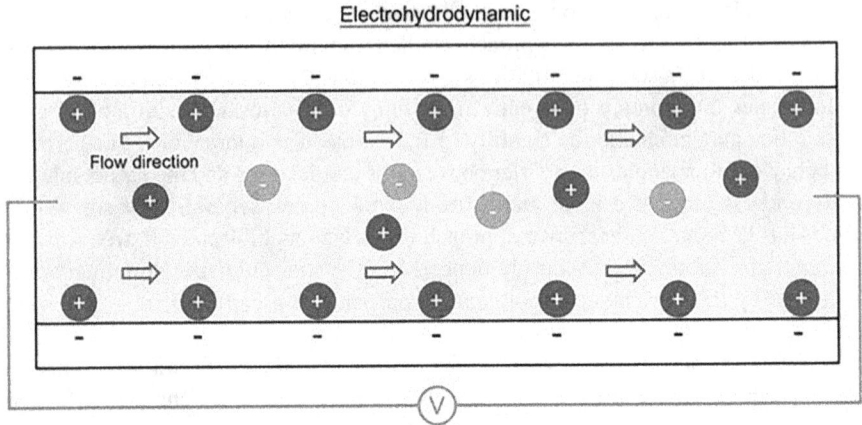

FIGURE 5.3 Schematic showing the electrohydrodynamic flow.

by a direct circuit (DC) or an alternative circuit (AC) electric filed (Figure 5.3). Electrohydrodynamics mainly studies the relative motions of ionic particles or molecules against the applied electric field, and the resulting fluid flow according to the interactions with the electric field and the surrounding particles. Based on the types of particles and fluid flow mechanism, electrohydrodynamics covers electrophoresis, electrokinesis, dielectrophoresis, electro-osmosis, and electrorotation.

Since no external hydraulic pressure is needed for the electrohydrodynamic separation of single molecules, the efficiency of separation can be improved mainly according to both the applied electric field and the associated channel arrangement. For example, the rapid separation, immobilization, and processing of nanoparticles (93 *versus* 216 nm diameter latex nanospheres), and DNA separation and enrichment were achieved through the incorporation of dielectrophoretic-induced trapping force and electrohydrodynamic-induced fluid flow (Chami, Socol, Manghi, & Bancaud, 2018; Green & Morgan, 1998; Ranchon et al., 2016).

A fluidic arrangement with nano-sized channels was further demonstrated to be able to separate and sort single DNA molecules (Yamamoto & Fujii, 2010). Combined with flow sensors and molecule switchers fabricated by microelectrodes with electrohydrodynamic activities, a critical DNA size of 300 bp can be separated and detected. Similarly, the single-molecule separations of DNA with small fragments, carbon-nanodots, and organic fluorophores in water were performed in nano-fabricated channels, cooperating with two-focus fluorescence correlation spectroscopy (2fFCS) (Ghosh, Karedla, & Gregor, 2020). The one-dimensional (1D) flows of 11-nm DNA fragments, 2-nm graphene quantum dots (GODs), and single organic fluorophores were performed using electro-osmosis flow induced in the nanochannels. The flow velocity of up to 300 μm/s can be achieved at a single-molecule detection level. In addition, Teillet et al. (2020) presented that the separation, concentration, and purification of DNA can be accomplished using the combination of hydraulic actuation of viscoelastic liquids with an opposite electrophoretic driving force. Based on the on-demand design of the fluidic channel with a Taylor dispersion geometry, the mechanism of DNA band broadening in electrohydrodynamic migration can be explored and the migration of a 600 bp DNA band can be also tracked following the matrix with different pressure-electric field interactions.

5.2.7 MICROFLUIDICS/NANOFLUIDICS-ASSISTED SEPARATION

In the beginning of the 1980s, the termed "microfluidics" emerged and was widely discussed to develop its potential applications in human life, such as micropump, microvalve, inkjet printing, and lab on a chip. At the critical scale of sub-millimeter, the control of fluids in such microchannels would be dominated by the surface forces instead of the volumetric forces. Thanks for the merits of multiplexing, automation, and high-throughput manipulation of microfluidics, it has been advanced for multidisciplinary fields, including physics, chemistry, engineering, biotechnology, nanotechnology, and medicine.

As the critical scale is down to nanometers (typically 1–100 nm), the thermodynamic properties and the associated chemical reactivity of reactants at the solid wall interface would be extremely different than that at the micro-scale. The term "nanofluidics" is the study of fluidic behavior in such scales. The increased viscosity near the channel wall and the differential solution ion-wall characteristics (e.g., electrical double layer, hydrodynamic radius, and electrohydrodynamic effect) would dominate the control of nanofluidics.

For the applications of single-molecule separation, most studies focus on the improved design of microchannels and/or nanochannels incorporated with hydrodynamic and/or electrohydrodynamic-based separations (Vicente, Plazl, Ventura, & Znidarsic-Plazl, 2020). For example, Haab and Mathies (1999) presented a microfabricated capillary electrophoresis chip for the single-molecule separation and detection of DNA. The chip was made with a Schott D263 glass, and the critical depth and width were 16 and 46 μm, respectively. The DNA separation limit using this chip was in the range of 100–1000 bp. Similarly, Tabuchi et al. (2003) demonstrated a microfluidic chip with a 12-channel array to be capable of 15-s protein separation from a human T lymphoblastic Jurkat cells. The called 2-D electrophoresis microchip was fabricated by poly(methyl methacrylate) (PMMA) having channels 30 μm depth and 100 μm width. The chip can achieve the protein separation criteria ranging from 5.7 to 200 kDa. The electrophoretic chip can be further integrated with laser-induced fluorescence (LIF) detection for single-molecule applications (Kim, Huang, & Zare, 2007). The chip was made of poly(dimethylsiloxane) (PDMS) with a microvalve for the sorting of sub-nanoliter (100–300 pL) analytes. They demonstrated that five A647-labeled streptavidin proteins can be separated during the period of 17.2–25.7 seconds. Another application using the microfluidics-based separation was developed to amplify nucleic acids in multiplexing (Dettloff, Yang, Rulison, Chow, & Farinas, 2008). The chip was made with 10 μm in depth, 180 μm in width, and 40 mm in separation, which is far enough to prevent neighboring DNA templates to contain discreet DNA molecules. It was successfully demonstrated using amplifying 2D6.6 CYP450 template to verify the wild-type and mutant sequences *in situ*.

Since the characteristic scale of nanofluidics resembles the biomolecules (e.g. DNA, RNA, protein), nanofluidics-based separation approach is widely adopted and be proven to have a superior separation performance of single-molecules than that using microfluidics (Napoli, Eijkel, & Pennathur, 2010). Jing et al. (2008) presented that a nanofluidic device with a critical depth of 500 nm can be used to separate single rabbit anti-HA polyclonal antibody and be verified fir the 2D photon burst analysis. Zrehen et al. (2020) presented an on-chip microfluidic chip for the separation of single proteins. The chip had a critical depth of 600 nm and was integrated with a sodium dodecyl sulfate–polyacrylamide gel electrophoresis (SDS-PAGE) enhancing the protein separation *in situ*. The migration of proteins inside the SDS-PAGE was achieved electrokinetically, and the low-profile channel concentrated the separated proteins and thus increased the detection sensitivity. An exponential relationship between the molecular weights of proteins and the measured mobilities was demonstrated by Atto647N-conjugated proteins (ranging from 14 to 70 kDa in size) in the chip.

The combination of microfluidics and nanofluidics to enhance single-molecule separation, concentration, and detection has also been widely discussed. Ghosh et al. (2020) used an electrokinetic 1D molecular mass flow incorporated with two-focus fluorescence correlation spectroscopy to *in situ* separate and detect single molecules. Following the significant improvement, the chip can be applied for the study of molecular shot noise as confined in a nano-environment. Varongchayakul et al. (2018) presented a microfluidic-nanopore chip for the single molecule detection of DNA PCR products. The chip was made of Zeonex plastic through micromachining and heat bonding. Through the incorporation of microchip and nanopore, the DNA sample can be purified and detected under the size range of 5 kbps.

5.3 REMARKS

Abnormal gene expression is typically caused by the conformational changes of DNA and its translated proteins, which are clinically correlated to most of the human diseases such as Alzheimer's disease (AD) and Parkinson's disease (PD). It highlights that technology that emerged for the separation and study of these single biomolecules should be eye-catching and invaluable against severe diseases. Based on these, single-molecule separation technology has been widely developed to investigate the intrinsic defects of DNA, proteins, and other critical biomolecules. The following analysis can be collected by profiling the "big data" from the separation of single molecules. Typical approaches for single-molecule separation include an aqueous two-phase system, capillary electrophoresis, protein crystallization, hydrodynamics and/or electrohydrodynamics-based technology, and micro/nanofluidics-assisted separations. The future step together with the single-molecule separation technology will be integrated achieving a monolithic analysis platform. Moreover, it will advance the study of clinical outcomes the personalized disease medicine.

ACKNOWLEDGEMENTS

Financial support from the Ministry of Science and Technology (MOST), Taiwan, under the grant 110-2221-E-110028 is gratefully acknowledged.

REFERENCES

Agrawal, A., Zhang, C., Byassee, T., Tripp, R. A., & Nie, S. (2006). Counting single native bio-molecules and intact viruses with color-coded nanoparticles. *Anal. Chem., 78*, 1061–1070.

Bishop, L. D. C., & Landes, C. F. (2018). From a protein's perspective: Elution at the single-molecule level. *Acc. Chem. Res., 51*, 2247–2254.

Bockelmann, U., Essevaz-Roulet, B., & Heslot, F. (1998). DNA strand separation studied by single molecule force measurements. *Phys. Rev. E, 58*(2), 2386.

Calabrase, W., Bishop, L. D. C., Dutta, C., Misiura, A., Landes, C. F., & Kisley, L. (2020). Transforming separation science with single-molecule methods. *Anal. Chem., 92*, 13622–13629.

Castro, A., & Shera, E. B. (1995). Single-molecule electrophoresis. *Anal. Chem., 67*, 3181–3186.

Chami, B., Socol, M., Manghi, M., & Bancaud, A. (2018). Modeling of DNA transport in viscoelastic electro-hydrodynamic flows for enhanced size separation. *Soft Matter, 14*, 5069–5079.

Chiesl, T. N., Forster, R. E., Root, B. E., Larkin, M., & Barron, A. E. (2007). Stochastic single-molecule videomicroscopy methods to measure electrophoretic DNA migration modalities in polymer solutions above and below entanglement. *Anal. Chem., 79*, 7740–7747.

Dettloff, R., Yang, E., Rulison, A., Chow, A., & Farinas, J. (2008). Nucleic acid amplification of individual molecules in a microfluidic device. *Anal. Chem., 80*, 4208–4213.

Fang, N., Zhang, H., Li, J., Li, H.-W., & Yeung, E. S. (2007). Mobility-based wall adsorption isotherms for comparing capillary electrophoresis with single-molecule observations. *Anal. Chem., 79*(16), 6047–6054.

Friedrich, S. M., Bang, R., Li, A., & Wang, T.-H. (2019). Versatile analysis of DNA–biomole-cule interactions in solution by hydrodynamic separation and single molecule detection. *Anal. Chem., 91*, 2822–2830.

Friedrich, S. M., Burke, J. M., Liu, K. J., Ivory, C. F., & Wang, T.-H. (2017). Molecular rheo-taxis directs DNA migration and concentration against a pressure-driven flow. *Nat. Commun., 8*, 1213.

Friedrich, S. M., Liu, K. J., & Wang, T.-H. (2016). Single molecule hydrodynamic separation allows sensitive and quantitative analysis of DNA conformation and binding interactions in free solution. *J. AM. Chem. Soc., 138*, 319–327.

Ghosh, S., Karedla, N., & Gregor, I. (2020). Single-molecule confinement with uniform elec-trodynamic nanofluidics. *Lab Chip, 20*, 3249–3257.

Gooding, J. J., & Gaus, K. (2016). Single molecule sensors: challenges and opportunities for quantitative analysis. *Angew. Chem. Int. Ed., 55*(38), 11354–11366.

Green, N. G., & Morgan, H. (1998). Separation of submicrometre particles using a combination of dielectrophoretic and electrohydrodynamic forces. *J. Phys. D: Appl. Phys., 31*, L25.

Grilo, A. L., Aires-Barros, M. R., & Azevedo, A. M. (2014). Partitioning in aqueous two-phase systems: Fundamentals, applications and trends. *Sep. Purif. Rev., 45*(1), 68–80. doi: 10.1080/15422119.2014.983128.

Haab, B. B., & Mathies, R. A. (1995). Single molecule fluorescence burst detection of DAN fragments separated by capillary electrophoresis. *Anal. Chem., 67*, 3253–3260.

Haab, B. B., & Mathies, R. A. (1996). *Single molecule fluorescence burst detection of DNA separated by capillary electrophoresis.* Paper presented at the Photonics West'96, San Jose, CA.

Haab, B. B., & Mathies, R. A. (1999). Single-molecule detection of DNA separations in micro-fabricated capillary electrophoresis chips employing focused molecular streams. *Anal. Chem., 71*, 5137–5145.

Hutti-Kaul, R. (2001). Aqueous two-phase systems. *Mol. Biotechnol., 19*, 269–277.

Iqbal, M., Tao, Y., Xie, S., Zhu, Y., Chen, D., Wang, X., … Yuan, Z. (2016). Aqueous two-phase system (ATPS): An overview and advances in its applications. *Biol. Proceed. Online, 18*, 18.

Jing, N., Kameoka, J., Su, C. B., Chou, C.-K., & Hung, M.-C. (2008). Nanofluidic devices for single molecule identification. *J. Photopolym. Sci. Technol., 21*(4), 531–536.

Kamagata, K. (2021). Single-molecule microscopy meets molecular dynamics simulations for characterizing the molecular action of proteins on DNA and in liquid condensates. *Front. Mol. Biosci., 8*, 795367.

Kim, S., Huang, B., & Zare, R. N. (2007). Microfluidic separation and capture of analytes for single-molecule spectroscopy. *Lab Chip, 7*, 1663–1665.

Kuo, C.-T., & Liu, C.-H. (2008a). A bubble-free AC electrokinetic micropump using the asymmetric capacitance-modulated microelectrode array for microfluidic flow control. *J. Microelectromech. Syst., 18*(1), 38–51.

Kuo, C.-T., & Liu, C.-H. (2008b). A novel microfluidic driver via AC electrokinetics. *Lab Chip, 8*(5), 725–733.

Laskowski, L., Kityk, I., Konieczny, P., Pastukh, O., Schabikowski, M., & Laskowska, M. (2019). The separation of the Mn_{12} single-molecule magnets onto spherical silica nanoparticles. *Nanomaterials, 9*(5), 764.

Leung, C.-H., Wu, K.-J., Li, G., Wu, C., Ko, C.-N., & Ma, D.-L. (2019). Application of label-free techniques in microfluidic for biomolecules detection and circulating tumor cells analysis. *Trends Anal. Chem., 117*, 78–83.

Li, A., Friedrich, S. M., & Wang, T.-H. (2018). Single molecule free solution hydrodynamic separation for size profiling of serum cell-free DNA. In: *Annual International Conference of the IEEE Engineering in Medicine and Biology Society (EMBC)*, 4476–4479.

Li, Y., Struwe, W. B., & Kukura, P. (2020). Single molecule mass photometry of nucleic acids. *Nucleic Acids Res., 48*(17), e97.

Liu, K. J., Rane, T. D., Zhang, Y., & Wang, T.-H. (2011). Single-molecule analysis enables free solution hydrodynamic separation using yoctomole levels of DNA. *J. AM. Chem. Soc., 133*, 6898–6901.

Lu, Y., & Bianco, P. (2021). High-yield purification of exceptional-quality, single molecule DNA substrates. *J. Biol. Methods, 8*(1), e145.

Lundgren, A., Fast, B. J., Block, S., Agnarsson, B., Reimhult, E., Gunnarsson, A., & Hook, F. (2018). Affinity purification and single-molecule analysis of integral membrane proteins from crude cell-membrane preparations. *Nano Lett., 18*, 381–385.

McPherson, A., & Gavira, J. A. (2014). Introduction to protein crystallization. *Acta Cryst., F70*, 2–20.

Mukherjee, S. K., Knop, J.-M., Mobitz, S., & Winter, R. H. A. (2020). Alteration of the conformational dynamics of a DNA hairpin by a-synuclein in the presence of aqueous two-phase systems. *Chem. Eur. J., 26*, 10987–10991.

Napoli, M., Eijkel, J. C. T., & Pennathur, S. (2010). Nanofluidic technology for biomolecule applications: a critical review. *Lab Chip, 10*, 957–985.

Navarro, A., Wu, H.-S., & Wang, S. S. (2009). Engineering problems in protein crystallization. *Sep. Purif. Technol., 68*, 129–137.

Pasti, L., Cavazzini, A., Felinger, A., Martin, M., & Dondi, F. (2005). Single-molecule observation and chromatography unified by lévy process representation. *Anal. Chem., 77*, 2524–2535.

Ranchon, H., Malbec, R., Picot, V., Boutonnet, A., Teerapanich, P., Joseph, P., … Bancaud, A. (2016). DNA separation and enrichment using electro-hydrodynamic bidirectional flows in viscoelastic liquids. *Lab Chip, 16*, 1243–1253.

Saadatpour, A., Lai, S., Guo, G., & Yuan, G.-C. (2015). Single-cell analysis in cancer genomics. *Trends Genet., 31*(10), 576–586.

Senavirathne, G., Lopez Jr., M. A., Messer, R., Fishel, R., & Yoder, K. E. (2018). Expression and purification of nuclease-free protocatechuate 3,4-dioxygenase for prolonged single-molecule fluorescence imaging. *Anal. Biochem., 556*, 78–84.

Shimizu, K., Mijiddorj, B., Usami, M., Mizoguchi, I., Yochida, S., Akayama, S., ... Kawwano, R. (2021). De novo design of a nanopore for single-molecule detection that incorporates a β-hairpin peptide. *Nat. Nanotechnol., 17*, 67–75.

Streitner, N., VoB, C., & Flaschel, E. (2007). Reverse extraction systems for the purification of pharmaceutical grade plasmid DNA. *J. Biotechnol., 131*, 188–196.

Suzuki, N., Hiraki, M., Yamada, Y., Matsugaki, N., Igarashi, N., Kato, R., ... Kawasaki, M. (2010). Crystallization of small proteins assisted by green fluorescent protein. *Acta Cryst., D66*, 1059–1066.

Swinney, K., & Bornhop, S. J. (2000). Detection in capillary electrophoresis. *Electrophoresis, 21*, 1239–1250.

Tabuchi, M., Kuramitsu, Y., Nakamura, K., & Baba, Y. (2003). A 15-s protein separation employing hydrodynamic force on a microchip. *Anal. Chem., 75*, 3799–3805.

Teillet, J., Martinez, Q., Tijunelyte, I., Chami, B., & Bancaud, A. (2020). Characterization and minimization of band broadening in DNA electrohydrodynamic migration for enhanced size separation. *Soft Matter, 16*, 5640–5649.

Varongchayakul, N., Hersey, J. S., Squires, A., Meller, A., & Grinstaff, M. W. (2018). A solid-state hard microfluidic–nanopore biosensor with multilayer fluidics and on-chip bioassay/purification chamber. *Adv. Funct. Mater., 28*, 1804182.

Vicente, F. A., Plazl, I., Ventura, S. P. M., & Znidarsic-Plazl, P. (2020). Separation and purification of biomacromolecules based on microfluidics. *Green Chem., 22*, 4391–4410.

Yamamoto, T., & Fujii, T. (2010). Nanofluidic single-molecule sorting of DNA: A new concept in separation and analysis of biomolecules towards ultimate level performance. *Nanotechnology, 21*, 395502.

Zhang, H., Shao, S., & Sun, Y. (2022). Characterization of liquid–liquid phase separation using super-resolution and single-molecule imaging. *Biophys. Rep.* doi:10.52601/bpr.2022.210043.

Zrehen, A., Ohayon, S., Huttner, D., & Meller, A. (2020). On-chip protein separation with single-molecule resolution. *Sci. Rep., 10*, 15313.

6 Single-Molecule and Single-Particle Tracking and Analysis in a Living Cell

Yen-Liang Liu
China Medical University

CONTENTS

6.1 BIOPHYSICAL INSIGHTS REVEALED BY SINGLE-PARTICLE TRACKING

Cells are highly controlled but complex systems, and cellular functionality is orchestrated through the biomolecule interactions, such as nucleic acids, proteins, and lipids. The direct observation of the dynamics of biomolecules inside a living cell was realized by the single-molecule and -particle tracking (SMT/SPT) techniques in the millisecond range and on spatial scales down to tens of nanometers [1,2]. With decades of efforts, researchers have forged SMT/SPT into indispensable tools to provide insight into subcellular processes, including motor protein kinetics [3,4], cell membrane dynamics [5–7], mRNA trafficking [8,9], virus infection [10,11], and even as a probe to measure mechanical properties of living cells [12–14]. The SMT can even zoom in on the most fundamental molecular processes, including enzymatic turnovers [15,16], gene expression [17,18], protein folding [19,20], and ligand-receptor interaction [21,22]. In particular, single-molecule detections have successfully unveiled intermediates during protein folding [23] and subpopulations of molecules

DOI: 10.1201/9781003409472-6

in a mixture [24], which could not be observed by conventional ensemble measurement techniques.

6.2   OVERVIEW OF SMT/SPT TECHNIQUES

The key to single-molecule detection lies in an extremely small detection volume (less than one femtoliter), significantly improving the signal-to-background ratio [25]. Two original techniques, confocal and total-internal-reflection fluorescence (TIRF) microscopy, provide small detection volumes for single-molecule detection. Confocal microscopy can detect single molecules as the molecule flows through the effective detection volume of about 0.2 femtoliter [26] in an aqueous solution. The collected photons generate burst signals in the single-photon-counting devices, and these photon bursts can provide photonic properties, including intensity [27], spectrum [28], anisotropy [29], and fluorescence lifetime [30], thereby offering information on molecular size, conformation, and stoichiometry. On the other hand, TIRF microscopy enables a higher-throughput data acquisition by employing a wide-field illumination scheme. Hundreds of single molecules can be imaged simultaneously but with the limitation of the shallow penetration depth of the evanescent wave field (~150 nm). The observation time of single molecules is still limited to a few minutes due to photobleaching [31,32]. TIRF is especially useful for *in vitro* observation of single-molecule behaviors on the surface. For example, the conformational changes of proteins, [33] DNA dynamics [34], and dynamics of transmembrane receptors on the plasma membrane [35] have been well characterized by TIRF microscopy at the single-molecule level.

Unlike the conventional single-molecule detection methods described above, SMT/SPT techniques can monitor the targets of interest and record their motion trajectories. The 1st generation SMT/SPT methods are based on TIRF and epifluorescence microscopy, with the additional capability to perform frame-by-frame video analysis. Single-target trajectories are generated by identifying the single targets in each frame and the calculation of displacements of the same targets in consecutive frames. As this frame-by-frame analysis can depict 2D trajectories of single targets within the evanescent wave field of a TIRF microscope [36–39] or the focal plane of an epifluorescence microscope [7,40], the 1st generation SMT/SPT methods can only investigate *in vitro* processes [4] or biomolecular processes on the plasma membrane [41]. To enable high-throughput SMT/SPT in living cells, HILO [42] and light-sheet microscopy [43] were developed based on a thin optical sectioning plane. We collectively term these two methods as the 2nd generation SMT/SPT methods. These methods allow direct observation of biomolecules inside living cells within the thin optical sectioning plane, which means that they are still 2D tracking techniques. For 3D tracking, the second-generation methods require a z-scan to capture the target motion in the third dimension in cells [44].

As cells exist in the three-dimensional space along with time flow, the 2D SMT/SPT may not be able to discover the critical information hidden in the third dimension entirely. Much more information can be obtained by extending to three-dimensional (3D) SPT: without this capability, direct observation of motions in the axial direction can easily be missed or confused, compromising scientific understanding [45,46]. To

TABLE 6.1

Overview of SMT/SPT Techniques [58–88]

Generation	Design	Feature	References
1st	Image-based tracking, non-feedback	Enables 2D SMT/SPT of single targets on cellular membranes or in *in-vitro* systems	[33,34,37,58–60]
2nd	Light-sheet microscopy and image-based tracking, non-feedback	Enables 2D/3D SMT/SPT in living cells but requires a z-scan to build 3D trajectories	[18,43,61–68]
3rd	3D, z-position encoded in the 2D image, non-feedback	Enables z-position localization within the imaging depth of objective lens	[69–74]
4th	Feedback-control 3D tracking microscopy	Enables high-resolution 3D tracking with a large z-tracking range and fluorescence lifetime detection. Multiple detectors are often required.	[52–54, 75–84]
5th	Feedback-control, multicolor and deep 3D tracking microscopy	Use one detector. Image depth up to 200 μm. Easy for multicolor detection.	[55, 85–88]

explore the research territory to the third dimension, researchers developed a variety of 3D-SPT methods, including the imaging of multiple focal planes [47,48], point-spread function engineering [49–55], and interferometric detection [56,57].

The 3rd generation tracking methods were designed to encode the z-position of the single targets in their 2D images. The multiple focal planes method captures images from multiple imaging planes using multiple cameras and estimates the z-position according to the size of the out-of-focus spot at each imaging plane[47,48]. The point-spread-function (PSF) engineering is the most significant breakthrough in developing 3rd generation tracking methods, in which the single emitter no longer appears as a single round spot at the imaging plane. Astigmatism-based 3D tracking technique encodes the z-position in the oval shape of an out-of-focus spot using only one camera [51]. Compared to cylindrical lenses and prisms, spatial light modulators provide more flexibility in PSF engineering and more control over localization accuracy. Spatial light modulators use liquid crystals to modulate the phase, amplitude, or polarization of incident light to generate a variety of excitation PSFs for 3D SMT/SPT, including double-helix [49,71,72], tetrapod PSF [73], self-bending PSF [74], corkscrew PSF [89], and bisected pupil PSF [90]. For instance, the double-helix PSF microscopy turns a single emitter into two spots at the imaging plane, and the z-position can be derived from the orientation and the separation distance of these two resulting spots.

From the 1st to the 3rd generation methods, the detection volume is either fixed or passively scanned. Once the targets diffuse away from the detection volume, systems lose track of targets and terminate recorded trajectories. To actively track single targets in 3D space, the 4th generation methods incorporated a feedback control system [83]. The idea of a feedback-control system was applied to track the motion of single bacteria by Howard C. Berg in 1971 [91]. The principle of feedback tracking is to keep the moving targets in the center of the detection volume by using an actuation mechanism. This can be achieved by iteratively bringing the moving targets back to the center of the detection volume, for example, by moving the whole sample using a 3D positioning stage and steering the laser beam to lock on the target. The motion path of the tracked target can be thus depicted according to the motion records of the 3D positioning stage or of the scanning mirrors of a laser beam [83]. Notably, in the 4th generation tracking methods, the target of interest is free to diffuse in the 3D space, while in particle trapping methods, such as optical tweezers [92] and electro-phoretic traps [93,94], the target is captured and spatially confined. We categorized the 1st, 2nd, and 3rd generation tracking microscopes as non-feedback-control SMT systems and the feedback SMT/SPT systems for the 4th and 5th generation tracking microscopes.

There are three major advantages of feedback tracking systems over non-feedback systems: (i) the greater axial tracking range (tens of μm vs. several μm); (ii) no need for complicated PSF design and calibration; (iii) simultaneous measurement of fluorescence lifetime of the tracer. The key difference between the non-feedback and the feedback tracking systems lies in the types of detectors (cameras vs. single-photon counting detectors). The feedback-control tracking system can be built based on single-pixel, single-photon-counting detectors, such as avalanche photodiode (APD) and photomultiplier tubes (PMT) [53,54]. The use of single-pixel detectors for SMT/SPT can significantly improve the temporal resolution and perform time-correlated single-photon counting (TCSPC), which enables the measurement of flu-orescence lifetime while tracking the targets. The 4th generation systems include circularly scanning laser tracking signals [82] and confocal-feedback tracking [79,83,95,96]. Four to five detectors focus on the excitation volume through pinholes and create a tetrahedral detection volume with axially offset for confocal-feedback tracking systems. A fluorescent molecule right in the center of the detection tetra-hedron will give equal photon counts in the four detectors, while any displacement from the center will lead to asymmetric photon count distribution. This asymme-try, known as the error signal, forms the basis for a feedback loop that drives the xyz piezo-stage to reposition the molecule to the center of the detection tetrahedron. However, the 4th generation feedback systems were limited by the low data through-put: only one target is actively tracked at a time.

Efficient multicolor detection remains a challenge for 3D-SPT – a task of the utmost importance for contextualizing biological data. For camera-based tracking techniques, multicolor detection has been made possible by various methods, includ-ing imaging with multiple cameras [97] or segmented dedicated fields of view [7], diffractive optics [98], or point-spread-function engineering [50]. However, using a camera limits the temporal resolution to ms [50] and the working depth in scattering samples [99]. In contrast, the single-element/photon counting detector-based 3D-SPT provides submillisecond temporal resolution, but for multicolor detection, the addition of a second image channel requires 3–5 extra detectors for spatial filtering [54,87,100].

In 2015, Dunn and Yeh groups demonstrated a 3D tracking microscope (Figure 6.1), termed TSUNAMI (Tracking of Single particles Using Nonlinear And Multiplexed Illumination) [55]. The 5th generation SMT/SPT method only requires one detector

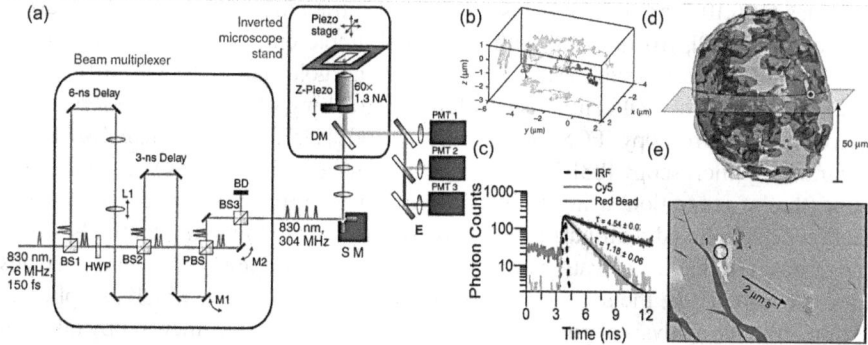

FIGURE 6.1 TSUNAMI tracking of EGFR in cancer spheroids. (a) Schematics of TSUNAMI 3D tracking microscope. (b) A representative trajectory of a free diffusing fluorescent bead in glycerol solution. (c) Fluorescence lifetime measurement during SPT. (d) 3D isocontour of a cancer spheroid taken with 2P-imaging. The white circle marks the exact location of the acquired EGFR trajectory on the spheroid. (e) Zoomed view of the trajectory. The trajectory began on the membrane with slow movement (mean velocity ~0.17 μm/s) for 250 seconds. The black circle marks the location where EGFR velocity suddenly increased to 2 μm/s and was sustained for 0.5 s unidirectionally. (These figures were adapted with permission from Perillo et al. *Nature Communications*, 2015. Copyright 2019 Springer Nature.)

(a photomultiplier) to achieve 3D SMT [55,85]. This approach is based on passive pulse splitters used for nonlinear microscopy. TSUNAMI discerns the 3D position of a molecule through spatiotemporally multiplexed two-photon excitation and temporally demultiplexed detection. A two-photon microscope by nature, TSUNAMI enables multicolor excitation with a single wavelength, which enabled the integration of multicolor 2P cellular imaging and 3D SPT on biomolecules using additional detectors [55,88]. As twice as many detectors and serious alignment effort may be required for multicolor detection of confocal 3D tracking systems [87]. A significant advantage of the 4th and 5th generation tracking microscope is their compatibility with a series of fluorescence techniques for detecting molecular interactions, including lifetime-based Förster resonance energy transfer (FRET) [31,84,101], multicolor imaging/spatiotemporal colocalization analysis [54,55,102], step counting [103,104], and MSD analysis [94,105].

6.3 3D ROTATIONAL TRACKING

While most of the SPT techniques today focus on monitoring the "translational dynamics," the "rotational dynamics" is able to shed light on the chemical and biological processes, such as how viral particles interact with their corresponding transmembrane receptors while the virus internalization. In fact, only a limited number of rotational tracking methods exist, and only a few are capable of 3D rotational tracking.

Within the long history of the development of SMT/SPT techniques, only seven methods have been proposed to measure the rotation of a particle: (i) tracking Janus particles [106,107] or colloidal clusters [108], (ii) tracking with digital holographic microscopy [109], (ii) imaging with defocused dipole emission [110], (iv) imaging with a dual-particle system [111–114], (v) tracking gold nanorods with differential interference contrast (DIC) microscopy [56]. (vi) tracking with polarized fluorescence correlation spectroscopy (FCS) on ellipsoids or rods [115], and (vii) *TSUNAMI*-based 5D tracking microscopy [88]. The first five systems are imaging-based tracking systems (the 1st generation) using optically asymmetric particles as probes. Both polarization detection and dual-particle system are based on 2D imaging methods with a major concern that the elevation angles θ and $-\theta$ is indistinguishable. Janus particles have the same issue, the uncertainty of the particle's orientation, which complicates the quantitative analysis of particle rotation. Digital holographic microscopy takes an approach similar to the 3rd generation SMT/SPT, generating a 2D holographic image that encodes the 3D information of the target in 6 modes (3 translational modes and 3 rotational modes). The complexity of processing and fitting holographic is a computationally expensive process. There is no simple SPT method that can probe both particle's translational and rotational dynamics in the 3D space.

In 2020, Yeh group developed a three-dimension two-color dual-particle-tracking (3D-2C-DPT) microscope [88] which is the multicolor extension of the TSUNAMI 3D-SPT technique (Figure 6.2). Using only one detector, TSUNAMI can perform 3D-SPT through spatiotemporally multiplexed two-photon excitation and temporally demultiplexed detection. This unique design enables simultaneous 3D -SPT of two spectrally-distinct targets within the tetrahedra excitation profile. The optically asymmetric tracer, a double-stranded DNA-linked dumbbell-like dimer, was composed of two fluorescent nanoparticles (green and red) connected by a single double-stranded DNA (dsDNA) linker. The dual-particle system is different from the one used in the traditional 2D rotational measurement – the coordinates of the green and red nanoparticles define the orientation of the tracer. Therefore, the analysis of DPT trajectories can reveal the translational and rotational movement of the dumbbell-like dimer and offer the information on coordinates (x, y, and z), rotation angles (θ and φ), separation distance (d), and photon-count rate ($B1$ and $B2$). The integration of the multicolor TSUNAMI and dual-particle probes lead to the invention of the 3D-2C-DPT. Since each particle can be individually localized in a 3D space by the two PMTs, a wealth of data can be extracted: 3D position, 2D rotation, separation distance, and the fluorescence lifetime of each target. In particular, this system provides unambiguous orientation determination of dual-particle systems with temporal resolution down to 5 ms and resolved separation distances from 33 to 250 nm with tracking precisions of around 15 nm.

6.4 APPLICATIONS OF SMT/SPT IN STUDYING CELL MEMBRANE DYNAMICS

In cell membrane biophysics, the SMT/SPT have significantly explored our understanding of cell membrane biophysics since the 1990s. The mosaic model of the plasma membrane was first challenged by the Kusumi group who proposed a membrane-skeleton mesh model in 1993 to explain the hop diffusion of epidermal

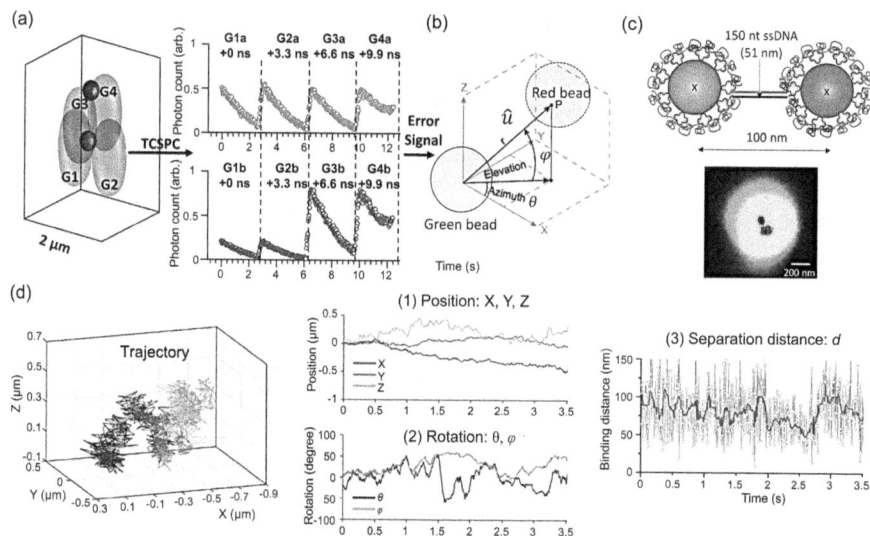

FIGURE 6.2 3D-2C-DPT for 3D rotational tracking. (a) Schematics of dual-particle tracking through spatiotemporally multiplexed two-photon excitation and temporally demultiplexed detection. (b) The dual particle creates a vector encoding the information of xyz coordinates and rotational angles. (c) Dual particle tracer with a single dsDNA linker. (d) A 3D trajectory offers data of 3D position, 2D rotation, and separation distance between two particles. (These figures were adapted with permission from Liu, et al. *ACS Nano*, 2020. Copyright 2020 American Chemical Society.)

growth factor receptors (EGFRs) on the cell membrane instead of Brownian diffusion [40]. Later, benefiting from the advance of SMT/SPT, this model further evolved to the compartmentalized fluid model, which accurately portraited the plasma membrane that is portioned into microdomains by the membrane-associated cytoskeleton (~40–300 nm in diameter) [1]. Later, the following studies also demonstrated that the cortical actin network did interact with the intracellular domains of transmembrane receptors, and the depolymerization of cortical actins released their constraints to transmembrane receptors and increased the diffusivity of transmembrane receptors (Figure 6.3), such as EGFR [40,116–118], G-protein coupled receptor [119], and B cell receptor [120,121]. Moreover, several groups have also reported that physical constraints enhanced the clustering of membrane receptors and their associated proteins [116,17,122] and then promoted cell signaling in compartments [119,123–125]. These results showed that the physical factors spatiotemporally manipulate the biochemical states of the cells.

Cell signaling is precisely regulated by the dynamics of transmembrane receptors. Their dynamics and binding kinetics can be affected by molecular factors such as the composition and structure of the plasma membrane, (1) cytoskeleton networks [126], and protein-protein interactions [59]. In particular, receptor tyrosine kinases (RTKs) play essential roles in many cell decision-making functions such as proliferation, survival, differentiation, and migration. Misregulated RTKs are the leading

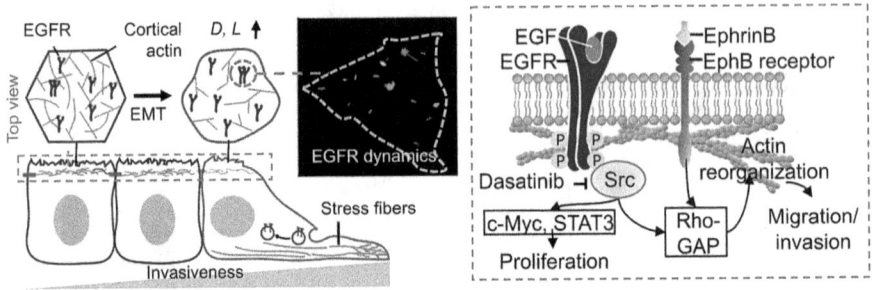

FIGURE 6.3 Cytoskeleton-associated transmembrane receptor dynamics and signaling. Cortical actin regulates the EGFR dynamics, and the epithelial-mesenchymal transition (EMT)-induced actin reorganization enhances cell invasiveness and increases the EGFR diffusivity. (These figures were adapted from Liu et al. *Cancers*, 2019. Copyright 2019 MDPI AG.)

cause of cancer [127]. With the help of advanced fluorescence techniques, emerging evidence indicates that the compromised spatial distribution and the derailed trafficking of RTKs could be hallmarks of carcinogenesis or even metastasis[128]. Grove group studied the spatial distribution of EphA2 receptors on the plasma membrane and demonstrated the clustering of EphA2 receptors is positively correlated with the increased malignancy of cancer cells [123]. Yeh group developed an SPT-based biophysical phenotyping assay named Transmembrane Receptor Dynamics (TReD) for quantifying metastatic potentials of the breast [126] and prostate cancer cells [129] based on the dynamics of EGFR on the plasma membrane (Figure 6.4). This study further investigated the changes of EGFR dynamics under the impact of epithelial-mesenchymal transition, indicating increased EGFR diffusivity is coupled with cell propagation, motility, and invasiveness of cancer cells.

6.5 TRAJECTORY ANALYSIS OF EGFR DYNAMICS ON THE PLASMA MEMBRANE

While SMT/SPT techniques visualize biomolecules' movements in tubes or living cells, unbiased and systematic analyses of recorded trajectories are needed to reveal the hidden features of biomolecular processes. For example, the tracking of transmembrane receptors explores our understanding of the dimerization and clustering of receptors [7,132] that regulate cell signaling. In addition, the transmembrane receptors can also serve as probes to detect the structures and compositions of the plasma membrane. (1) and even the cytoskeleton networks underneath the plasma membrane [126]. As a wealth of information about membrane structure, interior organization, and receptor biology can be derived from trajectories, trajectory analysis algorithms are needed.

The time-averaged mean-squared-displacement (MSD) is the most common approach for analyzing the SMT/SPT trajectory [2,130–133]. The epidermal growth factor receptors (EGFRs) have been comprehensively investigated for their dynamics on the plasma membrane and their intracellular trafficking. MSD plots can be used to

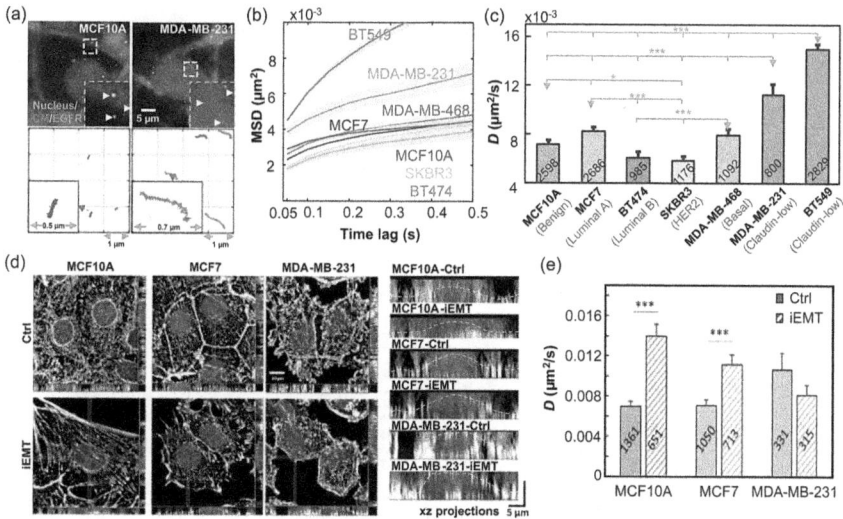

FIGURE 6.4 Transmembrane receptor dynamics reveals metastatic potentials. (a) Images of living cells and fluorescently labeled EGFRs. (b) The averaged MSD curves were derived from EGFR trajectories from these seven breast epithelial cell lines. (c) Characterization of EGFR diffusivity among cell lines. (d) Cortical actin is visualized by structured illumination microscopy. EMT induction transformed MCF10A and MCF7 cells and reorganized actin filaments. (e) EMT induction increases EGFR diffusivity of epithelial-like MCF10A and MCF7 cells. (These figures were adapted with permission from Liu et al. *Scientific Reports*, 2019. Copyright 2019 Springer Nature.)

estimate the EGFR diffusivity and the linear size of the compartment that constrains the movement of EGFRs in microdomains of the plasma membrane (Figure 6.5). While MSD analysis is widely used in numerous studies, much information is lost when processing the raw trajectories into the MSD plots [119]. Although new insights into biomolecular processes can still be revealed by carefully scrutinizing individual trajectories and developing novel anomalous diffusion models [58,105,134–137], an emerging deep learning-based algorithm has demonstrated its capability of extracting hidden features from a vast amount of receptor trajectories [138]. Liu group successfully translated the EGFR trajectories into a biomarker for the classification of breast cancer cells using a Residual neural Network (ResNet) trajectory analysis algorithm (Figure 6.5). With a well-trained classifier, breast cancer cell types can be differentiated based on hidden features extracted from the transmembrane receptor trajectories [138], which is significantly different from the previous studies that characterized diffusive states using the trajectory-trained machine learning or deep learning models [139–142].

6.6 PHASE TRANSITION OF EGFR TRAFFICKING

When EGFRs are membrane-bound, they show an interchange of Brownian diffusion and confined diffusion. When EGFRs are internalized, transitions from confined

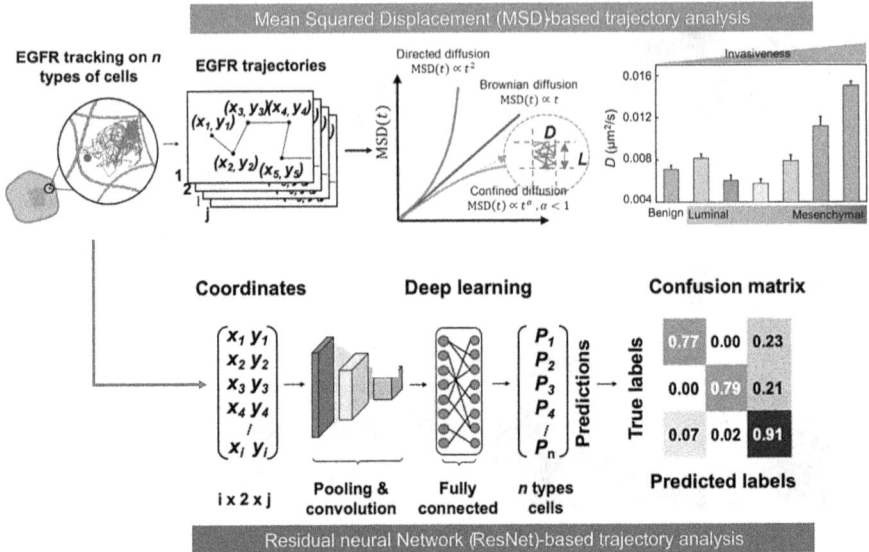

FIGURE 6.5 Comparison between MSD-based and Deep learning-based trajectory analyses. MSD analysis estimates EGFR diffusivity and the size of compartment, and these biophysical properties were correlated with metastatic potentials of breast cancer cells. The deep learning-based trajectory analysis identifies the hidden features from trajectories and creates a classifier for differentiating the subtype of breast cancer cells. (These figures were adapted from Kim et al. *Bioinformatics*, 2021. Copyright 2021 Oxford Academic.)

diffusion to directed diffusion and from directed diffusion back to confined diffusion are clearly seen using the TSUNAMI microscope (Figure 6.6). Based on the velocity, the trajectory was manually dissected into three phases: Phase I – Probably the formation of clathrin-coated pits. Actin-cytoskeleton-mediated movement. Confined diffusion in the vesicle (speed: 0.1~0.2 μm/s); Phase II – Fast, active transport on microtubules (speed: 0.5~2 μm/s); Phase III – Late endosome, confined diffusion (speed: 0.2~0.5 μm/s). As expected, the speed observed in active transport matches well with the typical molecular motor-mediated transport speed (0.5–2 μm/s) [143–146].

For trajectory analysis of long trafficking in living cells, recent studies have demonstrated that time-averaged MSD varies from molecule to molecule, so following a single representative particle over a long time does not provide the same information about an ensemble as monitoring the averaged itself [2,132,147]. As fourth and fifth generations of SMT/SPT can acquire 3D trajectories for several minutes, a sophisticated tool to segment and classify these trajectories according to their motion modes [40,148–150], extract physical parameters of the motion [55,151], and correlate that motion to the surrounding environment [133], all with the end goal of understanding the physical scenarios behind the observed motion [132,152]. Thus to discern the differences in diffusive behaviors among individual trajectories, several methods have also been developed to allow unbiased trajectory classification and exploit more of the information stored in a trajectory, including probability distribution of

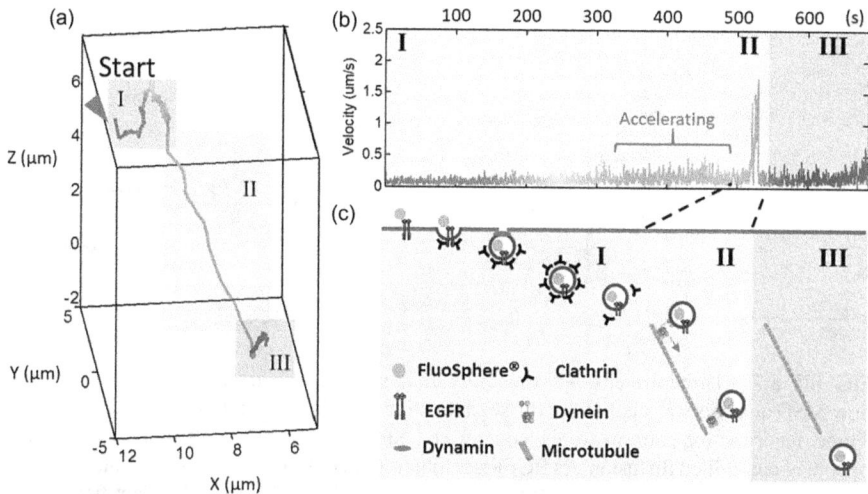

FIGURE 6.6 Motion transition of EGFR trafficking. (a) An example of EGF-stimulated EGFR trajectory started on the plasma membrane (Phase I), exhibited fast and linear movement in Phase II, and ended with confined diffusion (Phase III). (b) A plot shows the time trace of velocity. (c) Schematic shows the possible mechanism of EGFR internalization and trafficking.

square displacement (PDSD) analysis [153,154], moment scaling spectrum [155,156], Bayesian inference [157–160], distribution of directional changes [161], mean maximum excursion method [162], and fractionally integrated moving average (FIMA) [163], and correlation analysis [133,164]. In particular, considerable attention has lately been devoted to detecting transient changes of motion along the same trajectory and to visualizing spatial segments (also called subtrajectories) with different dynamic properties [2]. Trajectory segmentation has been obtained *via* a number of classification parameters calculated over the trajectory by rolling window analysis [148,149,165], supervised segmentation [166], maximum likelihood estimator [151], Bayesian methods [167], F-statistics [118], hidden Markov model [117] and wavelet analysis [64,147].

TSUNAMI microscope has been applied to track EGFR trafficking from the plasma membrane into the cytosol, and several long trajectories were acquired and characterized using a rolling-window MSD analysis [105] (Figure 6.7a and b). This algorithm is capable of trajectory segmentation and classification, adopting a combination of strategies [40,133,148–150] and a number of classification parameters (e.g. scaling exponent [148], directional persistence [149], and confinement index [150,168]). Figure 6.7c and d show that when EGFRs are membrane-bound, they show an interchange of Brownian diffusion and confined diffusion. When EGFRs are internalized, transitions from confined diffusion to directed diffusion and from directed diffusion back to confined diffusion are clearly seen. As expected, the speed observed in the directed diffusion matches well with the speed of the typical molecular motor-mediated transport (0.5–2 μm/s) [143–146].

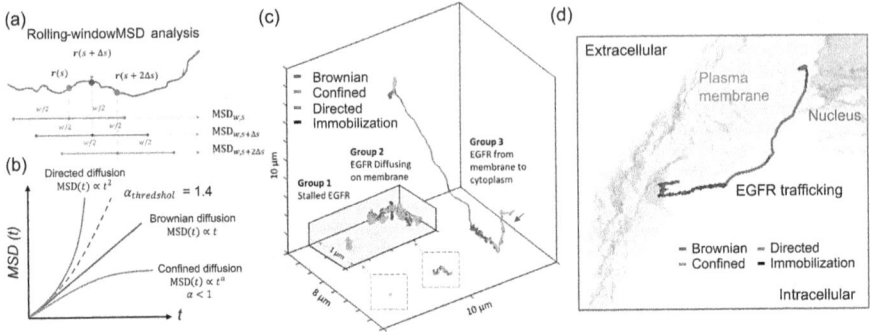

FIGURE 6.7 Three-dimensional trajectory analysis of EGFR trafficking. (a) Rolling window MSD analysis was applied for the segmentation and classification of long trajectories. (b) Three major motion patterns were characterized, including directed diffusion, Brownian diffusion, and confined diffusion. (c) EGFRs exhibit distinct motion patterns. (d) Colocalization of single-cell image with the EGFR trajectory that is color-coded with the identified motion patterns. (These figures were adapted with permission from Liu, et al. *Biophysical Journal*, 2016. Copyright 2016 Biophysical Society.)

6.7   ACTIVE TRANSPORT ON MICROTUBULE

The co-localization of the microtubule image and EGFR trajectory could be used to visualize the microtubule-mediated transportation (Figure 6.8). We tracked EGFR in live HeLa cells that were transfected with mKate2-tubulin-expressing plasmids. The 3D-stacked microtubule images were collected using 2-photon laser scanning microscopy in the TSUNAMI system, and then the single EGFR molecule was tracked with the TSUNAMI microscope. After tracking, the EGFR trajectories were co-registered to reconstructed microtubule 3D images. Figure 6.8a demonstrates the active transport of EGFR along microtubules in a live cell. In addition, a trajectory segmentation and classification algorithm was applied to the EGFR trajectory, enabling us to reveal the transitions of different motion modes along the trajectory (Figure 6.8b and c). The green, blue and red colors represent confined, Brownian, and directed diffusion, respectively. The fractions of free diffusion time, confinement time, and active transport time were 3.6%, 96.1%, and 0.3%, respectively. Thus, the EGFR trajectory could be clearly dissected into four phases: the confined diffusion on the plasma membrane in phase I, the short inward movement into the cell in phase II (endocytosis), the interchanges of confined and Brownian diffusion in phase III (in the cytoplasm), and the microtubule-mediated transport shown at phase IV. The time trace of instantaneous velocity also reflects the processes of EGFR internalization and active transport (Figure 6.8d).

6.8   APPLICATIONS OF 3D ROTATIONAL TRACKING

Nanoscopic optical rulers can be used to determine the separation distance of binding targets, binding kinetics, conformational change, or even the orientation of a targeting construct. While Förster Resonance Energy Transfer (FRET) has served as a critical

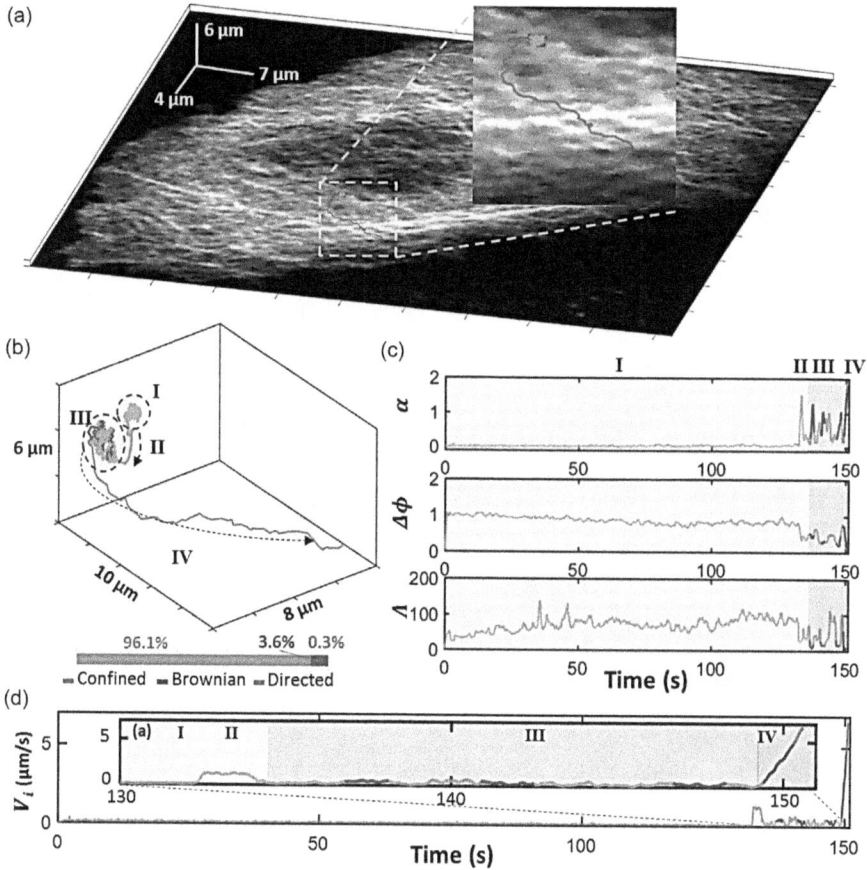

FIGURE 6.8 EGFR's active transport on microtubule. (a) A representative trajectory of a 40 nm fluorescent nanoparticle-tagged EGFR was measured in a mKate2-tubulin expressing HeLa cell. (b) This trajectory is segmented and classified using a segmentation algorithm. (c) Time traces of the three motion classification parameters (α, $\Delta\phi$, Λ). (d) Time trace of instantaneous velocity (V_i).

method for nanoscopic optical rulers for measuring conformational changes and intramolecular distances of single biomolecules [169–174], the FRET ruler is limited to a maximum of 10 nm within the FRET pair due to effective dipole-dipole interactions. Using plasmonic nanoparticle-based surface energy transfer, the accessible distance can be increased to 80 nm (depending on the size and coating of the nanoparticles) [175–177]. But the instrumentation of FRET and plasmonic resonance energy transfer relies on a slow camera-based approach without the capability of 3D tracking to freely diffusing targets, and they provide no rotational information. Using the 3D-2C-DPT technique, the dsDNA-linked dumbbell-like dimer was turned into a nanoscopic optical ruler extending the accessible distance up to 250 with 30 nm accuracy and providing direct observation of DNA flexing dynamics in free solution [88]. This technique

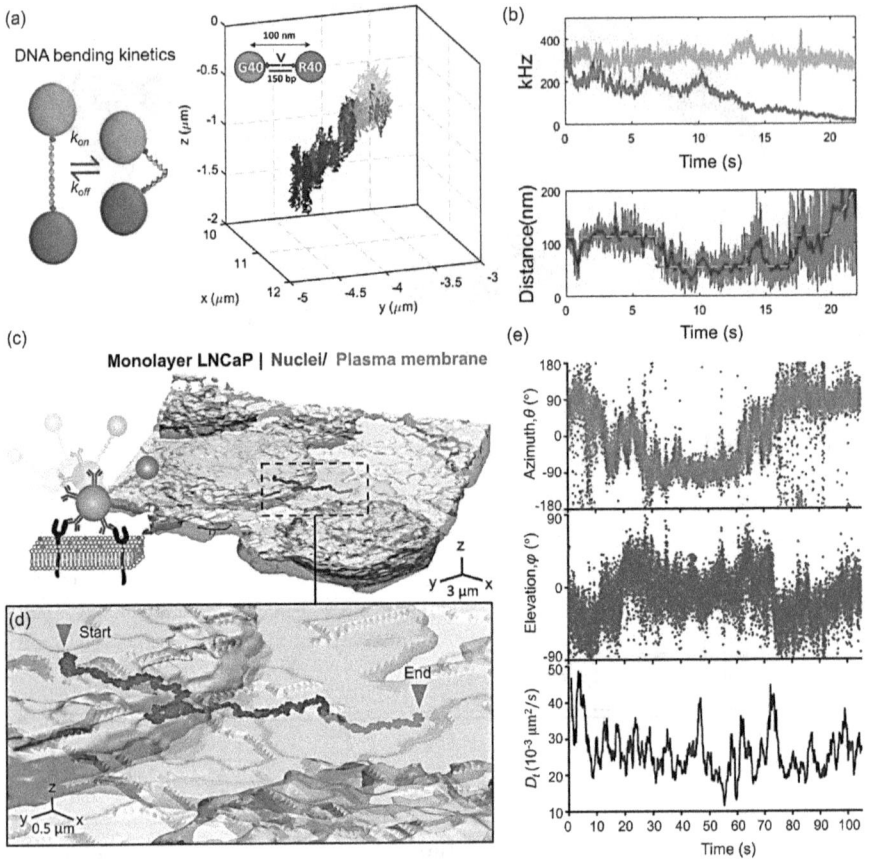

FIGURE 6.9 Three-dimension two-color dual-particle-tracking for investigating the DNA flexing dynamics and the landing of nanoparticles on the plasma membrane. (a) A trajectory of flexing dsDNA-linked dumbbell-like dimer in free solution and the changes of distances (b). (c) The landing of antibody-conjugated nanoparticles on a living cell and the information acquired during the entire process. (These figures were adapted with permission from Liu, et al. *ACS Nano*, 2020. Copyright 2020 American Chemical Society.)

enables the investigation of the bendability of nicked or gapped dsDNA molecules by manipulating the design of dsDNA linkers (Figure 6.9). In addition, its capability of rotational tracking can monitor the landing of antibody-conjugated nanoparticles on the plasma membrane of living cells (Figure 6.9). This demonstrates that the 3D-2C-DPT technique is a new tool to shed light on the conformational changes of biomolecules and the intermolecular interactions on the plasma membrane.

6.9 CONCLUSION

This chapter summarizes the development of five generations of SMT/SPT techniques and elaborates on their applications in cellular biology research. While SMT/SPT has made significant contributions to biophysical research, several challenges remain, such as no single solution that allows for super-resolution tracking of thousands of molecules in real-time in live tissues and limited tracking duration due to the photon budget of fluorescence tags. The next breakthroughs rely on advances in detector techniques, actuator techniques, objective techniques, laser, optical design, and fluorescence materials.

REFERENCES

1. Kusumi A., Tsunoyama T.A., Hirosawa K.M., Kasai R.S., Fujiwara T.K. Tracking single molecules at work in living cells. *Nature Chemical Biology.* 2014;10(7):524.
2. Manzo C., Garcia-Parajo M.F. A review of progress in single particle tracking: From methods to biophysical insights. *Reports on Progress in Physics.* 2015;78(12):124601.
3. Gelles J., Schnapp B.J., Sheetz M.P. Tracking kinesin-driven movements with nanometre-scale precision. *Nature.* 1988;331(6155):450.
4. Yildiz A., Forkey J.N., McKinney S.A., Ha T., Goldman Y.E., Selvin P.R. Myosin V walks hand-over-hand: Single fluorophore imaging with 1.5-nm localization. *Science.* 2003;300(5628):2061–5.
5. Fujiwara T., Ritchie K., Murakoshi H., Jacobson K., Kusumi A. Phospholipids undergo hop diffusion in compartmentalized cell membrane. *The Journal of Cell Biology.* 2002;157(6):1071–82.
6. Dahan M., Levi S., Luccardini C., Rostaing P., Riveau B., Triller A. Diffusion dynamics of glycine receptors revealed by single-quantum dot tracking. *Science.* 2003;302(5644): 442–5.
7. Andrews N.L., Lidke K.A., Pfeiffer J.R., Burns A.R., Wilson B.S., Oliver J.M., et al. Actin restricts FcεRI diffusion and facilitates antigen-induced receptor immobilization. *Nature Cell Biology.* 2008;10(8):955–63.
8. Shav-Tal Y., Darzacq X., Shenoy S.M., Fusco D., Janicki S.M., Spector D.L., et al. Dynamics of single mRNPs in nuclei of living cells. *Science.* 2004;304(5678):1797–800.
9. Park H.Y., Lim H., Yoon Y.J., Follenzi A., Nwokafor C., Lopez-Jones M., et al. Visualization of dynamics of single endogenous mRNA labeled in live mouse. *Science.* 2014;343(6169):422–4.
10. Lakadamyali M., Rust M.J., Babcock H.P., Zhuang X. Visualizing infection of individual influenza viruses. *Proceedings of the National Academy of Sciences.* 2003;100(16):9280–5.
11. Arhel N., Genovesio A., Kim K.-A., Miko S., Perret E., Olivo-Marin J.-C., et al. Quantitative four-dimensional tracking of cytoplasmic and nuclear HIV-1 complexes. *Nature Methods.* 2006;3(10):817.
12. Guo M., Ehrlicher A.J., Jensen M.H., Renz M., Moore J.R., Goldman R.D., et al. Probing the stochastic, motor-driven properties of the cytoplasm using force spectrum microscopy. *Cell.* 2014;158(4):822–32.
13. Wirtz D. Particle-tracking microrheology of living cells: Principles and applications. *Annual Review of Biophysics.* 2009;38:301–26.
14. Le V., Lee J., Chaterji S., Spencer A., Liu Y.-L., Kim P., et al. Syndecan-1 in mechanosensing of nanotopological cues in engineered materials. *Biomaterials.* 2018;155:13–24.
15. Lu H.P., Xun L., Xie X.S. Single-molecule enzymatic dynamics. *Science.* 1998;282(5395): 1877–82.
16. van Oijen A.M., Blainey P.C., Crampton D.J., Richardson C.C., Ellenberger T., Xie X.S. Single-molecule kinetics of λ exonuclease reveal base dependence and dynamic disorder. *Science.* 2003;301(5637):1235–8.
17. Cai L., Friedman N., Xie X.S. Stochastic protein expression in individual cells at the single molecule level. *Nature.* 2006;440(7082):358–62.

18. Elf J., Li G.-W., Xie X.S. Probing transcription factor dynamics at the single-molecule level in a living cell. *Science.* 2007;316(5828):1191–4.

19. Chung H.S., McHale K., Louis J.M., Eaton W.A. Single-molecule fluorescence experiments determine protein folding transition path times. *Science.* 2012;335(6071):981–4.

20. Chung H.S., Louis J.M., Eaton W.A. Experimental determination of upper bound for transition path times in protein folding from single-molecule photon-by-photon trajectories. *Proceedings of the National Academy of Sciences.* 2009;106(29):11837–44.

21. Sako Y., Minoghchi S., Yanagida T. Single-molecule imaging of EGFR signalling on the surface of living cells. *Nature Cell Biology.* 2000;2(3):168–72.

22. Teramura Y., Ichinose J., Takagi H., Nishida K., Yanagida T., Sako Y. Single-molecule analysis of epidermal growth factor binding on the surface of living cells. *The EMBO Journal.* 2006;25(18):4215–22.

23. Lipman E.A., Schuler B., Bakajin O., Eaton W.A. Single-molecule measurement of protein folding kinetics. *Science.* 2003;301(5637):1233–5.

24. Ha T., Ting A.Y., Liang J., Caldwell W.B., Deniz A.A., Chemla D.S., et al. Single-molecule fluorescence spectroscopy of enzyme conformational dynamics and cleavage mechanism. *Proceedings of the National Academy of Sciences.* 1999;96(3):893–8.

25. Zander C., Keller R.R., Enderlein J. *Single Molecule Detection in Solution.* Wiley Online Library; Hoboken, NJ, 2002.

26. Laurence T.A., Weiss S. How to detect weak pairs. *Science.* 2003;299(5607):667–8.

27. Kapanidis A.N., Lee N.K., Laurence T.A., Doose S., Margeat E., Weiss S. Fluorescence-aided molecule sorting: Analysis of structure and interactions by alternating-laser excitation of single molecules. *Proceedings of the National Academy of Sciences.* 2004;101(24):8936–41.

28. Prummer M., Hübner C.G., Sick B., Hecht B., Renn A., Wild U.P. Single-molecule identification by spectrally and time-resolved fluorescence detection. *Analytical Chemistry.* 2000;72(3):443–7.

29. Schaffer J., Volkmer A., Eggeling C., Subramaniam V., Striker G., Seidel C. Identification of single molecules in aqueous solution by time-resolved fluorescence anisotropy. *The Journal of Physical Chemistry A.* 1999;103(3):331–6.

30. Müller R., Zander C., Sauer M., Deimel M., Ko D.-S., Siebert S., et al. Time-resolved identification of single molecules in solution with a pulsed semiconductor diode laser. *Chemical Physics Letters.* 1996;262(6):716–22.

31. Myong S., Rasnik I., Joo C., Lohman T.M., Ha T. Repetitive shuttling of a motor protein on DNA. *Nature.* 2005;437(7063):1321–5.

32. Cisse I.I., Kim H., Ha T. A rule of seven in Watson-Crick base-pairing of mismatched sequences. *Nature Structural & Molecular Biology.* 2012;19(6):623–7.

33. Luo G., Wang M., Konigsberg W.H., Xie X.S. Single-molecule and ensemble fluorescence assays for a functionally important conformational change in T7 DNA polymerase. *Proceedings of the National Academy of Sciences.* 2007;104(31):12610–5.

34. McKinney S.A., Déclais A.-C., Lilley D.M., Ha T. Structural dynamics of individual Holliday junctions. *Nature Structural & Molecular Biology.* 2003;10(2):93–7.

35. Tsunoyama T.A., Watanabe Y., Goto J., Naito K., Kasai R.S., Suzuki K.G., et al. Super-long single-molecule tracking reveals dynamic-anchorage-induced integrin function. *Nature Chemical Biology.* 2018;14:497.

36. Kusumi A., Shirai Y.M., Koyama-Honda I., Suzuki K.G., Fujiwara T.K. Hierarchical organization of the plasma membrane: Investigations by single-molecule tracking vs. fluorescence correlation spectroscopy. *FEBS Letters.* 2010;584(9):1814–23.

37. Kasai R.S., Suzuki K.G., Prossnitz E.R., Koyama-Honda I., Nakada C., Fujiwara T.K., et al. Full characterization of GPCR monomer–dimer dynamic equilibrium by single molecule imaging. *The Journal of Cell Biology.* 2011;192(3):463–80.

38. Kusumi A., Suzuki K.G.N., Kasai R.S., Ritchie K., Fujiwara T.K. Hierarchical meso-scale domain organization of the plasma membrane. *Trends in Biochemical Sciences.* 2011;36(11):604–15.

39. Hern J.A., Baig A.H., Mashanov G.I., Birdsall B., Corrie J.E., Lazareno S., et al. Formation and dissociation of M1 muscarinic receptor dimers seen by total internal reflection fluorescence imaging of single molecules. *Proceedings of the National Academy of Sciences.* 2010;107(6):2693–8.

40. Kusumi A., Sako Y., Yamamoto M. Confined lateral diffusion of membrane receptors as studied by single particle tracking (nanovid microscopy). Effects of calcium-induced differentiation in cultured epithelial cells. *Biophysical Journal.* 1993;65(5):2021–40.

41. Lowe A.R., Siegel J.J., Kalab P., Siu M., Weis K., Liphardt J.T. Selectivity mechanism of the nuclear pore complex characterized by single cargo tracking. *Nature.* 2010;467(7315):600–3.

42. Tokunaga M., Imamoto N., Sakata-Sogawa K. Highly inclined thin illumination enables clear single-molecule imaging in cells. *Nature Methods.* 2008;5(2):159–61.

43. Gao L., Shao L., Chen B.-C., Betzig E. 3D live fluorescence imaging of cellular dynamics using Bessel beam plane illumination microscopy. *Nature Protocols.* 2014;9(5):1083–101.

44. Liu Z., Legant W.R., Chen B.-C., Li L., Grimm J.B., Lavis L.D., et al. 3D imaging of Sox_2 enhancer clusters in embryonic stem cells. *eLife.* 2014;3:e04236.

45. von Diezmann A., Shechtman Y., Moerner W. Three-dimensional localization of single molecules for super-resolution imaging and single-particle tracking. *Chemical Reviews.* 2017;117(11):7244–75.

46. Milenkovic L., Weiss L.E., Yoon J., Roth T.L., Su Y.S., Sahl S.J., et al. Single-molecule imaging of Hedgehog pathway protein Smoothened in primary cilia reveals binding events regulated by Patched1. *Proceedings of the National Academy of Sciences.* 2015;112(27):8320–5.

47. Toprak E., Balci H., Blehm B.H., Selvin P.R. Three-dimensional particle tracking via bifocal imaging. *Nano Letters.* 2007;7(7):2043–5.

48. Ram S., Prabhat P., Chao J., Ward E.S., Ober R.J. High accuracy 3D quantum dot tracking with multifocal plane microscopy for the study of fast intracellular dynamics in live cells. *Biophysical Journal.* 2008;95(12):6025–43.

49. Pavani S.R.P., Thompson M.A., Biteen J.S., Lord S.J., Liu N., Twieg R.J., et al. Three-dimensional, single-molecule fluorescence imaging beyond the diffraction limit by using a double-helix point spread function. *Proceedings of the National Academy of Sciences.* 2009;106(9):2995–9.

50. Shechtman Y., Weiss L.E., Backer A.S., Lee M.Y., Moerner W. Multicolour localization microscopy by point-spread-function engineering. *Nature Photonics.* 2016;10(9):590.

51. Huang B., Wang W., Bates M., Zhuang X. Three-dimensional super-resolution imaging by stochastic optical reconstruction microscopy. *Science.* 2008;319(5864):810–3.

52. McHale K., Berglund A.J., Mabuchi H. Quantum dot photon statistics measured by three-dimensional particle tracking. *Nano Letters.* 2007;7(11):3535–9.

53. Wells N.P., Lessard G.A., Goodwin P.M., Phipps M.E., Cutler P.J., Lidke D.S., et al. Time-resolved three-dimensional molecular tracking in live cells. *Nano Letters.* 2010;10(11):4732–7.

54. Welsher K., Yang H. Multi-resolution 3D visualization of the early stages of cellular uptake of peptide-coated nanoparticles. *Nature Nanotechnology.* 2014;9(3):198–203.

55. Perillo E.P., Liu Y.-L., Huynh K., Liu C., Chou C.-K., Hung M.-C., et al. Deep and high-resolution three-dimensional tracking of single particles using nonlinear and multiplexed illumination. *Nature Communications.* 2015;6:7874.

56. Chen K., Gu Y., Sun W., Wang G., Fan X., Xia T., et al. Characteristic rotational behaviors of rod-shaped cargo revealed by automated five-dimensional single particle tracking. *Nature Communications*. 2017;8(1):887.

57. Middendorff Cv., Egner A., Geisler C., Hell S., Schönle A. Isotropic 3D nanoscopy based on single emitter switching. *Optics Express*. 2008;16(25):20774–88.

58. Dietrich C., Yang B., Fujiwara T., Kusumi A., Jacobson K. Relationship of lipid rafts to transient confinement zones detected by single particle tracking. *Biophysical Journal*. 2002;82(1):274–84.

59. Kasai R.S., Kusumi A. Single-molecule imaging revealed dynamic GPCR dimerization. *Current Opinion in Cell Biology*. 2014;27:78–86.

60. Kusumi A., Ike H., Nakada C., Murase K., Fujiwara T., editors. *Single-Molecule Tracking of Membrane Molecules: Plasma Membrane Compartmentalization and Dynamic Assembly of Raft-Philic Signaling Molecules*. Seminars in Immunology; 2005: Elsevier, Amsterdam.

61. Ahrens M.B., Orger M.B., Robson D.N., Li J.M., Keller P.J. Whole-brain functional imaging at cellular resolution using light-sheet microscopy. *Nature Methods*. 2013;10(5):413–20.

62. Li Y., Hu Y., Cang H. Light sheet microscopy for tracking single molecules on the apical surface of living cells. *The Journal of Physical Chemistry B*. 2013;117(49):15503–11.

63. Ritter J.G., Veith R., Veenendaal A., Siebrasse J.P., Kubitscheck U. Light sheet microscopy for single molecule tracking in living tissue. *PLoS One*. 2010;5(7):e11639.

64. Chen K., Wang B., Granick S. Memoryless self-reinforcing directionality in endosomal active transport within living cells. *Nature Materials*. 2015;14(6):589–93.

65. Wang B., Kuo J., Granick S. Bursts of active transport in living cells. *Physical Review Letters*. 2013;111(20):208102.

66. Robin F.B., McFadden W.M., Yao B., Munro E.M. Single-molecule analysis of cell surface dynamics in Caenorhabditis elegans embryos. *Nature Methods*. 2014;11(6):677–82.

67. Ritter J.G., Veith R., Siebrasse J.-P., Kubitscheck U. High-contrast single-particle tracking by selective focal plane illumination microscopy. *Optics Express*. 2008;16(10):7142–52.

68. Planchon T.A., Gao L., Milkie D.E., Davidson M.W., Galbraith J.A., Galbraith C.G., et al. Rapid three-dimensional isotropic imaging of living cells using Bessel beam plane illumination. *Nature Methods*. 2011;8(5):417–23.

69. Kao H.P., Verkman A. Tracking of single fluorescent particles in three dimensions: Use of cylindrical optics to encode particle position. *Biophysical Journal*. 1994;67(3):1291–300.

70. Yajima J., Mizutani K., Nishizaka T. A torque component present in mitotic kinesin Eg5 revealed by three-dimensional tracking. *Nature Structural & Molecular Biology*. 2008;15(10):1119–21.

71. Thompson M.A., Casolari J.M., Badieirostami M., Brown P.O., Moerner W. Three-dimensional tracking of single mRNA particles in Saccharomyces cerevisiae using a double-helix point spread function. *Proceedings of the National Academy of Sciences*. 2010;107(42):17864–71.

72. Gahlmann A., Ptacin J.L., Grover G., Quirin S., von Diezmann A.R., Lee M.K., et al. Quantitative multicolor subdiffraction imaging of bacterial protein ultrastructures in three dimensions. *Nano Letters*. 2013;13(3):987–93.

73. Shechtman Y., Weiss L.E., Backer A.S., Sahl S.J., Moerner W. Precise three-dimensional scan-free multiple-particle tracking over large axial ranges with tetrapod point spread functions. *Nano Letters*. 2015;15(6):4194–9.

74. Jia S., Vaughan J.C., Zhuang X. Isotropic three-dimensional super-resolution imaging with a self-bending point spread function. *Nature Photonics*. 2014;8(4):302–6.

75. Levi V., Ruan Q., Kis-Petikova K., Gratton E. Scanning FCS, a novel method for three-dimensional particle tracking. *Biochemical Society Transactions*. 2003;31:997.

76. Levi V., Ruan Q., Gratton E. 3-D particle tracking in a two-photon microscope: Application to the study of molecular dynamics in cells. *Biophysical Journal*. 2005;88(4):2919–28.

77. Berglund A.J., Mabuchi H. Tracking-FCS: Fluorescence correlation spectroscopy of individual particles. *Optics Express*. 2005;13(20):8069–82.

78. Lessard G.A., Goodwin P.M., Werner J.H. Three-dimensional tracking of individual quantum dots. *Applied Physics Letters*. 2007;91(22):224106.

79. Xu C.S., Cang H., Montiel D., Yang H. Rapid and quantitative sizing of nanoparticles using three-dimensional single-particle tracking. *The Journal of Physical Chemistry C*. 2007;111(1):32–5.

80. Berglund A.J., McHale K., Mabuchi H. Feedback localization of freely diffusing fluorescent particles near the optical shot-noise limit. *Optics Letters*. 2007;32(2):145–7.

81. Levi V., Gratton E. Three-dimensional particle tracking in a laser scanning fluorescence microscope. *Single Particle Tracking and Single Molecule Energy Transfer*. 2009:1–24. doi: 10.1002/9783527628360.ch1.

82. Katayama Y., Burkacky O., Meyer M., Bräuchle C., Gratton E., Lamb D.C. Real-Time nanomicroscopy via three-dimensional single-particle tracking. *ChemPhysChem*. 2009;10(14):2458–64.

83. Juette M.F., Bewersdorf J. Three-dimensional tracking of single fluorescent particles with submillisecond temporal resolution. *Nano Letters*. 2010;10(11):4657–63.

84. Liu C., Obliosca J.M., Liu Y.-L., Chen Y.-A., Jiang N., Yeh H.-C. 3D single-molecule tracking enables direct hybridization kinetics measurement in solution. *Nanoscale*. 2017;9(17):5664–70.

85. Perillo E., Liu Y.-L., Liu C., Yeh H.-C., Dunn A.K., editors. Single particle tracking through highly scattering media with multiplexed two-photon excitation. SPIE BiOS; 2015: International Society for Optics and Photonics.

86. Liu C., Liu Y.-L., Perillo E., Jiang N., Dunn A., Yeh H.-C. Improving z-tracking accuracy in the two-photon single-particle tracking microscope. *Applied Physics Letters*. 2015;107(15):153701.

87. Keller A.M., DeVore M.S., Stich D.G., Vu D.M., Causgrove T., Werner J.H. Multicolor 3-dimensional tracking for single-molecule fluorescence resonance energy transfer measurements. *Analytical Chemistry*. 2018. doi: 10.1021/acs.analchem.8b00244.

88. Liu Y.-L., Perillo E.P., Ang P., Kim M., Nguyen D.T., Blocher K., et al. Three-dimensional two-color dual-particle tracking microscope for monitoring DNA conformational changes and nanoparticle landings on live cells. *ACS Nano*. 2020;14(7):7927–39.

89. Lew M.D., Lee S.F., Badieirostami M., Moerner W. Corkscrew point spread function for far-field three-dimensional nanoscale localization of pointlike objects. *Optics Letters*. 2011;36(2):202–4.

90. Backer A.S., Backlund M.P., von Diezmann A.R., Sahl S.J., Moerner W. A bisected pupil for studying single-molecule orientational dynamics and its application to three-dimensional super-resolution microscopy. *Applied Physics Letters*. 2014;104(19):193701.

91. Berg H.C. How to track bacteria. *Review of Scientific Instruments*. 1971;42(6):868–71.

92. Yang A.H., Moore S.D., Schmidt B.S., Klug M., Lipson M., Erickson D. Optical manipulation of nanoparticles and biomolecules in sub-wavelength slot waveguides. *Nature*. 2009;457(7225):71–5.

93. Cohen A.E., Moerner W. Suppressing Brownian motion of individual biomolecules in solution. *Proceedings of the National Academy of Sciences*. 2006;103(12):4362–5.

94. Wang Q., Moerner W. Single-molecule motions enable direct visualization of biomolecular interactions in solution. *Nature Methods*. 2014;11(5):555–8.

95. Cang H., Xu C.S., Montiel D., Yang H. Guiding a confocal microscope by single fluorescent nanoparticles. *Optics Letters*. 2007;32(18):2729–31.

96. Holtzer L., Meckel T., Schmidt T. Nanometric three-dimensional tracking of individual quantum dots in cells. *Applied Physics Letters*. 2007;90(5):053902.

97. Suzuki K.G., Fujiwara T.K., Sanematsu F., Iino R., Edidin M., Kusumi A. GPI-anchored receptor clusters transiently recruit Lyn and Gα for temporary cluster immobilization and Lyn activation: Single-molecule tracking study 1. *The Journal of Cell Biology.* 2007;177(4):717–30.

98. Broeken J., Rieger B., Stallinga S. Simultaneous measurement of position and color of single fluorescent emitters using diffractive optics. *Optics Letters.* 2014;39(11):3352–5.

99. Kim K.H., Buehler C., Bahlmann K., Ragan T., Lee W.-C.A., Nedivi E., et al. Multifocal multiphoton microscopy based on multianode photomultiplier tubes. *Optics Express.* 2007;15(18):11658–78.

100. Sahl S.J., Leutenegger M., Hilbert M., Hell S.W., Eggeling C. Fast molecular tracking maps nanoscale dynamics of plasma membrane lipids. *Proceedings of the National Academy of Sciences.* 2010;107(15):6829–34.

101. Chen Y.-I., Chang Y.-J., Nguyen T.D., Liu C., Phillion S., Kuo Y.-A., et al. Measuring DNA hybridization kinetics in live cells using a time-resolved 3D single-molecule tracking method. *Journal of the American Chemical Society.* 2019;141(40):15747–50.

102. Gebhardt J.C.M., Suter D.M., Roy R., Zhao Z.W., Chapman A.R., Basu S., et al. Single-molecule imaging of transcription factor binding to DNA in live mammalian cells. *Nature Methods.* 2013;10(5):421.

103. Ulbrich M.H., Isacoff E.Y. Subunit counting in membrane-bound proteins. *Nature Methods.* 2007;4(4):319.

104. Cocucci E., Aguet F., Boulant S., Kirchhausen T. The first five seconds in the life of a clathrin-coated pit. *Cell.* 2012;150(3):495–507.

105. Liu Y.-L., Perillo Evan P., Liu C., Yu P., Chou C.-K., Hung M.-C., et al. Segmentation of 3D trajectories acquired by TSUNAMI microscope: An application to EGFR trafficking. *Biophysical Journal.* 2016;111(10):2214–27.

106. Behrend C.J., Anker J.N., McNaughton B.H., Brasuel M., Philbert M.A., Kopelman R. Metal-capped brownian and magnetically modulated optical nanoprobes (MOONs): Micromechanics in chemical and biological microenvironments. *The Journal of Physical Chemistry B.* 2004;108(29):10408–14.

107. Anthony S.M., Hong L., Kim M., Granick S. Single-particle colloid tracking in four dimensions. *Langmuir.* 2006;22(24):9812–5.

108. Hong L., Anthony S.M., Granick S. Rotation in suspension of a rod-shaped colloid. *Langmuir.* 2006;22(17):7128–31.

109. Cheong F.C., Grier D.G. Rotational and translational diffusion of copper oxide nanorods measured with holographic video microscopy. *Optics Express.* 2010;18(7):6555–62.

110. Brokmann X., Ehrensperger M.-V., Hermier J.-P., Triller A., Dahan M. Orientational imaging and tracking of single CdSe nanocrystals by defocused microscopy. *Chemical Physics Letters.* 2005;406(1–3):210–4.

111. Hunter G.L., Edmond K.V., Elsesser M.T., Weeks E.R. Tracking rotational diffusion of colloidal clusters. *Optics Express.* 2011;19(18):17189–202.

112. Uspal W.E., Eral H.B., Doyle P.S. Engineering particle trajectories in microfluidic flows using particle shape. *Nature Communications.* 2013;4:2666.

113. Liu B., Böker A. Measuring rotational diffusion of colloidal spheres with confocal microscopy. *Soft Matter.* 2016;12(28):6033–7.

114. Yu Y., Gao Y., Yu Y. "Waltz" of cell membrane-coated nanoparticles on lipid bilayers: Tracking single particle rotation in ligand–receptor binding. *ACS Nano.* 2018;12(12):11871–80.

115. Tsay J.M., Doose S., Weiss S. Rotational and translational diffusion of peptide-coated CdSe/CdS/ZnS nanorods studied by fluorescence correlation spectroscopy. *Journal of the American Chemical Society.* 2006;128(5):1639–47.

116. Chung I., Akita R., Vandlen R., Toomre D., Schlessinger J., Mellman I. Spatial control of EGF receptor activation by reversible dimerization on living cells. *Nature.* 2010;464(7289):783–7.

117. Low-Nam S.T., Lidke K.A., Cutler P.J., Roovers R.C., van Bergen en Henegouwen P.M.P., Wilson B.S., et al. ErbB1 dimerization is promoted by domain co-confinement and stabilized by ligand binding. *Nature Structural & Molecular Biology*. 2011;18(11): 1244–9.

118. Clausen M.P., Lagerholm B.C. Visualization of plasma membrane compartmentalization by high-speed quantum dot tracking. *Nano Letters*. 2013;13(6):2332–7.

119. Kusumi A., Nakada C., Ritchie K., Murase K., Suzuki K., Murakoshi H., et al. Paradigm shift of the plasma membrane concept from the two-dimensional continuum fluid to the partitioned fluid: High-speed single-molecule tracking of membrane molecules. *Annual Review of Biophysics and Biomolecular Structure*. 2005;34:351–78.

120. Treanor B., Depoil D., Gonzalez-Granja A., Barral P., Weber M., Dushek O., et al. The membrane skeleton controls diffusion dynamics and signaling through the B cell receptor. *Immunity*. 2010;32(2):187–99.

121. Freeman S.A., Jaumouillé V., Choi K., Hsu B.E., Wong H.S., Abraham L., et al. Toll-like receptor ligands sensitize B-cell receptor signalling by reducing actin-dependent spatial confinement of the receptor. *Nature Communications*. 2015;6:6168.

122. Jaqaman K., Grinstein S. Regulation from within: The cytoskeleton in transmembrane signaling. *Trends in Cell Biology*. 2012;22(10):515–26.

123. Salaita K., Nair P.M., Petit R.S., Neve R.M., Das D., Gray J.W., et al. Restriction of receptor movement alters cellular response: Physical force sensing by EphA2. *Science*. 2010;327(5971):1380–5.

124. Pryor Meghan M., Low-Nam Shalini T., Halász Ádám M., Lidke Diane S., Wilson Bridget S., Edwards Jeremy S. Dynamic transition states of ErbB1 phosphorylation predicted by spatial stochastic modeling. *Biophysical Journal*. 2013;105(6):1533–43.

125. Ibach J., Radon Y., Gelléri M., Sonntag M.H., Brunsveld L., Bastiaens P.I., et al. Single particle tracking reveals that EGFR signaling activity is amplified in clathrin-coated pits. *PloS one*. 2015;10(11):e0143162.

126. Liu Y.-L., Chou C.-K., Kim M., Vasisht R., Kuo Y.-A., Ang P., et al. Assessing metastatic potential of breast cancer cells based on EGFR dynamics. *Scientific Reports*. 2019;9(1):3395.

127. Blume-Jensen P., Hunter T. Oncogenic kinase signalling. *Nature*. 2001;411(6835):355–65.

128. Casaletto J.B., McClatchey A.I. Spatial regulation of receptor tyrosine kinases in development and cancer. *Nature Reviews Cancer*. 2012;12(6):387–400.

129. Liu Y.-L., Horning A.M., Lieberman B., Kim M., Lin C.-K., Hung C.-N., et al. Spatial EGFR dynamics and metastatic phenotypes modulated by upregulated EphB2 and Src pathways in advanced prostate cancer. *Cancers*. 2019;11(12):1910.

130. Saxton M.J. Single-particle tracking - effects of corrals. *Biophysical Journal*. 1995; 69(2):389–98.

131. Saxton M.J. Single-particle tracking: The distribution of diffusion coefficients. *Biophysical Journal*. 1997;72(4):1744.

132. Meroz Y., Sokolov I.M. A toolbox for determining subdiffusive mechanisms. *Physics Reports*. 2015;573:1–29.

133. Di Rienzo C., Gratton E., Beltram F., Cardarelli F. Fast spatiotemporal correlation spectroscopy to determine protein lateral diffusion laws in live cell membranes. *Proceedings of the National Academy of Sciences*. 2013;110(30):12307–12.

134. Fujiwara T.K., Iwasawa K., Kalay Z., Tsunoyama T.A., Watanabe Y., Umemura Y.M., et al. Confined diffusion of transmembrane proteins and lipids induced by the same actin meshwork lining the plasma membrane. *Molecular Biology of the Cell*. 2016;27(7):1101–19.

135. Jin S., Haggie P.M., Verkman A. Single-particle tracking of membrane protein diffusion in a potential: Simulation, detection, and application to confined diffusion of CFTR Cl− channels. *Biophysical Journal*. 2007;93(3):1079–88.

136. Zhao R., Yuan J., Li N., Sun Y., Xia T., Fang X. Analysis of the diffusivity change from single-molecule trajectories on living cells. *Analytical Chemistry.* 2019;91(21):13390–7.

137. Ghosh R.P., Franklin J.M., Draper W.E., Shi Q., Beltran B., Spakowitz A.J., et al. A fluorogenic array for temporally unlimited single-molecule tracking. *Nature Chemical Biology.* 2019;15(4):401–9.

138. Kim M., Hong S., Yankeelov T.E., Yeh H.-C., Liu Y.-L. Deep learning-based classification of breast cancer cells using transmembrane receptor dynamics. *Bioinformatics.* 2021;38:243–249.

139. Matsuda Y., Hanasaki I., Iwao R., Yamaguchi H., Niimi T. Estimation of diffusive states from single-particle trajectory in heterogeneous medium using machine-learning methods. *Physical Chemistry Chemical Physics.* 2018;20(37):24099–108.

140. Granik N., Weiss L.E., Nehme E., Levin M., Chein M., Perlson E., et al. Single-particle diffusion characterization by deep learning. *Biophysical Journal.* 2019;117(2):185–92.

141. Wagner T., Kroll A., Haramagatti C.R., Lipinski H.-G., Wiemann M. Classification and segmentation of nanoparticle diffusion trajectories in cellular micro environments. *PloS one.* 2017;12(1):e0170165.

142. Dosset P., Rassam P., Fernandez L., Espenel C., Rubinstein E., Margeat E., et al. Automatic detection of diffusion modes within biological membranes using back-propagation neural network. *BMC Bioinformatics.* 2016;17(1):197.

143. Trinczek B., Ebneth A., Mandelkow E., Mandelkow E. Tau regulates the attachment/detachment but not the speed of motors in microtubule-dependent transport of single vesicles and organelles. *Journal of Cell Science.* 1999;112(14):2355–67.

144. de Bruin K., Ruthardt N., von Gersdorff K., Bausinger R., Wagner E., Ogris M., et al. Cellular dynamics of EGF receptor–targeted synthetic viruses. *Molecular Therapy.* 2007;15(7):1297–305.

145. Nielsen E., Severin F., Backer J.M., Hyman A.A., Zerial M. Rab5 regulates motility of early endosomes on microtubules. *Nature Cell Biology.* 1999;1(6):376–82.

146. Ichikawa T., Yamada M., Homma D., Cherry R.J., Morrison I.E., Kawato S. Digital fluorescence imaging of trafficking of endosomes containing low-density lipoprotein in brain astroglial cells. *Biochemical and Biophysical Research Communications.* 2000;269(1):25–30.

147. Ott D., Bendix P.M., Oddershede L.B. Revealing hidden dynamics within living soft matter. *ACS Nano.* 2013;7(10):8333–9.

148. Huet S., Karatekin E., Tran V.S., Fanget I., Cribier S., Henry J.-P. Analysis of transient behavior in complex trajectories: Application to secretory vesicle dynamics. *Biophysical Journal.* 2006;91(9):3542–59.

149. Arcizet D., Meier B., Sackmann E., Radler J.O,. Heinrich D. Temporal analysis of active and passive transport in living cells. *Physical Review Letters.* 2008;101(24):248103.

150. Meilhac N., Le Guyader L., Salome L., Destainville N. Detection of confinement and jumps in single-molecule membrane trajectories. *Physical Review E.* 2006;73(1):011915.

151. Montiel D., Cang H., Yang H. Quantitative characterization of changes in dynamical behavior for single-particle tracking studies. *The Journal of Physical Chemistry B.* 2006;110(40):19763–70.

152. Tabei S.M.A., Burov S., Kim H.Y., Kuznetsov A., Huynh T., Jureller J., et al. Intracellular transport of insulin granules is a subordinated random walk. *Proceedings of the National Academy of Sciences.* 2013;110(13):4911–6.

153. Schutz G.J., Schindler H., Schmidt T. Single-molecule microscopy on model membranes reveals anomalous diffusion. *Biophysical Journal.* 1997;73(2):1073–80.

154. Pinaud F., Michalet X., Iyer G., Margeat E., Moore H.P., Weiss S. Dynamic partitioning of a glycosyl-phosphatidylinositol-anchored protein in glycosphingolipid-rich microdomains imaged by single-quantum dot tracking. *Traffic.* 2009;10(6):691–712.

155. Ferrari R., Manfroi A.J., Young W.R. Strongly and weakly self-similar diffusion. *Physica D.* 2001;154(1–2):111–37.

156. Ewers H., Smith A.E., Sbalzarini I.F., Lilie H., Koumoutsakos P., Helenius A. Single-particle tracking of murine polyoma virus-like particles on live cells and artificial membranes. *Proceedings of the National Academy of Sciences.* 2005;102(42):15110–5.

157. Turkcan S., Masson J.B. Bayesian decision tree for the classification of the mode of motion in single-molecule trajectories. *PLoS One.* 2013;8(12):e82799.

158. Masson J.B., Dionne P., Salvatico C., Renner M., Specht C.G., Triller A., et al. Mapping the energy and diffusion landscapes of membrane proteins at the cell surface using high-density single-molecule imaging and Bayesian inference: Application to the multiscale dynamics of glycine receptors in the neuronal membrane. *Biophysical Journal.* 2014;106(1):74–83.

159. Turkcan S., Alexandrou A., Masson J.B. A Bayesian inference scheme to extract diffusivity and potential fields from confined single-molecule trajectories. *Biophysical Journal.* 2012;102(10):2288–98.

160. Voisinne G., Alexandrou A., Masson J.B. Quantifying biomolecule diffusivity using an optimal Bayesian method. *Biophysical Journal.* 2010;98(4):596–605.

161. Burov S., Tabei S.M.A., Huynh T., Murrell M.P., Philipson L.H., Rice S.A., et al. Distribution of directional change as a signature of complex dynamics. *Proceedings of the National Academy of Sciences.* 2013;110(49):19689–94.

162. Tejedor V., Benichou O., Voituriez R., Jungmann R., Simmel F., Selhuber-Unkel C., et al. Quantitative analysis of single particle trajectories: Mean maximal excursion method. *Biophysical Journal.* 2010;98(7):1364–72.

163. Burnecki K., Kepten E., Garini Y., Sikora G., Weron A. Estimating the anomalous diffusion exponent for single particle tracking data with measurement errors - an alternative approach. *Scientific Reports.* 2015;5:11306.

164. Westphal V., Rizzoli S.O., Lauterbach M.A., Kamin D., Jahn R., Hell S.W. Video-rate far-field optical nanoscopy dissects synaptic vesicle movement. *Science.* 2008;320(5873): 246–9.

165. Simson R., Sheets E.D., Jacobson K. Detection of temporary lateral confinement of membrane-proteins using single-particle tracking analysis. *Biophysical Journal.* 1995; 69(3):989–93.

166. Helmuth J.A., Burckhardt C.J., Koumoutsakos P., Greber U.F., Sbalzarini I.F. A novel supervised trajectory segmentation algorithm identifies distinct types of human adenovirus motion in host cells. *Journal of Structural Biology.* 2007;159(3):347–58.

167. Persson F., Lindén M., Unoson C., Elf J. Extracting intracellular diffusive states and transition rates from single-molecule tracking data. *Nature Methods.* 2013;10(3):265.

168. Serge A., Bertaux N., Rigneault H., Marguet D. Dynamic multiple-target tracing to probe spatiotemporal cartography of cell membranes. *Nature Methods.* 2008;5(8):687–94.

169. Förster T. Experimentelle und theoretische Untersuchung des zwischenmolekularen Übergangs von Elektronenanregungsenergie. *Zeitschrift für naturforschung A.* 1949;4(5): 321–7.

170. Stryer L., Haugland R.P. Energy transfer: A spectroscopic ruler. *Proceedings of the National Academy of Sciences.* 1967;58(2):719–26.

171. Stryer L. Fluorescence energy transfer as a spectroscopic ruler. *Annual Review of Biochemistry.* 1978;47(1):819–46.

172. Ha T., Enderle T., Ogletree D., Chemla D.S., Selvin P.R., Weiss S. Probing the interaction between two single molecules: Fluorescence resonance energy transfer between a single donor and a single acceptor. *Proceedings of the National Academy of Sciences.* 1996;93(13):6264–8.

173. Weiss S. Fluorescence spectroscopy of single biomolecules. *Science.* 1999;283(5408): 1676–83.

174. Zhuang X., Bartley L.E., Babcock H.P., Russell R., Ha T., Herschlag D., et al. A single-molecule study of RNA catalysis and folding. *Science.* 2000;288(5473):2048–51.

175. Sönnichsen C., Reinhard B.M., Liphardt J., Alivisatos A.P. A molecular ruler based on plasmon coupling of single gold and silver nanoparticles. *Nature Biotechnology.* 2005;23(6):741–5.
176. Reinhard B.M., Siu M., Agarwal H., Alivisatos A.P., Liphardt J. Calibration of dynamic molecular rulers based on plasmon coupling between gold nanoparticles. *Nano Letters.* 2005;5(11):2246–52.
177. Yun C., Javier A., Jennings T., Fisher M., Hira S., Peterson S., et al. Nanometal surface energy transfer in optical rulers, breaking the FRET barrier. *Journal of the American Chemical Society.* 2005;127(9):3115–9.

7 Determining the Location and Movement of Biomolecules and Biomolecular Complexes in Single Microbial Cells

E. O. Puchkov
Russian Academy of Sciences

CONTENTS

7.1   INTRODUCTION

Modern microbiology has acquired a toolbox of powerful quantitative analytical methods for studies of microorganisms at the level of single cells. The studies using quantitative analytical methods for single cells of microorganisms have opened up

DOI: 10.1201/9781003409472-7

prospects for solving such problems as genotypic and phenotypic heterogeneity of microbial populations, the nature of unculturable and persistent forms, the development of biofilms, the interaction of microbial pathogens and host cells, the relationship of structure and function in the metabolism, and others (Puchkov 2021a). Among these methods, computerized fluorescence microscopy (CFM) and atomic force microscopy (AFM) are of particular interest, since they allow for determining the location and movement of individual biomolecules and biomolecular complexes in single microbial cells. This Chapter presents a brief description of these methods and examples of their use in the studies of various microorganisms. These studies began relatively long ago, and the results obtained in this area of research were summarized in several reviews. Therefore, the corresponding information is presented below with the reference to the main ones.

7.2 COMPUTERIZED FLUORESCENCE MICROSCOPY

Over the past 20 years, fluorescence microscopy has evolved from a subjective visual method to an objective analytical approach, *computerized fluorescence microscopy* (CFM) (Puchkov 2021b). This was the result of the following combination of innovations. First, new fluorescence microscopes were created that provide highly sensitive recording of fluorescent images in *digital format* (Sanderson et al. 2014). Secondly, computers and software for processing and analyzing digital images with high speed were developed that allow *real-time processing and analysis of digital images* (Puchkov 2016a). Thirdly, an extensive set of *fluorescing molecules* with various properties has been produced (Johnson and Spence 2010; Spahn et al. 2018).

Compared to conventional fluorescence microscopy, CFM provides two important new possibilities. First, CFM makes it possible to improve the visual appearance of the entire image or its individual components through the procedure called *image deconvolution*. Secondly, the data encoded in the digital image can be used for *quantitative processing*. This approach is called *computer image analysis*. With the help of computer image analysis, the optical properties of objects are determined in conjunction with their spatial characteristics. CFM made it possible to use almost all quantitative analytical methods based on the phenomenon of fluorescence (Lakowicz 2006) to study single cells of microorganisms at the subcellular level. CFM can be used to combine quantitative fluorescence measurements with visual analysis. This makes it possible to relate the fluorescence parameters with the structural elements of the cells.

New research capabilities of CFM were opened up due to the development of special techniques based on the generation and use of *fluorescent proteins* (FPs).

7.2.1 FLUORESCENT PROTEINS (FPS)

The *first* technology for the generation of FPs was invented due to the finding of a natural protein of the jellyfish *Aequorea victoria*, which is known as *green fluorescent protein* (GFP) (Tsien 1998). Later, natural and artificial GFP homologues with various fluorescence properties were discovered and synthesized. These proteins can be *linked* (*fused*) to target proteins inside the cells as fluorescent tags by molecular biology methods to create *hybrid FPs*. The history of development and the principles of the technology for creating FPs based on GFP homologues were well described by Snapp (2005) and Campbell (2008).

The principle of the *second* technology is based on the creation of a hybrid of an intracellular target protein with a *special protein tag*, which has the ability to covalently attach low-molecular mass fluorophores. Due to this ability such proteins were called *"self-labeling proteins"*. They are enzymes that covalently bind fluorescent fragments attached to their substrates (Fernández-Suárez and Ting 2008; Hinner and Johnsson 2010). One of the popular proteins that can "label themselves" is SNAP-tag, which is currently commercially available as part of various expression vectors (Juillerat et al. 2003). There are other systems for labeling intracellular proteins with exogenous fluorophores using this technology, for example, CLIP-tag and Halo-tag (Gautier et al. 2008; Los et al. 2008).

Interesting experimental possibilities appeared after the discovery of so-called *photoswitchable FPs*. Some of such FPs acquire the ability to fluoresce only after being photoactivated with the light of certain wavelengths. The properties and possibilities of using such FPs in CFM were discussed in detail in the reviews of Chozinski et al. (2014), as well as in Minoshima and Kikuchi (2017).

One of the most important results of the combined application of CFM and the technologies for the generation of intracellular FPs was the development of the methods of *super-resolution microscopy* or *nanoscopy* (Hell 2007).

7.2.2   Super-Resolution Microscopy (Nanoscopy)

The resolution of standard optical microscopy is limited due to the diffraction properties of light. In practice, this means that it is impossible with a conventional optical microscope to distinguish two objects that are located closer than several hundred nanometers (Sanderson et al. 2014). Super-resolution microscopy methods made it possible to overcome this limit in various ways. To emphasize the possibility of studying objects below the "diffraction limit" with a resolution in the nanometer range, the term *nanoscopy* was introduced (Betzig and Chichester 1993).

The general principles of nanoscopy were well described in (Hell 2007; Requejo-Isidro 2013; Han et al. 2013; Lakadamyali 2014; Nienhaus and Nienhaus 2016). The specifics of application for the study of microorganisms were discussed (Coltharp and Xiao 2012; Sanderson et al. 2014; Yao and Carballido-López 2014; Gahlmann and Moerner 2014; Tuson and Biteen 2015; Kocaoglu and Carlson 2016; Tuson et al. 2016; Endesfelder 2019; Ho and Tirrell 2019; Cambré and Aertsen 2020; Singh and Kenney 2021).

Conventionally, nanoscopy methods can be divided into two categories.

7.2.2.1   Methods of Single Molecule Localization Microscopy

Single molecule localization microscopy (SMLM) (Lelek et al. 2021), such as fluorescence photoactivation localization microscopy (PALM) (Betzig et al. 2006; Hess et al. 2006) and stochastic optical reconstruction microscopy (STORM) (Rust et al. 2006; Bates et al. 2007). The general principle that underlies this category of methods is as follows. The fluorescent molecules in the test sample are excited in such a way that only a fraction of them can fluoresce at the same time. This procedure is carried out repeatedly with the imaging. The images are superimposed on each other. The resulting image includes all the fluorescent molecules, including those that are located at such a close distance from each other that they would be indistinguishable with conventional microscopy (Figure 7.1a). To implement this principle in the PALM methods, special photoswitchable FPs are used which can be "turned on/off" by certain combinations of excitation illumination (Dickson et al.1997; Lippincott-Schwartz and Patterson 2009; Heilemann et al. 2009). In STORM methods, specially selected pairs of fluorophores are used. One of these pairs acts as an "activator" and the other as an "acceptor" of excitation (Rust et al. 2006; Bates et al. 2007). The resolution of PALM/STORM methods in the lateral and axial directions is about 10–20 nm and 20–100 nm, respectively. Some other methods of this category can be found in (Lelek et al. 2021).

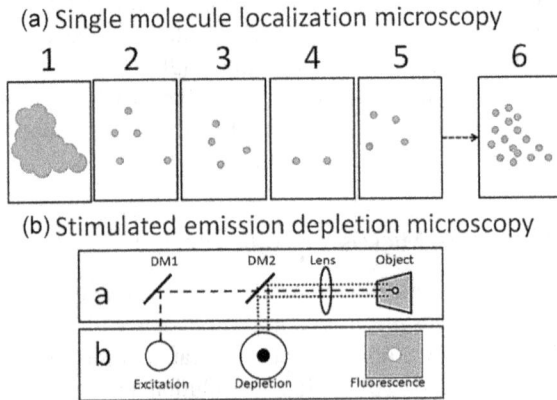

FIGURE 7.1 General principles of super-resolution microscopy. (a) Closely located to each other fluorescing molecules are undistinguishable if observed under conventional fluorescence microscope since their images overlap (1). However, image analysis techniques make it possible to localize precise positions of single molecules if they are separated from each other by some large enough distance. SMLM is based on collecting under a fluorescent microscope a large number of images each containing just a few isolated from each other fluorophores that fluoresce simultaneously (2–5). These images are overlaid which results in the image, where individual fluorescing molecules can be seen even at a close distance from each other (6). In this methodology, photoswitchable fluorophores are used. The spontaneously occurring phenomenon of photobleaching is exploited in fluorescence photoactivation localization microscopy (PALM), whereas reversible switching between a fluorescent on-state and a dark off-state of a dye is exploited in stochastic optical reconstruction microscopy (STORM). (b) Stimulated emission depletion microscopy (STED) is one of the methods based on special illumination. In STED microscopy, excitation with two laser pulses is used. The first pulse excites the photoswitchable fluorophores to a fluorescent state. The second pulse is used to suppress fluorescence in a fraction of the fluorophores in the area surrounding the narrow focus area of the objective lens. The sample is scanned under this excitation mode, and a fluorescent image of the entire sample is obtained. Simplified schemes of optical setup (a) (DM, dichroic mirror) and light beam profiles (b).

7.2.2.2   Methods Based on Special Illumination of Samples

Methods based on special illumination of samples such as stimulated emission depletion microscopy (STED) (Hell and Wichmann 1994; Klar et al. 2000; Hell 2007)
and structured illumination microscopy (SIM) (Gustafsson 2005). To implement this
category of methods, an ultra-high sensitivity for recording weak fluorescence of
individual molecules is not required. In STED microscopy, excitation with two laser
pulses is used. The first pulse excites the photoswitchable fluorophores to a fluorescent state. The second pulse is used to suppress fluorescence in a fraction of the
fluorophores in the area surrounding the narrow focus area of the objective lens. The
sample is scanned under this excitation mode, and a fluorescent image of the entire
sample is obtained (Figure 7.1b). The resolution of this method in the lateral direction
is of 30–60 nm. In the microscopes for SIM, the excitation light flux passes through a
special optical lattice. This leads to the formation of spatially structured illumination
in the focal plane of the sample due to the interference (Moire) effect. After processing a series of fluorescent images of the sample obtained in this way by special computer programs, the resolution can be improved to approximately 100 nm.

7.2.3   Examples of Studies with CFM

The examples below illustrate the capabilities of various CFM methods for determining the location and movement of biomolecules and biomolecular complexes in
single microbial cells.

7.2.3.1   Yeast

CFM can be performed on a standard fluorescence microscope with a digital photo/
video camera and ImageJ image analysis software available on the Internet. The
possibility of measurements of the Brownian motion of inorganic polyphosphate
complexes in the vacuoles of single *Saccharomyces cerevisiae* yeast cells was demonstrated with such a relatively simple set of experimental tools (Figure 7.2). This
approach made it possible to assess viscosity in the vacuoles (Puchkov 2016b).

Methods of visualization and quantification of single RNA molecules have been
developed (Femino et al. 1998), which were used for the *single-molecule fluorescence
in situ hybridization* (smFISH) on *S. cerevisiae* yeast. It was shown with smFISH
that two isoforms of mRNA of one of the kinetochore proteins were synthesized
at different stages of the cell cycle and exported from the nucleus to the cytoplasm
(Chen et al. 2018). The kinetics of expression and intracellular distribution of mRNA
of the STL1 and CTT1 genes in *S. cerevisiae* cells under osmotic shock were studied
(Li and Neuert 2019).

High-speed single-molecule tracking of functional Mig1 protein (Zn-containing
transcription factor) was used to study the *dynamics of its binding* to the DNA of live
S. cerevisiae cells. The obtained data made it possible to characterize the 3D distribution of the protein binding sites on DNA. Subsequent computer simulation allowed
for the prediction of the 3D architecture of the genome (Wollman et al. 2020).

Quantitative confocal microscopy (QCM) and nanoscopy were used to study the
localization of macromolecules, as well as the dynamics of intracellular processes

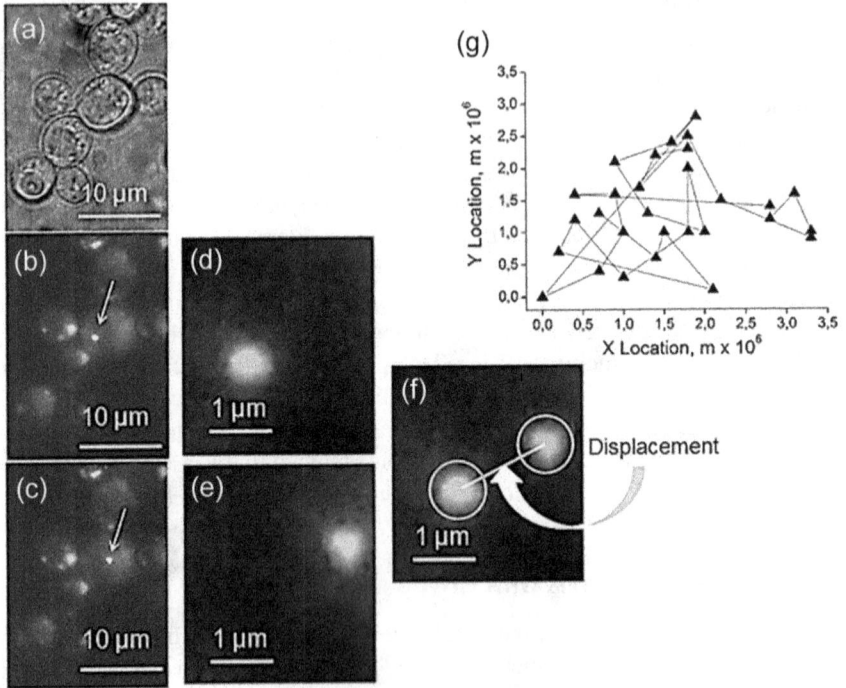

FIGURE 7.2 Characterization of the Brownian motion of inorganic polyphosphate complexes (IPCs) located in the vacuoles of *Saccharomyces cerevisiae* cells. Transmitted (a) and fluorescence (b, c) microscopy images of the cells stained with 4′, 6-diamidino-2-phenylindoleand (DAPI). The images (b) and (c) were captured with 2 s interval. Arrows indicate inorganic polyphosphate complexes (IPCs) fluorescing bright yellow. Other visible "spots" and "dots" fluorescing blue are the nuclei and mitochondria, respectively. (d) and (e) show the magnified images of the IPCs from (b) and (c), respectively. (f) shows how displacement of the IPC can be assessed with ImageJ software after the overlay of the images (d) and (e). (g) Locations of one IPC in Cartesian coordinates detected by the series of shots with 2 s interval (unpublished results). This approach was used for the viscosity measurements in the vacuoles of single yeast cells (Puchkov 2016b).

in *S. pombe*. For example, QCM was used to measure the distances between the cluster of Chr1 genes, the cytoplasmic membrane and the polar body of the spindle depending on the growth conditions of the cells (Bjerling et al. 2012). The QCM and PALM methods were used to study the structure and dynamics of assembly/ dissociation of interphase nodules, precursors of the cytokinetic contractile ring (Laplante et al. 2016; Akamatsu et al. 2017). The mechanism of polymerization of actin filaments during clathrin-induced endocytosis was studied with PALM (Arasada et al. 2018). SIM enabled to identify the spatial interaction of the protein components of the polar spindle bodies during their duplication (Burns et al. 2015; Bestul et al. 2017).

7.2.3.2 Bacteria

A number of reviews on the use of various CFM methods in the studies focused on determining the location and movement of biomolecules and biomolecular complexes in single bacterial calls have been published. In particular, there are reviews on the molecular structure of the cytoskeleton and nucleoid (Yao and Carballido-López 2014); the molecular dynamics of transcription (Stracy and Kapanidis 2017) and translation (Gahlmann and Moerner 2014); the dynamics of protein-DNA interaction (Uphoff 2016); DNA replication and repair (Li et al. 2018); intracellular signaling and gene expression dynamics (Kentner and Sourjik 2010); and prospects for using CFM in the study of intracellular biochemistry at the level of single molecules (Endesfelder 2019).

Some more examples of studies that were not considered in the reviews mentioned above are shown below.

To implement the STORM methodology on live *E. coli* cells, an original technique has been developed based on the creation of hybrid with "self-labeling" proteins using the eukaryotic enzyme N-myristoyl transferase. Spirolactam derivatives of rhodamine, which can be "turned on/off" under the light of specific wavelength, were linked to the Tar and CheA chemotoxis proteins and the FtsZ and FtsA division proteins as fluorescent tags. This made it possible to localize these proteins in the cells using STORM with an accuracy of 15 and 30 nm in the radial and axial directions, respectively (Ho and Tirrell 2019).

The application of the method of labeling target proteins of live *Caulobacter crescentus* cells by attaching the so-called *fluorogen-activating peptides* was described (Szent-Gyorgyi et al. 2008). The dye malachite green was used as a fluorogen, and the peptide dL5 was used as a fluorogen-activating peptide (Szent-Gyorgyi et al. 2013). The complex of this peptide and dye had a well-pronounced fluorescence in the red region of the spectrum, photostability and high affinity of the components. Fluorescent images of three proteins labeled in this way in bacterial cells were obtained using STED. It was shown that the developed technique allows for a four-fold improvement in resolution compared to standard microscopy. In addition, it was noted that the addition of dL5 to one of the proteins (CreS) significantly less affected its structure compared to FPs (Saurabh et al. 2016).

The possibility of improving the brightness and stability of FP fluorescence in *Vibrio cholerae* cells using *plasmon resonance* phenomenon was investigated. For this purpose, a special substrate of gold nanoparticles was created, on which the cells with the TcpP virulence regulator protein labeled with mCherry FP were immobilized. The excitation of the fluorescence of hybrid proteins in such a system made it possible to increase the number of recorded fluorescence photons before the onset of fluorescence decay. The conclusion was made about the possibility of using plasmon-enhanced fluorescence for the localization of individual protein molecules in live bacterial cells in general and in *V. cholerae*, in particular (Flynn et al. 2016).

Another CFM technique based on PALM was developed to study the mutual disposition of proteins of the polar complex associated with chemotaxis and flagella in *V. cholerae* cells. It allowed for more precise localization of target proteins labeled with FPs. This was achieved by simultaneous recording of the fluorescent image of both the cell contours and target proteins, due to the labeling of the cytoplasmic membrane, periplasm and polar complex proteins with specific photoswitchable FPs. Special software has been created that enabled the processing of PALM data simultaneously on the structure of at least 100 cells and on more than 10,000 molecules of target proteins. It was found with this approach, in particular, that one of the proteins of the polar complex tends to be strictly localized in the site close to the internal bend of the cytoplasmic membrane (Altinoglu et al. 2019).

Spatiotemporal *image correlation spectroscopy* (Hebert et al. 2005) was used to characterize the movement of protein molecules in live *E. coli* cells. This technique made it possible to overcome the problem of blurring of the fluorescent image caused by the difference in the speeds of image recording and the movement of molecules. The approach was based on the exact determination of the width of the correlation function by direct calculation of the variance of the correlation function instead of the commonly used Gaussian fitting procedure. The effectiveness of the technique was confirmed by assessing the diffusion parameters of the photoswitchable mMaple3 fluorescent protein in live *E. coli* cells (Rowland et al. 2016).

7.3 ATOMIC FORCE MICROSCOPY

Atomic force microscopy (AFM) has been developed in 1986 (Binnig et al. 1986). It is based on the recording of the interaction of a miniature mechanical probing tip mounted at the end of a flexible cantilever with the surface of an investigated object during its scanning (Figure 7.3a). Importantly, the size of the probe is so small that it is possible to characterize the physical and chemical properties of the surface (to obtain "physicochemical images") with a resolution of the size of single molecules (Sun 2018) and under certain conditions it is possible of resolving even the chemical structure of a single molecule on a surface (Iwata et al. 2015). AFM is one of the numerous methods of the so-called scanning probe microscopy family (Bhushan and Marti 2011). However, according to available literature, only AFM was used to study biomolecules and molecular complexes located on the surface of microbial cells.

7.3.1 AFM IMAGING

Initially, AFM was used to characterize the topography of biomolecules and biomolecular complexes on the surface of native microbial cells by *AFM imaging* (Figure 7.3b). Two imaging modes are mainly used in AFM analysis, namely, the *contact* and *non-contact* or *tapping* modes. The first is the foremost imaging mode of operation, where the tip remains in continuous contact with the surface during scanning. In this mode of operation, two types of images can be obtained, namely, the *height* and *deflection* images. The height image is obtained by recording the cantilever deflection as the tip is scanned over the sample surface (Figure 7.3b, a). The deflection image is generated by adjusting the sample height using a feedback loop to keep the deflection of the cantilever constant while scanning the AFM tip over the sample surface (Figure 7.3b, b). In tapping mode, the AFM probe is excited externally near the resonance frequency of the cantilever and the amplitude and phase of the cantilever are monitored while the tip is scanned over the sample surface

(Figure 7.3b, c). In this case, the tip of the probe interacts with the surface intermittently at the end of its downward movement. This mode minimizes potential damage to the sample by the probe. AFM imaging has a resolution comparable to that of *scanning electron microscopy* (SEM). The cells can be examined in an aqueous medium that provides the advantage of AFM over SEM, as the latter could be carried out only with dry or frozen preparations under vacuum conditions (Dorobantu and Gray 2010; Dorobantu et al. 2012; Dufrêne 2014; Sun 2018; Touhami 2020). The use of AFM for obtaining images of the structure of polysaccharides, peptidoglycan, teichoic acids, pili, flagella, and many other components on the surfaces of various microbial cells under native conditions has been demonstrated in numerous studies. In addition, AFM imaging made it possible to investigate structural changes of molecular complexes on the surface of living microbial cells during growth and under various stresses, e.g. antibiotic treatment (Dufrêne 2014; Touhami 2020).

7.3.2 Single-Molecule Force Spectroscopy (SMFS)

AFM can be used to probe the physical properties of the molecules on the surface of microbial cells using so-called *single-molecule force spectroscopy* (SMFS) (Figure 7.3c, a) (Angeloni et al. 2016; Touhami 2020). In this technique, force-distance curves are recorded by monitoring the cantilever deflection as a function of the vertical displacement at a given location in Cartesian coordinates. By using appropriate corrections, these data are converted into a force-versus-separation distance curve (Figure 7.3c, b). The different parts of this curve can provide the following information. The approach portion of the force-distance curve can characterize surface forces, including van der Waals and electrostatic, solvation, hydration, and steric/bridging forces. The shape of the curve in the region of contact of the probe with the sample may provide information on the elasticity of the sample. When the probe is retracted from the surface, the curve often shows a hysteresis, which can be used to estimate the binding forces between the probe and the studied surface. In the presence of long, flexible molecules, an attractive force, referred to as an elongation force, may develop nonlinearly due to macromolecular stretching (Dorobantu et al. 2012; Dufrêne 2014; Tang et al. 2018; Touhami 2020).

7.3.3 AFM with Functionalized Tip

Modification (functionalization) of the AFM tip with cognate ligands made it possible to specifically locate various biomolecules and molecular complexes relevant to the ligands on the outer layers of cells (Figure 7.3c, a) (Wang et al. 2015; Beaussart and El-Kirat-Chatel 2019; Touhami 2020). In addition, the measurement of specific receptor-ligand forces either on model substrates or on living cells has provided new research possibilities for SMFS (Tang et al. 2018; Wang et al. 2020). This approach has been used to unravel the binding mechanisms of microbial adhesion. Direct information has been obtained on binding strength, affinity, and specificity of the adhesin molecules of several bacterial species. Single-molecule experiments have highlighted multifunctional properties of adhesion proteins that could play a role in pathogen-host interactions. SMFS has been used in experiments to design molecules

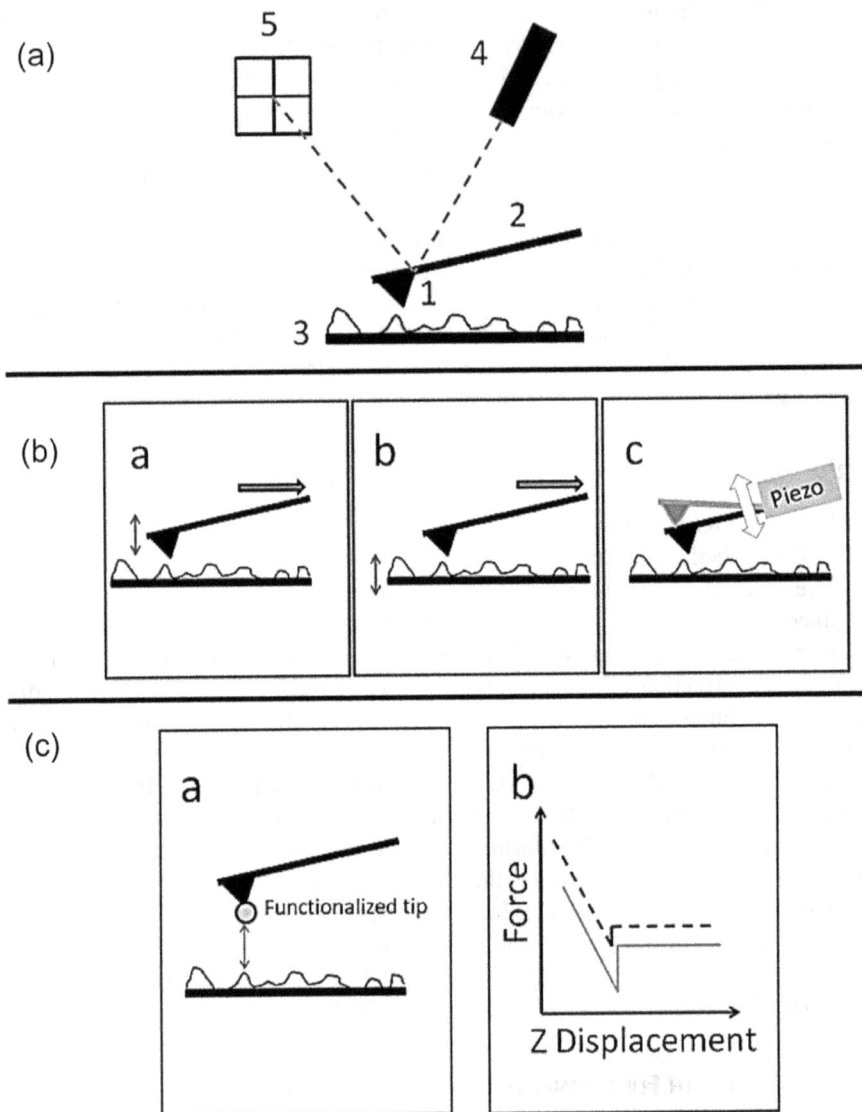

FIGURE 7.3 General principles of atomic force microscopy (AFM). (A) AFM is based on the recording of the interaction of a miniature mechanical probing tip (1) mounted at the end of a flexible cantilever (2) with the surface of an investigated object (3) during its scanning. Deflections of the cantilever (2) are recorded by the optical system consisting of a laser (4) and a 4 quadrant photodetector (5). Importantly, the size of the probe is so small that it is possible to characterize the physical and chemical properties of the surface with a resolution of the size of single molecules. (B) AFM imaging. Two imaging modes are used in AFM analysis, namely, the contact and non-contact or tapping modes. There are two contact modes: height mode, the cantilever deflection is recorded as the tip is scanned over the sample surface (a); and deflection mode, the sample height is adjusted to keep the deflection of the cantilever constant while scanning the AFM tip over the sample surface using a feedback loop (b).

that promote (probiotics) or inhibit (pathogens) bacterial adhesion. Unique possibilities of SMFS made it possible to tackle the problem of mechanical properties of the surface proteins involved in adhesion and mechanical signal transduction (El-Kirat-Chatel et al. 2013; El-Kirat-Chatel and Dufrêne 2018; Alsteens et al. 2013; Dufrêne 2014; Angeloni et al. 2016; Mathelié-Guinlet et al. 2019; Yang et al. 2020; Touhami 2020).

7.3.4 AFM of Isolated Microbial Supramolecular Complexes

The resolution of AFM enables to study the structural arrangement of single protein molecules in isolated microbial supramolecular complexes. One of the examples of the realization of this unique possibility of AFM is the characterization of spatial self-assembly of protein molecules in the envelope of bacterial microcompartments of *Haliangium ochraceum* (Sutter et al. 2016). High-resolution AFM can also be used to record the dynamics of the formation of supramolecular protein complexes. As an example, structural features of the assembly and activation of MotPS stator component of the Na$^+$-type flagella motor of *Bacillus subtilis* have been studied with a special technique of high-speed AFM (Terahara et al. 2017).

7.3.5 Combined Use of AFM

New experimental platforms based on the combined use of AFM with other physical methods for probing molecular features of single microbial cells have been demonstrated. One example of the use of this approach is the study of molecular changes in the cell wall of single-living *Staphylococcus aureus* bacteria by atomic force microscopy combined with infrared spectroscopy (AFM–IR). The measurements were performed with a NanoIR2 system (Anasys Instruments Inc., Santa Barbara, CA, USA) that allowed for the high lateral resolution of AFM–IR, relative to that achievable by conventional IR microspectroscopy. Specific changes in the IR spectra in the region of the septum during division were found (Kochan et al. 2018). Another example is the study aimed at mapping the synthesis of the bacterial

---

In tapping mode, the tip of the probe interacts with the surface intermittently. The AFM probe is excited externally near the resonance frequency of the cantilever and the amplitude and phase of the cantilever are monitored while the tip is scanned over the sample surface (c). (C) Single-molecule force spectroscopy. This AFM technique is based on monitoring the cantilever deflection as a function of the vertical displacement at a given location in Cartesian coordinates. Modification (functionalization) of the AFM tip with cognate ligands makes it possible to specifically locate various biomolecules and molecular complexes relevant to the ligands on the outer layers of cells (a). The data on cantilever deflection as a function of the vertical displacement are converted into force-versus-separation distance curves, which may differ during approach (dashed line) and retract (gray line) of the tip to the target molecule (hysteresis) (b). The different parts of these curves can provide information on surface forces and elasticity of relevant biomolecules (see text).

cell wall structural macromolecule, peptidoglycan, during growth and division in the rod-shaped bacterium *Bacillus subtilis*. In this study, AFM was combined with super-resolution optical microscopy, namely, STORM and SIM (see section B). The synthesis of fluorescently labeled peptidoglycan was correlated with cell wall architecture recorded with high-resolution AFM. It was found that during division, septal synthesis occurred across its developing surface, suggesting a two-stage process. During cell growth, the elongation occurred through bands of synthesis, spaced by ~300 nm (Cadby et al. 2021).

7.4   LIMITATIONS AND PERSPECTIVES

The first limitation of CFM is due to the fact that it makes possible to study only molecules that fluoresce. There are very few natural molecules in the cells of microorganisms that can fluoresce. To study the target molecules, they must be labeled in one way or another. However, such procedures are not developed for all molecules. In addition, labeling molecules with fluorophores inevitably change their properties, and it is not always possible to determine to what extent the properties of such molecules are violated. The second limitation is that the light of fluorescence excitation of fluorophores can be damaging to cellular structures. This imposes restrictions on the choice of fluorophores and the corresponding optical system. Finally, many fluorophores are prone to photobleaching during prolonged exposure to exciting light. On the one hand, it is not always possible to take this effect into account when processing the results. On the other hand, because of this effect, the duration of some experiments is limited.

The main limitation of using AFM in the studies of microorganisms is that it can provide information only about biomolecules and molecular complexes, which are located at their surfaces. The quality of the AFM images is highly dependent on probe geometry and possibility of the violation of "delicate" molecular surfaces of the cells. Although AFM provides the advantage of AFM over SEM because the cells can be examined in a native state in an aqueous medium, its scanning speed is comparatively lower than that of SEM. Potential improvement of AFM in this respect might be possible by what is being termed video-AFM (Fleming et al. 2010).

CFM and AFM are in constant development in various ways both technically and methodologically. The most promising perspectives of using CFM and AFM in the studies of location and movement of biomolecules and biomolecular complexes in single microbial cells would be in their combined use as well as with other physicochemical methods and methods of genetics and molecular biology.

7.5   CONCLUSIONS

CFM and AFM are mighty tools for studying the location and movement of biomolecules and biomolecular complexes in single microbial cells. CFM made it possible to quantitatively measure the fluorescence parameters of the target fluorescent molecules in association with certain structures of individual cells. It expanded the boundaries of visualization/resolution of intracellular components beyond the "diffraction limit" of light microscopy into the nanometer range. This enabled to obtain unique information about the localization and dynamics of intracellular processes at the molecular level.

CFM and AFM combine quantitative measurements with visual analysis to link physical parameters to structural elements of cells. It is also important that many CFM and AFM techniques can be implemented on live cells. To date, many CFM and AFM methods have been developed to solve a wide range of problems related to the structural and functional organization of microbial cells.

All this determines a special place for CFM and AFM in the modern *analytical toolbox for single cell studies* of microorganisms (Puchkov 2021a).

FUNDING

This research has received funding from the Ministry of Science and Higher Education of the Russian Federation under grant agreement № 075-15-2021-1051.

REFERENCES

Akamatsu, M., Lin, Y., Bewersdorf, J. et al. 2017. Analysis of interphase node proteins in fission yeast by quantitative and superresolution fluorescence microscopy. *Mol Biol Cell* 28: 3203–14.

Alsteens, D., Beaussart, A., El-Kirat-Chatel, S., Sullan, R. M., Dufrêne, Y.F. 2013. Atomic force microscopy: a new look at pathogens. *PLoS Pathog.* 9(9): e1003516. doi: 10.1371/journal. ppat.1003516.

Altinoglu, I., Merrifield, C. J., Yamaichi, Y. 2019. Single molecule super-resolution imaging of bacterial cell pole proteins with high-throughput quantitative analysis pipeline. *Sci Rep* 9: 6680.

Angeloni, L., Passeri, D., Reggente, M. et al. 2016. Microbial cells force spectroscopy by atomic force microscopy: A review. *Nanoscience and Nanometrology* 2(1), 30–40. doi: 10.11648/j. nsnm.20160201.13.

Arasada, R., Sayyad, W. A., Berro, J. et al. 2018. High-speed superresolution imaging of the proteins in fission yeast clathrin-mediated endocytic actin patches. *Mol Biol Cell* 29: 295–303.

Bates, M., Huang, B., Dempsey, G. T. et al. 2007. Multicolor super-resolution imaging with photoswitchable fluorescent probes. *Science* 317: 1749–53.

Beaussart, A., El-Kirat-Chatel, S. 2019. Microbial adhesion and ultrastructure from the single-molecule to the single-cell levels by Atomic Force Microscopy. *Cell Surf* 5:100031. doi: 10.1016/j.tcsw.2019.100031.

Bestul, A. J., Yu, Z., Unruh, J. R. et al. 2017 Molecular model of fission yeast centrosome assembly determined by superresolution imaging. *J Cell Biol* 216: 2409–24.

Betzig, E., Chichester, R. J. 1993. Single molecules observed by near-field scanning optical microscopy. *Science* 262: 1422–25.

Betzig, E., Patterson, G.H., Sougrat, R. et al. 2006. Imaging intracellular fluorescent proteins at nanometer resolution. *Science* 313: 1642–45.

Bhushan, B., Marti, O. 2011. Scanning probe microscopy – principle of operation, instrumentation and probes. In *Nanotribology and nanomechanics*, ed. B. Bhushan, 37–110. Berlin/Heidelberg: Springer. doi:10.1007/978-3-642-15283-2_2.

Binnig, G., Quate, C. F., Gerber, C. 1986. Atomic force microscope. *Phys Rev Lett* 56: 930–33.

Bjerling, P., Olsson, I., Meng, X. 2012. Quantitative live cell fluorescence-microscopy analysis of fission yeast. *J Vis Exp* 59: e3454.

Burns, S., Avena, J.S., Unruh, J. R. et al. 2015. Structured illumination with particle averaging reveals novel roles for yeast centrosome components during duplication. *eLife* 4: e08586.

Cadby, A., Foster, S. J., Hobbs, J. K. 2021. Correlative super-resolution optical and atomic force microscopy reveals relationships between bacterial cell wall architecture and synthesis in *Bacillus subtilis*. *ACS Nano* 15(10):16011–18. doi: 10.1021/acsnano.1c04375.

Cambré, A., Aertsen, A. 2020. Bacterial vivisection, how fluorescence-based imaging techniques shed a light on the inner workings of bacteria. *Microbiol Mol Biol Rev* 84: e00008–20.

Campbell, R. E. 2008. Fluorescent proteins. *Scholarpedia* 3: 5410.

Chen, J., McSwiggen, D., Ünal, E. 2018. Single molecule fluorescence in situ hybridization (smFISH) analysis in budding yeast vegetative growth and meiosis. *J Vis Exp* 135: e57774.

Chozinski, T.J., Gagnon, L. A., Vaughan, J. C. 2014. Twinkle, twinkle little star, photoswitchable fluorophores for super-resolution imaging. *FEBS Lett* 588: 3603–12.

Coltharp, C., Xiao, J. 2012. Superresolution microscopy for microbiology. *Cell Microbiol* 14: 1808–18.

Dickson, R. M., Cubitt, A. B., Tsien, R. Y. et al. 1997. On/off blinking and switching behavior of single molecules of green fluorescent protein. *Nature* 388: 355–58.

Dorobantu, L. S., Gray, M. R. 2010. Application of atomic force microscopy in bacterial research. *Scanning* 32: 74–96.

Dorobantu, L. S., Goss, G. G., Burrell, R. E. 2012. Atomic force microscopy: a nanoscopic view of microbial cell surfaces. *Micron* 43(12):1312–22. doi: 10.1016/j.micron.2012.05.005.

Dufrêne, Y. F. 2014. Atomic force microscopy in microbiology: new structural and functional insights into the microbial cell surface. *MBio* 5(4): e01363. doi:10.1128/mBio.01363-14.

El-Kirat-Chatel, S., Beaussart, A., Alsteens, D., et al. 2013. Single-molecule analysis of the major glycopolymers of pathogenic and non-pathogenic yeast cells. *Nanoscale* 5(11):4855–63. doi: 10.1039/c3nr00813d.

El-Kirat-Chatel, S., Beaussart, A. 2018. Probing bacterial adhesion at the single-molecule and single-cell levels by AFM-based force spectroscopy. *Methods Mol Biol* 1814: 403–14. doi: 10.1007/978-1-4939-8591-3_24.

Endesfelder, U. 2019. From single bacterial cell imaging towards in vivo single-molecule biochemistry studies. *Essays in Biochemistry* 63: 187–96.

Femino, A. M., Fay, F. S., Fogarty, K. et al. 1998. Visualization of single RNA transcripts in situ. *Science* 280: 585–90.

Fernández-Suárez, M., Ting, A. Y. 2008. Fluorescent probes for superresolution imaging in living cells. *Nat Rev Mol Cell Biol* 9: 929–43.

Fleming, A. J., Kenton, B. J., Leang, K. K. 2010. Bridging the gap between conventional and video-speed scanning probe microscopes. *Ultramicroscopy* 110(9):1205–14.

Flynn, J. D., Haas, B. L., Biteen, J. S. 2016. Plasmon-enhanced fluorescence from single proteins in living bacteria. *J Phys Chem C* 120: 20512–17.

Gahlmann, A., Moerner, W. E. 2014. Exploring bacterial cell biology with single-molecule tracking and super-resolution imaging. *Nat Rev Microbiol* 12: 9–22.

Gautier, A., Juillerat, A., Heinis, C. et al. 2008. An engineered protein tag for multiprotein labeling in living cells. *Chem Biol* 15: 128–36.

Gustafsson, M. G. L. 2005. Nonlinear structured-illumination microscopy: Wide-field fluorescence imaging with theoretically unlimited resolution. *Proc Natl Acad Sci USA* 102: 13081–6.

Han, R., Li, Z., Fan, Y., Jiang, Y. 2013. Recent advances in super-resolution fluorescence imaging and its applications in biology. *J Genet Genom* 40: 583–95.

Hebert, B., Costantino, S., Wiseman, P. W. 2005. Spatiotemporal image correlation spectroscopy (STICS) theory, verification, and application to protein velocity mapping in living CHO cells. *Biophys J* 88: 3601–14.

Heilemann, M., Dedecker, P., Hofkens, J. et al. 2009. Photoswitches: Key molecules for sub-diffraction resolution fluorescence imaging and molecular quantification. *Laser Photon Rev* 3:180–202.

Hell, S. W. 2007. Far-field optical nanoscopy. *Science* 316: 1153–58.

Hell, S. W., Wichmann, J. 1994. Breaking the diffraction resolution limit by stimulated emission: stimulated emission-depletion fluorescence microscopy. *Opt Lett* 19: 780–2.

Hess, S. T., Girirajan, T. P., Mason, M. D. 2006 Ultra-high resolution imaging by fluorescence photoactivation localization microscopy. *Biophys J* 91: 4258–72.

Hinner, M. J., Johnsson, K. 2010. How to obtain labeled proteins and what to do with them. Curr *Opin Biotechnol* 21: 766–76.

Ho, S. H., Tirrell, D. A. 2019. Enzymatic labeling of bacterial proteins for super-resolution imaging in live cells. *ACS Central Science* 5: 1911–19.

Iwata, K., Yamazaki, S., Mutombo, P., et al. 2015. Chemical structure imaging of a single molecule by atomic force microscopy at room temperature. *Nat Commun* 6: 7766. doi: 10.1038/ncomms8766.

Johnson, I., Spence, M. 2010. *The Molecular Probes Handbook*. A Guide to Fluorescent Probes and Labeling Technologies. 11th edn., Life Technologies, Carlsbad, CA.

Juillerat, A., Gronemeyer, T., Keppler, A. et al. 2003. Directed evolution of O^6-alkylguanine-DNA alkyltransferase for efficient labeling of fusion proteins with small molecules in vivo. *Chemistry and Biology* 10: 313–17.

Kentner, D., Sourjik, V. 2010. Use of fluorescence microscopy to study intracellular signaling in bacteria. *Annu Rev Microbiol* 64: 373–90.

Klar, T. A., Jakobs, S., Dyba, M. et al. 2000. Fluorescence microscopy with diffraction resolution barrier broken by stimulated emission. *Proc Natl Acad Sci USA* 97: 8206–10.

Kocaoglu, O., Carlson, E. E. 2016. Progress and prospects for small-molecule probes of bacterial imaging. *Nature Chem Biol* 12: 472–78.

Kochan, K., Perez-Guaita, D., Pissang, J., et al. 2018. In vivo atomic force microscopy–infrared spectroscopy of bacteria. *J R Soc Interface* 15: 20180115. doi: 10.1098/rsif.2018.0115.

Lakadamyali, M. 2014. Super-resolution microscopy, going live and going fast. *Chemphyschem* 15: 630–36.

Lakowicz, J. R. 2006. *Principles of Fluorescence Spectroscopy*, 3rd edn. Berlin: Springer Science+Business Media.

Laplante, C., Huang, F., Tebbs, I. R. et al. 2016. Molecular organization of cytokinesis nodes and contractile rings by super-resolution fluorescence microscopy of live fission yeast. *Proc Natl Acad Sci USA* 113: E5876–E5885.

Lelek, M., Gyparaki, M. T., Beliu, G. et al. 2021. Single-molecule localization microscopy. *Nat Rev Methods Primers* 1, 39. doi: 10.1038/s43586-021-00038-x.

Li, G., Neuert, G. 2019. Multiplex RNA single molecule FISH of inducible mRNAs in single yeast cells. *Sci Data* 6: 94.

Li, Y., Schroeder, J. W., Simmons, L. A. et al. 2018. Visualizing bacterial DNA replication and repair with molecular resolution. *Curr Opin Microbiol* 43: 38–45.

Lippincott-Schwartz, J., Patterson, G. H. 2009. Photoactivatable fluorescent proteins for diffraction-limited and super-resolution imaging. *Trends Cell Biol* 19: 555–65.

Los, G. V., Encell, L. P., McDougall, M. G. et al. 2008. HaloTag, a novel protein labeling technology for cell imaging and protein analysis. *ACS Chem Biol* 3: 373–82.

Mathelié-Guinlet, M., Viela, F., Viljoen, A., Dehullu, J., Dufrêne, Y. F. 2019. Single-molecule atomic force microscopy studies of microbial pathogens. *Current Opinion in Biomedical Engineering* 12: 1–7.

Minoshima, M., Kikuchi, K. 2017. Photostable and photoswitching fluorescent dyes for super-resolution imaging. *J Biol Inorg Chem* 22: 639–52.

Nienhaus, K., Nienhaus, G. U. 2016. Where do we stand with super-resolution optical microscopy? *J Mol Biol* 428(2 Pt A): 308–22.

Puchkov, E. 2016a. Image analysis in microbiology: A review. *Journal of Computer and Communications* 4: 8–32.

Puchkov, E. 2016b. Microfluorimetry of single yeast cells by fluorescence microscopy combined with digital photography and computer image analysis. In *Advances in Medicine and Biology*. vol. 98. ed. L. V. Berhardt, 69–90. New York: Nova Science Publishers Inc.

Puchkov, E. 2021a. Analytical techniques for single-cell studies in microbiology. In *Handbook of Single Cell Technologies*, eds. T. Santra and F. G. Tseng, 1–32. Singapore: Springer. doi: 10.1007/978-981-10-4857-9_17-3.

Puchkov, E. O. 2021b. Computerized fluorescence microscopy of microbial cells. *World J Microbiol Biotechnol* 37: 189. doi:10.1007/s11274-021-03159-3.

Requejo-Isidro, J. 2013. Fluorescence nanoscopy. Methods and applications. *J Chem Biol* 6: 97–120.

Rowland, D. J., Tuson, H. H., Biteen, J. S. 2016. Resolving fast, confined diffusion in bacteria with image correlation spectroscopy. *Biophys J* 110: 2241–51.

Rust, M. J., Bates, M., Zhuang, X. 2006. Sub-diffraction-limit imaging by stochastic optical reconstruction microscopy (STORM). *Nat Methods* 3: 793–95.

Sanderson, M. J., Smith, I., Parker, I. et al. 2014. Fluorescence microscopy. *Cold Spring Harb Protoc*. doi: 10.1101/pdb.top071795.

Sutter, M., Faulkner, M., Aussignargues, C. et al. 2016. Visualization of bacterial microcompartment facet assembly using high-speed atomic force microscopy. *Nano Lett* 16 (3):1590–95. doi: 10.1021/acs.nanolett.5b04259.

Saurabh, S., Perez, A. M., Comerci, C. J. et al. 2016. Super-resolution imaging of live bacteria cells using a genetically directed, highly photostable fluoromodule. *J Am Chem Soc* 138: 10398–401.

Singh, M. K., Kenney, L. J. 2021. Super-resolution imaging of bacterial pathogens and visualization of their secreted effectors. *FEMS Microbiol Rev* 45. doi: 10.1093/femsre/fuaa050.

Snapp, E. 2005. Design and use of fluorescent fusion proteins in cell biology. *Current Protocols in Cell Biology* 21: 21.4.1.

Stracy, M., Kapanidis, A. N. 2017. Single-molecule and super-resolution imaging of transcription in living bacteria. *Methods* 120: 103–14.

Spahn, C. K., Glaesmann, M., Grimm, J. B. et al. 2018. A toolbox for multiplexed super-resolution imaging of the *E. coli* nucleoid and membrane using novel PAINT labels. *Sci Rep*. 8: 14768.

Sun, W. 2018. Principles of atomic force microscopy. In *Atomic Force Microscopy in Molecular and Cell Biology*, ed. J. Cai, 1–28. Springer Nature Singapore Pte Ltd. doi: 10.1007/978-981-13-1510-7_1.

Szent-Gyorgyi, C., Schmidt, B. F., Creeger, Y., et al. 2008. Fluorogen-activating single-chain antibodies for imaging cell surface proteins. *Nat Biotechnol* 26: 235–40.

Szent-Gyorgyi, C., Stanfield, R. L., Andreko, S., et al. 2013. Malachite green mediates homodimerization of antibody VL domains to form a fluorescent ternary complex with singular symmetric interfaces. *J Mol Biol* 425: 4595–613.

Tang, C., Fan, Y., Lu, J. 2018. Atomic force microscopy-based single molecule force spectroscopy for biological application. In *Atomic Force Microscopy in Molecular and Cell Biology*, ed. J. Cai, 29–40. Springer Nature Singapore Pte Ltd. doi:10.1007/978-981-13-1510-7_1.

Terahara, N., Kodera, N., Uchihashi, T. et al. 2017. Na $^+$-induced structural transition of MotPS for stator assembly of the Bacillus flagellar motor. *Sci Adv* 3(11): eaao4119. doi: 10.1126/sciadv.aao4119.

Touhami, A. 2020. Atomic force microscopy: A new look at microbes. *Synthesis Lectures on Materials and Optics*. Morgan & Claypool Publ. doi: 10.2200/S01005ED1V01Y2020 04MOP003.

Tsien, R. Y. 1998. The green fluorescent protein. *Annu Rev Biochem* 67: 509–44.

Tuson, H.H., Biteen, J.S. 2015. Unveiling the inner workings of live bacteria using super-resolution microscopy. *Anal Chem* 87: 42–63.

Tuson, H. H., Aliaj, A., Brandes, E. R. et al. 2016. Addressing the requirements of high-sensitivity single-molecule imaging of low-copy-number proteins in bacteria. *Chemphyschem* 17: 1435–40.

Uphoff, S. 2016. Super-resolution microscopy and tracking of DNA-binding proteins in bacterial cells. *Methods Mol Biol* 1431: 221–34.

Wang, C., Ehrhardt, C. J., Yadavalli, V. K. 2015. Single cell profiling of surface carbohydrates on Bacillus cereus. *J R Soc. Interface* 12. doi: 10.1098/rsif.2014.1109.

Wang, L., Yuhong, Q., Yantao, S., Bin, L., Gang, W. 2020. Single-molecule force spectroscopy: A facile technique for studying the interactions between biomolecules and materials interfaces *Rev Anal Chem* 39: 116–29. doi: 10.1515/revac-2020-0115.

Wollman, A., Hedlund, E. G., Shashkova, S. et al. 2020. Towards mapping the 3D genome through high speed single-molecule tracking of functional transcription factors in single living cells. *Methods* 170: 82–89.

Yang, B., Liu, Z., Liu, H., Nash, M. A. 2020. Next generation methods for single-molecule force spectroscopy on polyproteins and receptor-ligand complexes. *Front Mol Biosci* 7: 85. doi: 10.3389/fmolb.2020.00085.

Yao, Z., Carballido-López, R. 2014. Fluorescence imaging for bacterial cell biology: from localization to dynamics, from ensembles to single molecules. *Annu Rev Microbiol* 68: 459–76.

8 Pull Down Assay-Based Protein Analysis

Amandeep Kaur
Central Scientific Instruments Organisation (CSIR-CSIO)

Satish Pandey
Mizoram University

Suman Singh
Central Scientific Instruments Organisation (CSIR-CSIO)

CONTENTS

DOI: 10.1201/9781003409472-8

8.1 ENZYME LINKED IMMUNO-SORBENT IMMUNOASSAY

The enzyme-linked immunosorbent assay, often known as ELISA, is an analytical biochemical assay used to determine the concentration of antigen, peptide, and protein in a patient's blood sample. The antibody generated against a certain antigen is used to detect it. Engvall and Perlmann invented the ELISA test in 1971. The radioimmunoassay was used to measure analytes before the advent of the ELISA technique. A radioimmunoassay uses a radioactively labeled antigen or antibody to quantify the signal generated by their interaction, showing the presence of the antigen or antibody in the sample. Because radioactive materials pose a health risk, ELISA is the best alternative to using radioactive signals. ELISA uses a chromogenic substrate, which causes a visible colour shift to indicate the presence of a substance. This alteration occurs as a result of reactions between enzymes and their substrates, such as hoarse peroxide and TMB or ABTS. To begin, unknown antigens were immobilised on a solid surface platform, such as a microtiter plate made of polystyrene. After the antigen has been immobilised, a detection antibody is added, which binds to the antigen's specific epitope non-covalently (electrostatic, hydrogen bonding Waals forces, and hydrophobic effects, among other things). This antigen-antibody complex is identified by a secondary antibody that has been conjugated with enzymes. To remove non-specially bound proteins or antibodies, each stage takes 5–6 washes with a mild non-ionic detergent. The colour formed as a result of the reaction between the enzyme and the antigen-substrate complex is measured calorimetrically at a given wavelength in the final step. There are numerous ELISA forms available depending on the antigen-antibody combination, including direct, indirect, and sandwich ELISA, as well as competitive ELISA. Each ELISA format has its own set of benefits and drawbacks. The first and most important step was to immobilise the antigen of interest. This antigen is identified either directly (with a labeled primary antibody) or indirectly (with an unlabeled secondary antibody) (labeled secondary antibody.

8.1.1 Direct ELISA

A protein of interest or antigen is immobilized on a plate in a direct ELISA, and an enzyme-bound detection antibody is added, which directly binds to the antigen. A spectrophotometer was used to measure the enzyme activity at a specified wavelength. The steps of direct ELISA (Lequin, 2005) involved:

i. Antigen to be tested is prepared in buffer solution and added to each 96-well microplate (it will take time to adhere to the microplate through charge interaction).
ii. Nonspecific binding sites are blocked with casein or Bovine serum albumin is added to each well.
iii. The primary antibody (coupled with enzyme-conjugated) specific to antigen to be tested is added.
iv. A substrate (Example: TMB or ABTS) is added to the plate well. Often substrates change the color upon reaction with enzyme.
v. Color change signal measured with spectrophotometer at specific wavelength (Table 8.1).

8.1.2 Indirect ELISA

A target protein is immobilized on a microtiter plate's surface, and then a primary antibody is applied to the specific antigen, followed by an enzyme-linked secondary antibody that binds to the antigen-antibody combination (Lin, 2015). Steps involved in Indirect ELISA assay:

i. Micro-plate wells coated with test antigen
ii. Unbounded antigen is removed by thrice washing with buffer.
iii. The plate wells surface is blocked with BSA/Casein.
iv. Primary antibody is added to each well in order to bind the specific antigen. Excess antibody is removed followed by washing.

TABLE 8.1

Enlist Advantages and Disadvantages of Direct ELISA

Advantages	Simple and quick protocol
	Cross-reactivity is no longer an issue.
Disadvantages	Because only one main antibody was used, the results were less specific.
	Primary antibody immunoreactivity may be harmed as a result of labeling or tagging.
	It's time-consuming and costly.

TABLE 8.2

Advantages and Disadvantages of Indirect ELISA

Advantages	Commercially available secondary antibodies that have been labeled are available.
	Primary antibody immunoreactivity at its peak.
	Amplification of the signal.
Disadvantages	Due to the use of a secondary antibody, cross-reactivity may be hampered.
	Time-consuming; many incubation processes are required.

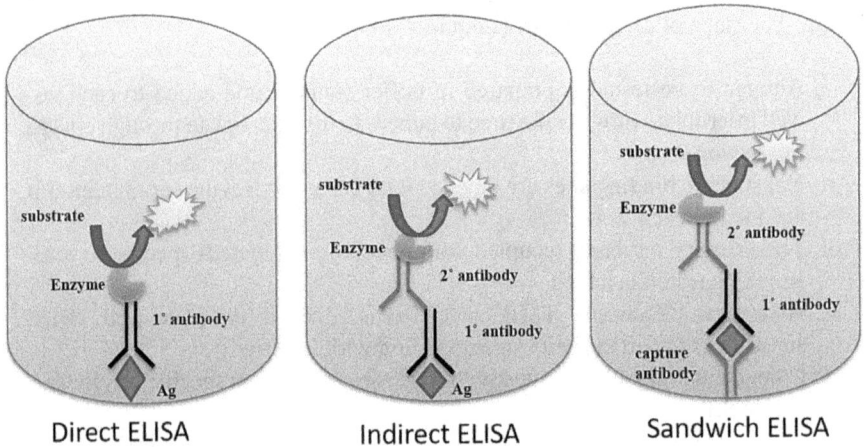

Direct ELISA   Indirect ELISA   Sandwich ELISA

FIGURE 8.1   Types of Enzyme-linked immuno-sorbent immunoassay (ELISA).

v. Secondary antibody enzyme labeled is added, it will bind to the antigen-antibody complex.

vi. Add substrate, enzyme on the antibody elicit a noticeable color change due to enzyme-substrate reaction.

vii. Observe the color change and measured the signal by spectrophotometer (Table 8.2).

8.1.3 SANDWICH ELISA

The most preferable platform used is sandwich ELISA. In this, the analyte to be measured is bound or sandwiched between two antibodies i.e. capture and detection antibody. Each capture antibody detects the different epitopes of target antigen. Capture antibody were immobilized on the surface of plate whereas conjugated - detection antibody bind to additional epitope on the targeted protein (Gonzalez et al., 2008) (Figure 8.1). Sandwich ELISA follow the following steps:

i. Capture antibody is coated on the surface of microtitre plate.

ii. Nonspecific binding sites are blocked with a suitable blocking agent.

TABLE 8.3

Advantages and Disadvantages of Sandwich ELISA

Advantages	Highly specific and sensitive
	Different detection methods can be used with the same capture antibody
Disadvantages	Long protocol,
	Time-consuming as multiple incubation steps are required
	More optimization requires to limited cross-reactivity

iii. Test antigen is applied to the plate wells in order to capture by the capturing antibody.

iv. The plate is washed to remove the unbounded antigen.

v. Primary antibody specific to antigen is added and binds to antigen.

vi. Enzyme-linked secondary antibody is added as a detection antibody.

vii. A substrate is added to be converted by enzyme into color signals.

viii. Color signal is recorded by spectrophotometer to check the presence of antigen in the sample (Table 8.3).

8.1.4 COMPETITIVE ELISA

Competitive ELISA is commonly used for small molecules or peptides. Competitive ELISA is also known as inhibition ELISA where sample antigen and reference antigen compete for the binding site to the specific antibody. Unlabeled primary antibodies are incubated with sample antigen; these antigen-antibody complexes are added to 96-well microtiter plates which are pre-coated with the reference antigen. Afterward, wash the microtiter plates to remove the unbound antibody. Then, a secondary antibody is conjugated with an enzyme that is specific to the primary antibody and finally a specific substrate is added, and the remaining enzymes elicit an optical signal. Both free antigen in the sample and pre-coated antigen will compete for binding to the primary antibody. The more antigen in the sample, the fewer reference antigen will bind to the primary antibody and the weaker the signal (Eva Engvall & Peter Perlmann, 1971) (Figure 8.2).

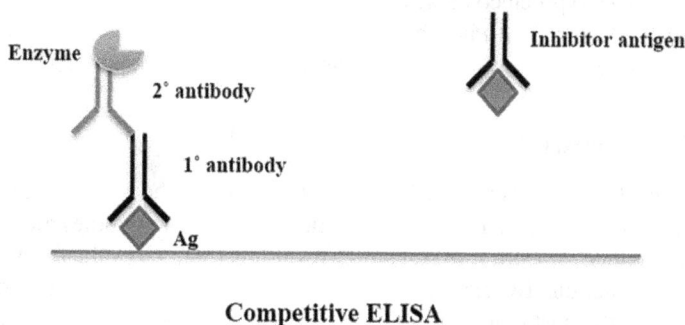

Competitive ELISA

FIGURE 8.2 Principal of competitive ELISA.

TABLE 8.4

Advantages and Disadvantages of Competitive Assay

Advantages	Ability to quantify small-size molecules
	More robust, no sample processing required
	Less variation between duplicates samples
Disadvantages	Complex protocol processing
	Required inhibition antigen

The steps involved in Competitive ELISA (Eva Engvall & Peter Perlmann, 1971) are as follows:

    i. Bounded antigen/antibody complex added to antigen-coated wells.
    ii. Unbound antibodies are removed by washing.
    iii. Enzyme labeled secondary antibody specific to primary antibody is added to each well.
    iv. A substrate is added, elicit chromogenic signals.
    v. Signal recorded spectrophotometry.
    vi. Reactions stopped to prevent saturation of the signal (Table 8.4).

8.1.5 COMMONLY USED SUBSTRATE AND ENZYMATIC MARKERS

A substrate plays an important role in various ELISA procedures. On Substrate – enzymatic reactions generate a chromogenic color change during assay performance. There is a list of enzymatic markers commonly used in ELISA assay:

    a. OPD (*o*-phenylenediamine dihydrochloride) turns amber on HRP detection (Clark, Lister, & Bar-Joseph, 1986).
    b. TMB (3,3',5,5'-tetramethylbenzidine) turns blue upon reaction with HRP and yellow after the addition of sulphuric or phosphoric acid (Madersbacher & Berger, 1991).
    c. ABTS (2,2'-Azinobis [3-ethylbenzothiazoline-6-sulfonic acid]-diammonium salt) change to green color with HRP detection (Matsuda et al., 1984).
    d. PNPP (*p*-Nitrophenyl Phosphate, Disodium Salt turns yellow while detecting alkaline phosphasatse (Huang, Olson, You, & Haugland, 1992).

8.1.6 CHECKERBOARD ELISA

The purpose of the checkerboard ELISA is to optimize the sample and antibody concentration for detection. The sample dilution factor and antibodies must be tuned in order to produce the optimum results. Two variables can be evaluated at the same time in this situation. By experimenting with different sample-to-antibody ratio concentrations, the best concentration and sample-to-antibody ratio for detecting a

1:1 1:2 1:4 1:8 1:16.....

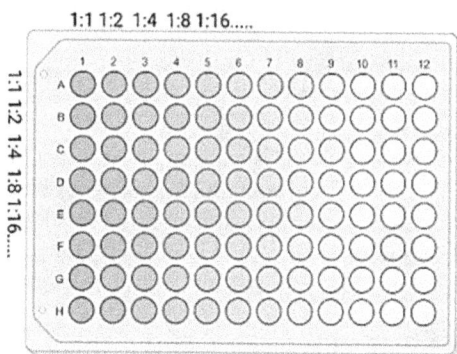

FIGURE 8.3 Checkboard ELISA.

certain target molecule can be identified. The results of the checkerboard test are utilized to perform ELISA for a more widespread application of the choice with better results (Skjerve, Bos, & van der Gaag, 1991). An example of checkerboard ELISA is shown below. Each of column 1–12 represent antigen dilution factor whereas rows A-H denotes antibody dilution factors (Figure 8.3).

A recommended starting concentration of antigen would be in the range of 1–20μg/mL whereas detection antibody concentration starts from 500 ng/mL.

Direct checkerboard titration assay followed as:

Stage 1: Dilute the specific antigen from columns 1–12. Incubate the plate at the required temperature and wash the wells with buffer to remove unbound antigen.

Stage 2: Similarly dilute the antibody starting from row A to row G (note: antibody is diluted in same blocking buffer). Incubate the plate and wash it thrice with a specific buffer to avoid non-specific binding.

Stage 3: Add chromophore/substrate (Mostly 3, 3′, 5, 5′-Tetramethylbenzidine or TMB is used as a chromogenic substrate for the staining procedure in immunohistochemistry and ELISA as visualizing reagent).

Stage 4: Record absorbance at a specific wavelength using spectrophotometer.

8.1.6.1 Data Analysis

The difference in color between columns and rows reflects the effect of altering the concentration of antigen and detection antibody. Rows accessed for a maximum color development reflects the highest concentration of antibody. However, the rows where there is no decrease in color upon diluting the antigen tell about the maximum antigen saturation is coated to the plates. Thus, we can get the information from the checkerboard about the region having excess antigen/antibody. Low color signal exhibit detection antibody specific of antigen is very low in concentration or vice-versa or do not bind due to a limited amount of antigen/antibody.

8.1.6.2 Troubleshooting

High background: Possible source is insufficient washing of wells; this causes the non-specific binding resulting strong signal. Substrate exposed to light prior to use.

No signal: No signal when a signal is expected because of many reasons such as insufficient reagent addition, contaminations in buffers, HRP, not enough antibody is used, detection antibody did not bind to plate.

Low signals: The plate wells with very low color exhibit poor signals because of very weak reactions. The concentration of binding antigen or antibody is very low.

Poor reproducibility: Reproducibility is affected by multiple factors like insufficient washing, variation in incubation temperature, protocol variations, buffer contaminations, plate sealer reused and so on.

Poor standard curve: This was achieved because of incorrect dilution prepared, capture antibody didn't bind to plate, longer incubation times than recommended.

8.1.7 APPLICATIONS OF ELISA

ELISA has a wide variety of applications in the field of science such as food industry, vaccine development, toxicology, drug monitoring, pharmaceutical industry and transplantation etc. are reviewed.

8.1.7.1 Food Industry

ELISA plays a major role in identifying food allergens in the food industry such as those present in milk products, peanuts, eggs and almonds (Peng et al., 2012). Peng et al. developed a sandwich ELISA assay for the authenticity of the food products. They develop monoclonal antibodies for the detection of ovalbumin in food, a major cause of food allergy (Nayak, Li, Ahmed, & Lin, 2017; Peng et al., 2012). ELISA has proven to be a reliable technique to ensure quality control of fish, milk-based products, and other harmful components that can be transferred to humans such as bovine spongiform encephalopathy (Asensio, Gonzalez Alonso, Garcia, & Martín, 2008).

8.1.7.2 Immunology

Measuring and monitoring the change occur in the immune system when encountered with foreign particles or antigen. Systematically the sera sample from immunized persons can be tested for the presence of antibodies produced against the specific antigen (Pizza et al., 2000). An ELISA can also be used to determine the viral load in the testing sample such as Lentivirus determination by p24 ELISA. In this protocol microtiter plate (12×8 well strips) coated with anti-HIV-1 p24 capture antibodies which bind with p24 (HIV core protein) in the targeted test sample. Specifically bounded p24 is detected through sandwich ELISA format using biotinylated labeled anti-p24 secondary antibody along with streptavidin-HRP conjugated. Color-producing substrate reaction with antigen-antibody complex determined color intensity which is measured spectrophotometrically to indicate the level of p24 viral load in the tested sample. While some efforts were made to optimize the ELISA protocol to enhance the accuracy, sensitivity and selectivity of the test for further clinical studies.

8.1.7.3 Diagnosis

ELISA has been proven for detecting a wide variety of diseases such as HIV (Nandi, Maity, Bhunia, & Saha, 2014), Dengue fever (Hunsperger et al., 2014), influenza (Tarigan, Indriani, Durr, & Ignjatovic, 2015; Welch, Chang, & Litwin, 2014), Lyme disease (Hinckley et al., 2014), Ebola (Schieffelin, Moses, Shaffer, Goba, & Grant, 2016) and many more. Specific ELISA pregnancy kits were developed based on the detection of human chorionic gonadotropin (hCG) protein one month after fertilization (Chard, 1992). Sometimes, another estriol (E3) pregnancy hormone is associated with and confirmed by the ELISA technique in the saliva at the 6 weeks of pregnancy. Advancement in ELISA technology has provided its applications in cancer detection based on the cancer biomarker, which is the most challenging biomolecule as well as a crucial early diagnosis for patient survival. For example, CA125 is present in the serum, which is an important biomarker for the timely detection of ovarian cancer (Scholler et al., 2006). However infectious diseases caused by subcellular entities such as prokaryotic bacteria, eukaryotic fungi and some protozoa invade and colonize the host tissues, inducing immune response to generate specific antibodies against the organism. These antibodies can be detected and measured by the immunoassay method such as ELISA to quantify the infection load in the tested samples for diagnosis. Creative diagnosis offered a wide range of detection of human IgM, IgA, IgG antibodies to bacterial, fungal, viral antigens using commercially Scholler et al. (2006) available ELISA kits. Recently ELISA kits were used for the diagnosis of the SARS-CoV-2 virus (Gong, Wei, Li, Liu, & Li, 2021). Various research group is trying hard to make antibody-based diagnostic test which includes Enzyme immunosorbent assay (ELISA), Lateral flow immunoassay (LFIA), fluorescence immunoassay, etc. (Adekanmbi & Olude, 2021). These ELISA kits are ready-to-use, simple, less time-consuming, robust method features with high selectivity and specificity for various microbial and infectious disease research applications.

8.1.7.4 Cell Cytotoxicity

Cell cytotoxicity measures the loss of integrity of the cell membrane after cell death. One gold standard protocol to measure cytotoxicity is based on the measurement of activity of LDH (Lactate dehydrogenase) because lactate enzymes are located in the cytosolic membrane of all cells and are released into supernatant during plasma membrane damage. Alternatively, solid-phase ELISA has been developed to determine the selective & specific detection of apoptotic cells (Frankfurt & Krishan, 2001). In this assay, the cell binds with 96-well microtiter plates, denaturation of DNA involves cell treatment with formamide to convert denatured DNA in apoptotic cells followed by one step staining procedure of denatured DNA with a mixture of anti-ssDNA MAb and conjugated mouse IgM secondary antibody. Untreated, cells with broken ss-DNA induced by H2O2 and necrotic cells do not produce signals (Frankfurt & Krishan, 2001).

8.1.7.5 Pharmaceutical Industry

ELISA is a molecular technique that utilizes the specific antibody as well as the sensitivity of enzyme assay for the determination of molecules such as hormones, peptides, antibodies and proteins. Identification of cancer markers such as ovarian cancer and breast cancer at an early stage is made easy using ELISA-based techniques. Furthermore, monitoring the concentration level of various drugs such as amphetamines, cannabinoids, cocaine and methadone, can be achieved by using ELISA in urine samples. This approach can be used for therapeutic drug monitoring (TDM) monitoring drug levels in plasma samples (Fraaij, Rakhmanina, Burger, & de Groot, 2004). Currently, most of the commercially available methods for quantification of drug infliximab (IFX) are ELISA based with an average recovery of IFX 92% and having a correlation coefficient with these established methods were 0.939. This strategy reported in the literature, cost-effective method for the accurate analysis of antidrug antibodies for monitoring therapeutic drug levels (Hock et al., 2016; Iwamoto et al., 2018).

8.2 DIGITAL ELISA

Newly developed digital ELISA is a powerful technique that exhibits 1000-fold more sensitivity than conventional ELISA. This technology works for the quantification of suitable proteins such as biological markers, and cytokines concentration in a wide variety of different matrices such as serum, plasma, cerebrospinal fluid, urine, cell extracts, etc. at femtogram (fg) level. It exhibits the following features:

- Ultra-sensitivity -1000 times more improvement than existing immunoassay.
- Detect single protein molecule
- Enhanced data reproducibility
- Wide range provides >4 logs of dynamic range.
- Developed customized assay
- Measure biomarkers associated with serum, plasma, CSF
- Capability of tested four different analytes from single sample

Firstly, antibody capture agents are attached to the surface of the bead containing approximately 250,000 attachment sites, each bead having its lawn of capture molecule. Adding so many beads exhibit its own advantages having labeled immunocomplex follows a Poisson distribution. At low concentrations of protein, bead will capture either single immunocomplex or none. Secondly, with so many beads, diffusion of targeted analytes molecule occur quickly because molecule has multiple collision with multiple beads. Non-specifically bound proteins were eluted out through washing and incubated with biotin-labeled antibodies, each bead capturing a single molecule with an enzyme labeled. Beads are added into femtoliter-sized wells (4.25 μm width, 3.25 μm depth). Beads were added in the presence of substrate, if targeted analytes capture the resulting substrate will convert into the fluorescent product by the captured enzyme. The protein concentration of unknown targeted analytes molecule was determined by counting the ratio of the number of wells containing both a bead and fluorescent product relative to the total number of wells containing beads.

8.3 PULL-DOWN ASSAY

Pull-down assay is mainly used to determine the physical interaction between protein-protein or protein-ligand interaction. This method involves the use of binding partner proteins or other ligand molecules for direct examining the affinity extent between two purified proteins or identifying novel interacting partners. The basic principle of pull-down assay was to identify the component of the complex. For this firstly to isolate the protein complexes and immobilization of affinity tag protein such as His, biotin and GST labeled tag as a bait protein on glutathione affinity resin. Capture protein (prey protein) interacts and binds with the specific bait protein and pulls down after cell lysate flows through. Unbound proteins are removed by subsequently washing. The method of elution depends upon the affinity between bait and prey protein (Louche, Salcedo, & Bigot, 2017).

8.3.1 BASIC COMPONENTS NEEDED FOR PULL DOWN ASSAY

8.3.1.1 Binding Ligands and Fusion Tags

Bait protein can be developed by linking an affinity tag to a protein or either by expressing recombinant fusion protein. The purified proteins are tags with affinity tags commonly used for labeling whereas if the gene is cloned then a molecular approach is used to clone the gene of our interest in a selective construct along with a fusion protein expression in the expression vector. Purified bait protein, which is further useful for pull-down assay (Table 8.5).

8.3.1.2 Binding Parameter

Studying protein–protein interaction mainly depends upon the binding affinity developed between bait and prey protein. It can be of two types: stable and transient interactions. Stable interaction is easy to isolate by physical method for further use pull-down assay. Whereas transient interactions are temporal interactions mostly found in enzymatic mechanisms, are challenging to isolate due to their interaction with other proteins and often require cofactors and energy source for hydrolysis.

TABLE 8.5
Commonly Used Fusion Tag with Affinity Binding Ligands

Fusion Tag	Affinity Ligand
Glutathione S-transferase (GST)	Glutathione
Poly-histidine (polyHis or 6xHis)	Nickel or cobalt chelate complexes
Biotin	Streptavidin

Source: https://www.thermofisher.com/in/en/home/life-science/protein-biology/protein-biology-learning-center/protein-biology-resource-library/pierce-protein-methods/pull-down-assays.html.

8.3.1.3   Elution of Protein Complexes

The entire complex of bait-prey protein elution from affinity support was achieved by using sodium dodecyl sulfate–polyacrylamide gel electrophoresis (SDS-PAGE) or competitive elution. Sodium dodecyl sulfate (SDS) an anionic detergent may interfere with analysis, so recommended to avoid denaturing agent as it distorts the structure of the protein complex. The competitive elution method was a non-denaturing agent, suitable for specific bait-prey interaction. Thus, it can elute biologically active molecules which could be useful for further study. Alternatively, the stepwise gradient elution method employed by increasing salt concentration and decreasing pH, which allowed selective elution of prey protein without disturbing bait protein, remained immobilized to the supporting matrix.

8.3.1.4   Detection of Bait-Prey Complex

Finally, the obtained eluted protein complex can be determined by using associated methods such as SDS, gel staining, western blotting and S-35 radio isotopic detection. The final determination of protein-protein interaction was carried out by digestion of peptide, and detection of the fragmented peptide by mass spectroscopy.

8.3.2   ADVANTAGES

- Screening and identification of novel protein partners
- Confirmation of protein–protein interaction discovered by previously existing methods
- Additional quantitative analysis about the amount of protein interacting to an affinity bead

8.4   SIGNIFICANCE OF BIOMOLECULE INTERACTION

Living cells are made up of diverse biomolecules that process life in a systematic order. Biomolecule functions by changing the structure or either establishing the physical interaction between the molecules. These interactions are important for enzymatic activity, as the enzyme fitted to the specific target molecule carries out the catalysis of enzymatic reaction. Biomolecule interaction enables a single protein to affect various cellular behavior of connected partners through a cell-signaling cascade. This interaction plays a fundamental role in determining multiple disease states.

Biologist tends to use functional biomolecular machines such as X-rays, diffraction crystallography, and nuclear magnetic resonance, to study the molecular interactions in protein complexes. But they were limited by the analyte complex size. Gold standard method to study large protein complex with respect to size and interaction is co-immuno-precpitation with western –blotting. However, the limitation of Co-IP and western blotting motivates a researcher to design the new easiest method for studying protein-protein interactions. SiMPull is the best alternative valuable tool to characterize single protein molecule interaction with high selectivity and specificity.

8.4.1   Single Molecule Pull Down Assay (SiMPull Assay)

Analysis of protein interaction is important to understand cell function and its regulation. This can be achieved by pull-down assay. Molecular architecture of protein assemblies can be understood by single-molecule experiment, which is a combination of two techniques -classical pull-down assay and single-molecule fluorescence microscopy, that enables direct visualization of a cellular protein complex (Jain et al., 2011). The assay required an internal total internal reflection fluorescence. Firstly, antibodies are immobilized on a polymer passivated flow chamber against bait protein. When cell extract is added to the chamber, antibodies capture this bait protein present in the complex with protein or nucleic acid. Other cellular components or unbound proteins were removed after washing. The composition of the protein complex is determined by using multicolor fluorescence imaging with single fluorophore sensitivity (Hoskins et al., 2011; Jain et al., 2011; Yeom et al., 2011).

8.4.2   Concepts Behind SiMpull

In the SiMpull assay firstly antibody against known protein is immobilized on the surface of the microscopic slide. Microscopic slides are passivated with biotin-PEG, which prevents stickiness of protein and increase the specificity of the assay. Streptavidin is used as an affinity ligand to attach a biotin-labeled antibody. This PEG-biotin-coated microscopic slide is placed in a multi-flow chamber. After successful assembly of the flow chamber, streptavidin is added which performs binding sites for biotin-labeled antibodies. Cell lysate is prepared after antibody immobilization, to carry out pull-down assay for specific known bait protein which brings to study all protein–protein interaction forming protein complex. Prey protein visualized by immunofluorescence labeling technique (Aggarwal & Ha, 2014). SiMpull is a revolutionary technique in two different ways, firstly it allows the detection of each protein complex along with the quantification of protein interaction associated with biomolecule protein complex, secondly for studying the protein dynamics by monitoring the fluorescence intensity (Aggarwal & Ha, 2014).

8.4.3   Steps Involved During Preparation

8.4.3.1   Preparation of Flow Chamber

Components required for the construction of the flow chamber are quartz microscope slide, passivated with methoxy polyethylene glycol PEG (prevent non-specific adsorption of cell extract) side facing upward and cover slip for rapid exchange of contents. Edges were sealed with epoxy. The slide holes are used for flowing solutions introduced with the help of pipette. Evanescent excitation field created using the prism on top of the slide at the quartz-aqueous buffer interface.

8.4.3.2   Sample Preparation

SiMPull assay preparation required using purified protein, tissue and cell lysate. Sample extracts were prepared by using lysis cell cultured or tissue with detergent. Try to use non-denaturing detergent in order to preserve the psychological structure and interaction of protein. Avoid the use of SDS or any other ionic detergent for lysis which may hinder, distort the original structure of the protein complex.

8.4.3.3 Antibody Immobilization

Antibody immobilization was achieved for the selective capture of bait protein. Doped with biotinylated labeled antibodies against bait protein, immobilized on the surface at a concentration of 10–20nM. Biotinylated secondary antibodies recruit the primary antibody against bait protein. When cell extracts are fused in the chamber, labeled antibodies capture the bait protein along with the interacting partner.

8.4.3.4 Single Molecule Pull Down Assay

To obtain a single molecule image, the required quantity of interested protein was immobilized at sufficiently low density, so that the bait protein got detached from the slide surface. Dilution of cell lysates are prepared to obtain low-density cells followed by incubation of lysate protein with an antibody-coated surface. Note the density of coated antibody should be 100-fold higher than immobilize molecules so that, the protein capture is not limited by the availability of binding sites

8.4.3.5 Fluorophore Labeling

Two approaches have been used to visualize prey protein either through fluorescent tag protein or immunofluorescence labeling. Firstly, the protein of our interest can be expressed as a fusion protein in an expression construct and imaging visualized under total internal reflection fluorescence microscopy with single-molecule sensitivity. This approach is mainly used to determine the stoichiometry of protein complexes (Reyes-Lamothe, Sherratt, & Leake, 2010). Yellow fluorescent protein (YFP), Green fluorescent protein (GFP) and RFP variants are recommended probes used to tag protein.

Alternatively, the immunofluorescence labeling technique was used to detect the prey protein. This approach involves the use of antibody against the prey protein and the corresponding labeled secondary antibody with fluorophore. Care must be taken while using multiple antibodies to avoid cross-reactivity that leads to false positives.

8.4.4   Advantages of SiMPull Assay

- Discriminate protein complexes.
- Provide quantitative data on protein.
- Determine stoichiometry of the complexes using photobleaching.
- Study functional activity of protein complexes.
- Improvement in terms of cost, time and sensitivity over the conventional western blot.

8.4.5   Limitations

- This method does not resolve protein biomolecules based on sizes
- Relies solely on fluorophore detection
- Necessary requirement of appropriate control for interpretation of data

- SiMPull assay is applicable only when binding partner parameter is known
- Dilution of surfactant is required to obtain a low concentration of protein for a single-cell imaging.
- Weak interaction with dissociation constant $>0.05\,s^{-1}$ may not be accessible by this method

8.4.6 APPLICATIONS OF SiMPULL ASSAY

Applicability of SiMPull to examine protein complexes using different capture and detection configuration from various cellular environments.

8.4.6.1 Receptor Pull Down

It is difficult to analyze membrane protein complexes with conventional methods [26], (Schwarzenbacher et al., 2008). SiMPull assay was applied to study the b2-adrenergic receptor (b2AR). This b2-adrenergic receptor tag with fusion protein YFP resulted in Flag–YFP–b2AR transfected with HEK293 cells and membrane proteins were solubilized. Pull down the receptor b2AR using antibodies against YFP or flag.

8.4.6.2 Mitochondrial Protein Pull Down

Mitochondrial outer membrane protein examined by SiMPull technique. Isolated mitochondrial fraction from cell overexpressed in YFP-MAVS construct [28]. Pull-down mitochondrial antiviral signaling (MAVS) proteins using anti-YFP and observed fluorescent structure in insolubilized preparations are due to immune-precipitated mitochondrial patches indicating the presence of several YFP molecules. On pre-solubilized the mitochondrial outer membrane protein using mild detergent, shown single YFP spot to which 86% indicates the presence of monomeric state of solubilized MAVS after single-step photobleaching.

8.4.6.3 Pull Down of Endogenous Complex in Native Tissues

Pull down of endogenous complex although desirable, is challenging owing to lack of high-affinity antibodies, low abundance and high interfered background interaction with other cellular proteins. So the SiMPull assay was used to study the interaction between proteins. The antibody against the bait protein i.e. protein kinase (PKA) immobilized on a surface using biotin labeled secondary antibody. Probing was done for prey protein (AKAP-150) using a primary antibody and fluorescent labeled secondary antibody resulting in a 10-fold increase in a fluorescent spot in the protein kinase antibody-associated channel in comparison to the control one. Due to its high sensitivity, the required 20-fold lower sample volume allowed the detection of PKA-AKAP binding from mouse heart tissue under below detection limit as compared to conventional western blot.

8.4.6.4 Single Molecule Localization and Tracking

Single molecules have potential applications with respect to studies of molecular tracking, molecule interaction and dynamics, and super-resolution imaging. Additionally, single molecules have the capability of visualizing protein complexes on the image surface by pulling down the protein from the cell lysate. This technique provides fast avenue quantification to analyze protein-protein interaction, stoichiometry and the activity of biomolecular complexes. Strategy involved the use of a fabricated flow chamber (1 0µm height) where cells were immobilized on one side of the inner surface, however antibodies against the specific protein target coated on the other inner side of the chamber. Targeted protein molecules released upon cell lysate were arrested by the antibody surface. This design uses a spacer of 10µm which significantly reduces capture radius and allows to increase the protein-antibody interaction within which around 70% of targeted protein can capture to below 30µm. The developed platform does not require microfabrication, and provides high compatibility through protein analysis at cell-level single-molecule imaging based on total internal reflection microscopy.

8.5 SIMPULL AS BIOSENSOR

SiMPull work like a biosensor because biosensor can detect different analyst such as DNA (Teles & Fonseca, 2008), protein (Anish Kumar, Jung, & Ji, 2011), Lipids (Wills, Goulden, & Hammond, 2018), Sugars (Yoo & Lee, 2010), heavy metals (Stocker et al., 2003), etc. based on physiochemical signal developed between them. The specificity of analytes is increased by using special interactions such as antibodies, enzymes, RNA etc. Similar way SiMPull is an analytical method used to detect biomolecules based on fluorescence.

8.5.1 SiMPull in Contrast to ELISA

As discussed previously enzyme-linked immunosorbent assay was performed to determine the antigen or antibody concertation present in the sample. It is the oldest technique based on the affinity binding between two molecules (antigen or antibody), detected by a linked enzyme on either side of the binding partners (Eva Engvall & Peter Perlmann, 1971; Van Weemen & Schuurs, 1971). It becomes a very important tool in various pharmaceutical, clinical diagnostic and food industry. The detection limits depend upon the test molecules (Zhang, Garcia-D'Angeli, Brennan, & Huo, 2014). SiMpull has very satisfactory detection limits of a single molecule in picomolar and has additional advantages to perform biochemical measurements of the analytes, which is absent in case of ELISA. Therefore recent efforts made to develop digital ELISA for single molecule (Chen, Svedendahl, Van Duyne, & Käll, 2011; Eyer, Stratz, Kuhn, Kuster, & Dittrich, 2013; Rissin et al., 2010).

8.5.2 SiMPull in Contrast to Surface Plasmon Resonance (SPR)

Surface Plasmon resonance (SPR) is a powerful technique to determine biomolecular interactions, biochemical detection and immunoassay (Homola, 2003; Jason-Moller, Murphy, & Bruno, 2006). Surface Plasmon resonance (SPR) works on the principle of excitation of electromagnetic waves. The ultimate performance of the SPR-biosensor is in that one detects the change in refractive index upon analyte binding to the surface (Abbas, Linman, & Cheng, 2011; Boozer, Kim, Cong, Guan, & Londergan, 2006). This makes the SPR a label-free optical biosensor detection method (Campbell &

TABLE 8.6
Comparison of SiMPull, ELISA Detection Parameter

Methods	Specificity (%)	Sensitivity	LOD	Biocompatibility	Time
SiMPull	90–95	pM-nM (Berggård, Linse, & James, 2007)	Detect single protein and complexes	Proteins from cultured cell, nucleic acid, lipids	Real-time
ELISA	95–98	<fM-pM (Engvall & Perlmann, 1971)	<pg-few ng/mL	Proteins from antibodies, serum, blood	2–3 hours

Kim, 2007). SPR employ the use of total internal reflection to achieve high selectivity by exciting surface plasmon at a definite angle. Substrate immobilization to which analyte binds is conducted by absorption on the gold surface, where in case SiMpull assay polyethyalglycol (PEG) and streptavidin-biotin is used for antibody immobilization (Dostálek & Knoll, 2008; Jain et al., 2011) (Table 8.6).

SiMpull assay biomolecule detection is fluorescence-based. It provides advantages to determining biochemical and functional measurement on analyte by measuring fluorescence intensity (Dostálek & Knoll, 2008; Mayer, Hao, Lee, Nordlander, & Hafner, 2010; P. Zijlstra, Paulo, & Orrit, 2012). In addition, reliance on fluorescence for identification enables SiMPull to detect analyses directly from contaminated extracts by filtering nonspecific binding from various biomolecules present in the extraction. As a non-labeled method, SPR cannot distinguish confidently between certain binding events of the analyte and other biomolecules obtained from the extraction of contaminated cells or serum (Jason-Moller et al., 2006; Kyo, Usui-Aoki, & Koga, 2005; Aubé, Breault-Turcot, Chaurand, Pelletier, & Masson, 2013). Researchers have developed fluorescence SPR, called SPFS (surface- Plasmon-enhanced fluorescence spectroscopy) (Dostálek & Knoll, 2008; Yu, Persson, Löfås, & Knoll, 2004), but suffered from multiple complications and did not extend the use of serum and cell extracts. A major obstacle is the high level of indirect adsorption of lipids and other biomolecules from crude serum to the gold field of SPR (Aubé et al., 2013). SiMPull solves this hurdle by performing glass passivation with polyethylene glycol (PEG) polymer, which improves the clarity of analyte scanning.

8.6 ALPHA-SYNUCLEIN (α-SYN) PROTEIN AND OLIGOMER DETECTION

Alpha-synuclein (α-SYN) is a presynaptic neuron protein encoded by the SNCA gene. This human protein is made up of 120 amino acids. This protein is associated with Parkinson's disease pathogenesis. The characteristics of alpha-synuclein protein and its aggregation state from human brain tissue are mainly studied by using recombinant protein (Chen et al., 2015; Tosatto et al., 2015; Tuttle et al., 2016; Weinreb, Zhen, Poon, Conway, & Lansbury, 1996; N. Zijlstra, Blum, Segers-Nolten,

Claessens, & Subramaniam, 2012). However, the analysis of protein has been challenging, therefore the pull-down approach has been used to quantify alpha-synuclein and its aggregation state. For this experiment mainly four specific antibodies were required such as biotinylated labeled secondary antibody, primary monoclonal antibodies (capturing and detection antibody) recognized different specific epitopes of α-SYN, and Alexa 647-labeled secondary antibody. α-SYN were detected by single molecule fluorescence microscopy.

Similarly, alpha-synuclein oligomerization strongly affects the detection of α-SYN. So it is important to understand the oligomerization state associated with α-SYN for better diagnosis and progression of Parkinson disease. For this firstly, Alexa 647-labeled secondary fragmented antibodies were used instead of full-length IgG to avoid steric hindrances occurring between two antibodies. Then a single molecule pull-down assay was performed for oligomer alpha-SYN detection. Fluorescence intensity of molecule, proportional to the number of α-SYN molecules in single pull-down assay.

8.7   CONCLUSION AND FUTURE PROSPECT

Biomolecule functions have been studied by the combination of molecular and genetic approaches and action of mechanism is empowered by biochemical activities. A set of techniques involved isolation, purification and precipitation of biomolecules out from their cellular environment. However, the isolation of various sets of intrinsic protein assembly is yet extremely challenging because of their stability inside the cell. SiMpull technique builds up as a bridge to study these complexes by eliminating the purification requirements. It provides all stoichiometric and dynamic information to be studied for any biologically complex in its environment. Involvement of various factors such as antibody immobilization, surface passivation tools, fluoroscencent ligands for labeling bound to strengthen the technique. Additionally, in this chapter, we discussed Single molecule pull-down analysis and various ELISA tools for quantification of antigen-antibody complex and its potential applications in widespread areas.

REFERENCES

Abbas, A., Linman, M. J., & Cheng, Q. (2011). New trends in instrumental design for surface plasmon resonance-based biosensors. *Biosensors and Bioelectronics, 26*(5), 1815–1824.

Adekanmbi, A. J., & Olude, M. A. (2021). An alternate prospect in detecting presymptomatic and asymptomatic COVID-19 carriers through odor differentiation by HeroRATs. *Journal of Veterinary Behavior: Clinical Applications and Research: Official Journal of Australian Veterinary Behaviour Interest Group, International Working Dog Breeding Association, 42*, 26–29. doi: 10.1016/j.jveb.2020.12.001.

Aggarwal, V., & Ha, T. (2014). Single-molecule pull-down (SiMPull) for new-age biochemistry: methodology and biochemical applications of single-molecule pull-down (SiMPull) for probing biomolecular interactions in crude cell extracts. *Bioessays, 36*(11), 1109–1119. doi: 10.1002/bies.201400090.

Anish Kumar, M., Jung, S., & Ji, T. (2011). Protein biosensors based on polymer nanowires, carbon nanotubes and zinc oxide nanorods. *Sensors (Basel, Switzerland), 11*(5), 5087–5111. doi: 10.3390/s110505087.

Asensio, L., Gonzalez Alonso, I., Garcia, T., & Martín, R. (2008). Determination of food authenticity by enzyme-linked immunosorbent assay (ELISA). *Food Control, 19*, 1–8. doi: 10.1016/j.foodcont.2007.02.010.

Aubé, A., Breault-Turcot, J., Chaurand, P., Pelletier, J. N., & Masson, J.-F. (2013). Non-specific adsorption of crude cell lysate on surface plasmon resonance sensors. *Langmuir, 29*(32), 10141–10148.

Aydin, S. (2015). A short history, principles, and types of ELISA, and our laboratory experience with peptide/protein analyses using ELISA. *Peptides, 72*, 4–15. doi: 10.1016/j.peptides.2015.04.012.

Berggård, T., Linse, S., & James, P. (2007). Methods for the detection and analysis of protein-protein interactions. *Proteomics, 7*(16), 2833–2842. doi: 10.1002/pmic.200700131.

Boozer, C., Kim, G., Cong, S., Guan, H., & Londergan, T. (2006). Looking towards label-free biomolecular interaction analysis in a high-throughput format: A review of new surface plasmon resonance technologies. *Current Opinion in Biotechnology, 17*(4), 400–405.

Campbell, C. T., & Kim, G. (2007). SPR microscopy and its applications to high-throughput analyses of biomolecular binding events and their kinetics. *Biomaterials, 28*(15), 2380–2392.

Chard, T. (1992). Pregnancy tests: A review. *Human Reproduction, 7*(5), 701–710. doi: 10.1093/oxfordjournals.humrep.a137722.

Chen, S., Svedendahl, M., Van Duyne, R. P., & Käll, M. (2011). Plasmon-enhanced colorimetric ELISA with single molecule sensitivity. *Nano Letters, 11*(4), 1826–1830.

Chen, S. W., Drakulic, S., Deas, E., Ouberai, M., Aprile, F. A., Arranz, R., ... Cremades, N. (2015). Structural characterization of toxic oligomers that are kinetically trapped during α-synuclein fibril formation. *Proceedings of the National Academy of Sciences of the United States of America, 112*(16), E1994. doi: 10.1073/pnas.1421204112.

Clark, M. F., Lister, R. M., & Bar-Joseph, M. (1986). ELISA techniques. *Methods in Enzymology* (Vol. 118, pp. 742–766): Academic Press, Cambridge, MA.

Dostálek, J., & Knoll, W. (2008). Biosensors based on surface plasmon-enhanced fluorescence spectroscopy. *Biointerphases, 3*(3), FD12–FD22.

Engvall, E., & Perlmann, P. (1971). Enzyme-linked immunosorbent assay (ELISA). quantitative assay of immunoglobulin G. *Immunochemistry, 8*(9), 871–874. doi: 10.1016/0019-2791(71) 90454-x.

Eyer, K., Stratz, S., Kuhn, P., Kuster, S., & Dittrich, P. S. (2013). Implementing enzyme-linked immunosorbent assays on a microfluidic chip to quantify intracellular molecules in single cells. *Analytical Chemistry, 85*(6), 3280–3287.

Fraaij, P. L., Rakhmanina, N., Burger, D. M., & de Groot, R. (2004). Therapeutic drug monitoring in children with HIV/AIDS. *Therapeutic Drug Monitoring, 26*(2), 122–126. doi: 10.1097/00007691-200404000-00006.

Frankfurt, O. S., & Krishan, A. (2001). Enzyme-linked immunosorbent assay (ELISA) for the specific detection of apoptotic cells and its application to rapid drug screening. *The Journal of Immunological Methods, 253*(1–2), 133–144. doi: 10.1016/s0022-1759(01)00387-8.

Gong, F., Wei, H., Li, Q., Liu, L., & Li, B. (2021). Evaluation and comparison of serological methods for COVID-19 diagnosis. *Frontiers in Molecular Biosciences, 8*. doi: 10.3389/fmolb.2021.682405.

Gonzalez, R. M., Seurynck-Servoss, S. L., Crowley, S. A., Brown, M., Omenn, G. S., Hayes, D. F., & Zangar, R. C. (2008). Development and validation of sandwich ELISA microarrays with minimal assay interference. *Journal of Proteome Research, 7*(6), 2406–2414. doi: 10.1021/pr700822t.

Hinckley, A. F., Connally, N. P., Meek, J. I., Johnson, B. J., Kemperman, M. M., Feldman, K. A., ... Mead, P. S. (2014). Lyme disease testing by large commercial laboratories in the United States. *Clinical Infectious Diseases, 59*(5), 676–681. doi: 10.1093/cid/ciu397.

Hock, B. D., Stamp, L. K., Hayman, M. W., Keating, P. E., Helms, E. T., & Barclay, M. L. (2016). Development of an ELISA-based competitive binding assay for the analysis of drug concentration and antidrug antibody levels in patients receiving adalimumab or infliximab. *Therapeutic Drug Monitoring, 38*(1), 32–41. doi: 10.1097/ftd.0000000000000229.

Homola, J. (2003). Present and future of surface plasmon resonance biosensors. *Analytical and Bioanalytical Chemistry, 377*(3), 528–539.

Hoskins, A. A., Friedman, L. J., Gallagher, S. S., Crawford, D. J., Anderson, E. G., Wombacher, R., ... Moore, M. J. (2011). Ordered and dynamic assembly of single spliceosomes. *Science (New York, N.Y.), 331*(6022), 1289–1295. doi: 10.1126/science.1198830.

Huang, Z., Olson, N. A., You, W., & Haugland, R. P. (1992). A sensitive competitive ELISA for 2,4-dinitrophenol using 3,6-fluorescein diphosphate as a fluorogenic substrate. *Journal of Immunological Methods, 149*(2), 261–266. doi: 10.1016/0022-1759(92)90258-U.

Hunsperger, E. A., Yoksan, S., Buchy, P., Nguyen, V. C., Sekaran, S. D., Enria, D. A., ... Margolis, H. S. (2014). Evaluation of commercially available diagnostic tests for the detection of dengue virus NS1 antigen and anti-dengue virus IgM antibody. *PLoS Neglected Tropical Diseases, 8*(10), e3171. doi: 10.1371/journal.pntd.0003171.

Iwamoto, N., Yokoyama, K., Takanashi, M., Yonezawa, A., Matsubara, K., & Shimada, T. (2018). Verification between original and biosimilar therapeutic antibody infliximab using nSMOL Coupled LC-MS bioanalysis in human serum. *Current Pharmaceutical Biotechnology, 19*(6), 495–505. doi: 10.2174/1389201019666180703093517.

Jain, A., Liu, R., Ramani, B., Arauz, E., Ishitsuka, Y., Ragunathan, K., ... Ha, T. (2011). Probing cellular protein complexes using single-molecule pull-down. *Nature, 473*(7348), 484–488. doi: 10.1038/nature10016.

Jason-Moller, L., Murphy, M., & Bruno, J. (2006). Overview of Biacore systems and their applications. *Current Protocols in Protein Science, 45*(1), 19.13. 11–19.13. 14.

Kyo, M., Usui-Aoki, K., & Koga, H. (2005). Label-free detection of proteins in crude cell lysate with antibody arrays by a surface plasmon resonance imaging technique. *Analytical Chemistry, 77*(22), 7115–7121.

Lequin, R. M. (2005). Enzyme Immunoassay (EIA)/Enzyme-Linked Immunosorbent Assay (ELISA). *Clinical Chemistry, 51*(12), 2415–2418. doi: 10.1373/clinchem.2005.051532.

Lin, A. V. (2015). Indirect ELISA. *Methods in Molecular Biology, 1318*, 51–59. doi: 10.1007/978-1-4939-2742-5_5.

Louche, A., Salcedo, S. P., & Bigot, S. (2017). Protein-protein interactions: Pull-down assays. *Methods in Molecular Biology, 1615*, 247–255. doi: 10.1007/978-1-4939-7033-9_20.

Madersbacher, S., & Berger, P. (1991). Double wavelength measurement of 3,3′, 5,5′-tetramethylbenzidine (TMB) provides a three-fold enhancement of the ELISA measuring range. *Journal of Immunological Methods, 138*(1), 121–124. doi: 10.1016/0022-1759(91)90071-M.

Matsuda, H., Tanaka, H., Blas, B., Nosenas, J., Tokawa, T., & Ohsawa, S. (1984). Evaluation of ELISA with ABTS, 2-2'-azino-di-(3-ethylbenzthiazoline sulfonic acid), as the substrate of peroxidase and its application to the diagnosis of schistosomiasis. *The Japanese Journal of Experimental Medicine, 54*(3), 131–138.

Mayer, K. M., Hao, F., Lee, S., Nordlander, P., & Hafner, J. H. (2010). A single molecule immunoassay by localized surface plasmon resonance. *Nanotechnology, 21*(25), 255503.

Nandi, S., Maity, S., Bhunia, S. C., & Saha, M. K. (2014). Comparative assessment of commercial ELISA kits for detection of HIV in India. *BMC Research Notes, 7*(1), 436. doi: 10.1186/1756-0500-7-436.

Nayak, B., Li, Z., Ahmed, I., & Lin, H. (2017). Chapter 11- Removal of allergens in some food products using ultrasound. In D. Bermudez-Aguirre (Ed.), *Ultrasound: Advances for Food Processing and Preservation* (pp. 267–292): Academic Press, Cambridge, MA.

Peng, J., Meng, X., Deng, X., Zhu, J., Kuang, H., & Xu, C. (2012). Development of a monoclonal antibody-based sandwich ELISA for the detection of ovalbumin in foods. *Food and Agricultural Immunology, 25*, 1–8. doi: 10.1080/09540105.2012.716398.

Pizza, M., Scarlato, V., Masignani, V., Giuliani, M. M., Aricò, B., Comanducci, M., ... Rappuoli, R. (2000). Identification of vaccine candidates against serogroup B meningococcus by whole-genome sequencing. *Science (New York, N.Y.), 287*(5459), 1816–1820. doi: 10.1126/science.287.5459.1816.

Reyes-Lamothe, R., Sherratt, D. J., & Leake, M. C. (2010). Stoichiometry and architecture of active DNA replication machinery in Escherichia coli. *Science (New York, N.Y.), 328*(-5977), 498–501. doi: 10.1126/science.1185757.

Rissin, D. M., Kan, C. W., Campbell, T. G., Howes, S. C., Fournier, D. R., Song, L., ... Rivnak, A. J. (2010). Single-molecule enzyme-linked immunosorbent assay detects serum proteins at subfemtomolar concentrations. *Nature Biotechnology, 28*(6), 595–599.

Schieffelin, J., Moses, L. M., Shaffer, J., Goba, A., & Grant, D. S. (2016). Clinical validation trial of a diagnostic for Ebola Zaire antigen detection: Design rationale and challenges to implementation. *Clinical Trials (London, England), 13*(1), 66–72. doi: 10.1177/1740774515621013.

Scholler, N., Crawford, M., Sato, A., Drescher, C. W., O'Briant, K. C., Kiviat, N., ... Urban, N. (2006). Bead-based ELISA for validation of ovarian cancer early detection markers. *Clinical Cancer Research, 12*(7 Pt 1), 2117–2124. doi: 10.1158/1078-0432.ccr-05-2007.

Schwarzenbacher, M., Kaltenbrunner, M., Brameshuber, M., Hesch, C., Paster, W., Weghuber, J., ... Schütz, G. J. (2008). Micropatterning for quantitative analysis of protein-protein interactions in living cells. *Nature Methods, 5*(12), 1053–1060. doi: 10.1038/nmeth.1268.

Skjerve, E., Bos, W., & van der Gaag, B. (1991). Evaluation of monoclonal antibodies to Listeria monocytogenes flagella by checkerboard ELISA and cluster analysis. *Journal of Immunological Methods, 144*(1), 11–17.

Stocker, J., Balluch, D., Gsell, M., Harms, H., Feliciano, J., Daunert, S., ... van der Meer, J. R. (2003). Development of a set of simple bacterial biosensors for quantitative and rapid measurements of arsenite and arsenate in potable water. *Environmental Science & Technology, 37*(20), 4743–4750. doi: 10.1021/es034258b.

Tarigan, S., Indriani, R., Durr, P. A., & Ignjatovic, J. (2015). Characterization of the M2e antibody response following highly pathogenic H5N1 avian influenza virus infection and reliability of M2e ELISA for identifying infected among vaccinated chickens. *Avian Pathology, 44*(4), 259–268. doi: 10.1080/03079457.2015.1042428.

Teles, F. R. R., & Fonseca, L. P. (2008). Trends in DNA biosensors. *Talanta, 77*(2), 606–623. doi: 10.1016/j.talanta.2008.07.024.

Tosatto, L., Horrocks, M. H., Dear, A. J., Knowles, T. P. J., Dalla Serra, M., Cremades, N., ... Klenerman, D. (2015). Single-molecule FRET studies on alpha-synuclein oligomerization of Parkinson's disease genetically related mutants. *Scientific Reports, 5*, 16696. http://europepmc.org/abstract/MED/26582456.

Tuttle, M. D., Comellas, G., Nieuwkoop, A. J., Covell, D. J., Berthold, D. A., Kloepper, K. D., ... Rienstra, C. M. (2016). Solid-state NMR structure of a pathogenic fibril of full-length human α-synuclein. *Nature Structural & Molecular Biology, 23*(5), 409–415. doi: 10.1038/nsmb.3194.

Van Weemen, B., & Schuurs, A. (1971). Immunoassay using antigen—enzyme conjugates. *FEBS Letters, 15*(3), 232–236.

Weinreb, P. H., Zhen, W., Poon, A. W., Conway, K. A., & Lansbury Jr., P. T. (1996). NACP, a protein implicated in Alzheimer's disease and learning, is natively unfolded. *Biochemistry, 35*(43), 13709–13715. doi: 10.1021/bi961799n.

Welch, R. J., Chang, G. J., & Litwin, C. M. (2014). Comparison of a commercial dengue IgM capture ELISA with dengue antigen focus reduction microneutralization test and the Centers for Disease Control dengue IgM capture-ELISA. *Journal of Virological Methods, 195*, 247–249. doi: 10.1016/j.jviromet.2013.10.019.

Wills, R. C., Goulden, B. D., & Hammond, G. R. V. (2018). Genetically encoded lipid biosensors. *Molecular Biology of the Cell, 29*(13), 1526–1532. doi: 10.1091/mbc.E17-12-0738.

Yeom, K.-H., Heo, I., Lee, J., Hohng, S., Kim, V. N., & Joo, C. (2011). Single-molecule approach to immunoprecipitated protein complexes: insights into miRNA uridylation. *EMBO Reports, 12*(7), 690–696. doi: 10.1038/embor.2011.100.

Yoo, E.-H., & Lee, S.-Y. (2010). Glucose biosensors: An overview of use in clinical practice. *Sensors (Basel, Switzerland), 10*(5), 4558–4576. doi: 10.3390/s100504558.

Yu, F., Persson, B., Löfås, S., & Knoll, W. (2004). Attomolar sensitivity in bioassays based on surface plasmon fluorescence spectroscopy. *Journal of the American Chemical Society, 126*(29), 8902–8903.

Zhang, S., Garcia-D'Angeli, A., Brennan, J. P., & Huo, Q. (2014). Predicting detection limits of enzyme-linked immunosorbent assay (ELISA) and bioanalytical techniques in general. *Analyst, 139*(2), 439–445.

Zijlstra, N., Blum, C., Segers-Nolten, I. M. J., Claessens, M. M. A. E., & Subramaniam, V. (2012). Molecular composition of sub-stoichiometrically labeled α-synuclein oligomers determined by single-molecule photobleaching. *Angewandte Chemie International Edition, 51*(35), 8821–8824. doi: 10.1002/anie.201200813.

Zijlstra, P., Paulo, P. M., & Orrit, M. (2012). Optical detection of single non-absorbing molecules using the surface plasmon resonance of a gold nanorod. *Nature Nanotechnology, 7*(6), 379–382.

9 Atomic Force Microscopy for Single Molecule Detection and Analysis

Kavitha Illath and Ashwini Shinde
Indian Institute of Technology

Moeto Nagai
Toyohashi University of Technology

Tuhin Subhra Santra
Indian Institute of Technology

CONTENTS

DOI: 10.1201/9781003409472-9

9.1 INTRODUCTION

The first single-molecule experiment was carried out 52 years ago; thereafter, tremendous developments evolved in this field (Schroeder, Hedrich, and Fernandez 1984). As already explained in this book in a couple of chapters, the single-molecule approach enables us to understand the basic biomolecular mechanisms such as DNA-RNA transcription, translation, rotary motor F1-ATPase, cytoskeletal motors behavior, etc. Due to the technological advancement in microscopy and crystallographic procedures, it can observe the structure of biomolecules at the atomic level. For example, cytoskeletal motors at atomic scale level structure under different conformations have been successfully viewed under X-ray crystallographic techniques using crystals of motors' catalytic domains (Veigel and Schmidt 2011). With electron microscopy, it is able to image a single cytoskeletal motor molecule that is fixed or frozen under specific physiological conditions. However, electron microscopy is a static technique, which hinders the study of live cells, dynamics of intramolecular motion, folding, unfolding, etc. (Jadavi et al. 2021). Hence, there is a need for dynamic and single-molecule studying techniques and evolved techniques such as atomic force microscopy (AFM), optical tweezers and single-molecule fluorescence microscopy (Veigel and Schmidt 2011).

In 1986, the first atomic force microscopy (AFM) was presented which changed the complete scenario in the field of cell biology to molecular resolution (Binnig, Quate, and Gerber 1986). In general, AFM is a mechanical microscope that visualizes the topography of sample surface (Katan and Dekker 2011). It reconstructs the 3-D morphology of the sample placed on a flat surface by recording the force curves varying with the distance between the sample and the sharp tip of the cantilever probe (Ruggeri et al. 2019). It was first developed in 1986, and shortly thereafter, its potential in studying the structure of biological samples at the micro and nanoscale level has been implemented (Fang and Hu 2021; Schön 2018). The widespread usage of AFM in biology was initiated after the invention of the optical lever detection principle, which is considered a breakthrough event in AFM design (Schön 2018). It allowed the imaging of moving biomolecules under buffer solutions with a prominent signal to noise ratio, which is an essential requirement to maintain the actual physiological environment (Müller and Dufrêne 2008; Schön 2018). Further, it paved the way in analyzing the biological samples without the usage of any labelling or staining (Schön 2018; Veigel and Schmidt 2011). It totally replaced the conventional X-ray crystallography and optical crystallographic techniques (Müller and Dufrêne 2008). Thereafter, AFM has been treated as a unique probing method that provides imaging and force probing of biological samples under the force resolution of piconewton (pN) or topographical distance of nanometer (nm) (Schön 2018). It provides imaging (micrograph is the more technical terminology than image) of biological samples by working under various modes to characterize and manipulate samples under the 'nm' scale (Müller and Dufrêne 2008). Originally AFM used as an imaging tool, and it has been exploited as single-molecule force spectroscopy, single-cell force spectroscopy, molecular recognition microscopy, etc. (Müller and Dufrêne 2008).

This chapter focuses on various AFM techniques used in single-molecule studies. Different AFM approaches and their working principles are explained in detail with explanatory schematics, followed by the requirements for single-molecule AFM imaging. The following section describes the illustrative examples in single-molecule imaging and force kinetics using AFM. Later, it discusses the limitations associated with the current AFM-based single-molecule study and provides summary and future prospects.

9.2 AFM WORKING PRINCIPLE AND IMAGING MODES

AFM works on the principle of force interaction between the sample and a sharp tip while scanning the sample (Demir-Yilmaz, Guiraud, and Formosa-Dague 2021). Figure 9.1a shows the schematic of the AFM setup that works under multimode. It essentially consists of a soft mechanical lever arm (cantilever) whose one end is fixed, and the other end has a sharp tip (or probe) to sample the height profile of an immobilized surface kept on a substrate (Li et al. 2019). Mica, glass, gold, and highly ordered pyrolytic graphite are the commonly used substrate materials (Ruggeri et al. 2019). The piezoelectric actuator controls the probe tip movement to maintain the required interaction force between tip atoms and sample surface atoms (Li et al. 2019). The sample to be imaged should be placed on a substrate attached to a piezoelectric tube scanner that can move in 3-D space with sub-nm precision (Hodel and Hammond 2021; Ruggeri et al. 2019). For imaging, as the cantilever perform raster scans over the sample, due to height variation of the sample surface, the cantilever tip bend or deflects proportionally to the occurring force (Figure 9.1b). Laser gets reflected upon tip deflection and recorded by the photodiode (sensitive to position) (Demir-Yilmaz, Guiraud, and Formosa-Dague 2021; Hodel and Hammond 2021; Ruggeri et al. 2019). It records the tip-sample interaction force and distance (deflection) over time whenever the tip engages and retract from the sample, as shown in Figure 9.1c (Hodel and Hammond 2021). The user predefines the scanning surface in raster pixel for imaging purposes. An example of a raster scanning pattern is schematically drawn in Figure 9.1d (Hodel and Hammond 2021). Each pixel stores the recorded force-distance value, and once the scanning is performed, it reconstructs the sample surface into its 3-D representation, which is illustrated as a 2-D map (XY) corresponding to each sample height (Z) (Ruggeri et al. 2019). The raster scan performs line-by-line, and each line is scanned twice in the opposite direction. Hence, it generates separate images for trace and retraces that are useful in analyzing images for artefact (Hodel and Hammond 2021).

Every image is recorded based on the setpoint value defined by the user for sample-tip interaction. This setpoint can be provided in 'Volts' in terms of relative deflection value or spring constant or deflection sensitivity if it is already known as an absolute force value (Hodel and Hammond 2021). An important parameter that dictates the image resolution is the choice of probe and its features (Hodel and Hammond 2021; Veigel and Schmidt 2011). Some probes can only be operated with a specific machine and/or specific imaging mode. The bending elastic constant of the cantilever varies from 10 to 10,000 $pNnm^{-1}$, and the common material choice is silicon nitride cantilever with silicon tips (Hodel and Hammond 2021; Veigel and Schmidt 2011). Each material operates on different resonance frequency (9–45 kHz) and possess various

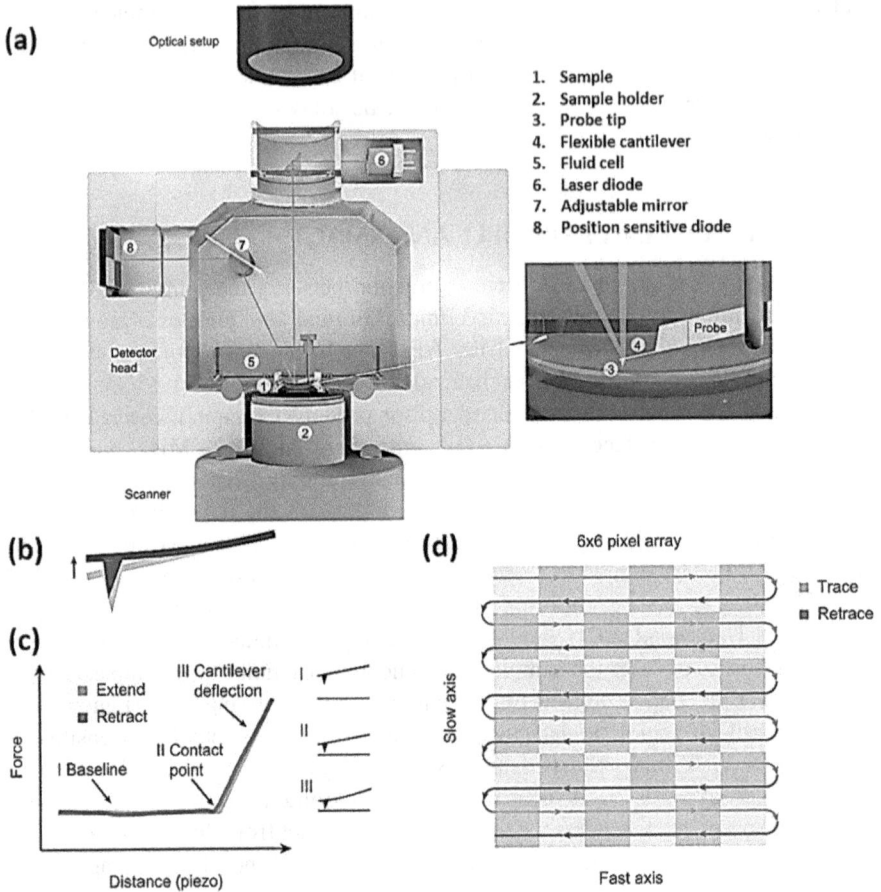

FIGURE 9.1 AFM working principle; (a) schematic drawing of AFM machine with multimode setup. Its main parts are a scanner, detector head and optical setup. Sample (1) is placed on a sample holder (2). A probe tip (3) extends from a flexible vibrating cantilever (4) above sample is placed via Fluid cell (5). The laser emitted from the diode (6) focuses on using an adjustable mirror and detecting using position-sensitive diode (8). (b) By Hooke's law, cantilever tip deflection 'x' is proportional to the force exerted on the probe tip. (c) force-distance curve recorded by AFM when the tip moves towards the sample surface. (d) Raster scanning performed line by line on 6×6 pixel array (Reprinted with permission from Hodel and Hammond 2021.)

tip radius (1–50 nm) (Hodel and Hammond 2021; Ruggeri et al. 2019). The probe can be either pyramidal or conical shaped (Ruggeri et al. 2019). For specific biological applications, the probe materials vary. For example, the probe used in the protein pore formation study includes a micro sharp nitride lever (MSNL-E, MSNL-F) and AC-40 (Hodel and Hammond 2021). The cantilever possesses various shapes, with triangular or rectangular being the most common having 10–200 μm in length (Ruggeri et al. 2019).

Conventional AFM captures static images of biomolecules, and it takes approximately 1 minute to capture only a single image. However, acquisition time has been improved over the past decade, and it is able to capture ten images per second (Katan and Dekker 2011). For bio applications, tips used for AFM should be functionalized with the inert molecules, biomolecules or cells, which will be discussed in detail in later sections.

Apart from imaging, AFM is a powerful tool in studying the forces; it is able to record the forces down to 20 pN. It records the force-distance curve generated by the interaction between the cantilever tip and sample, plot the force generated in the probe as a function of distance or separation between sample and probe (Demir-Yilmaz, Guiraud, and Formosa-Dague 2021). Depending on the origin of the force that develops between tip and sample, it can be divided into two, repulsive and attractive interaction. Repulsive short-range (Pauli repulsion) interaction comes if there are no external stimuli such as electrical or magnetic field, while at a distance of above a few nm, attractive force (Van der walls and capillary) dominates. Further, adhesive and viscoelastic interactions may also be present in both cases (Ruggeri et al. 2019). In the case of cells, these curves can be interpreted with the help of various physical models to arrive at the nanomechanical and nano adhesive features of cell surfaces (Demir-Yilmaz, Guiraud, and Formosa-Dague 2021). Hence, AFM can be used to study the details of biological processes involving cells and molecules as it has the capabilities in analyzing mechanical properties (Katan and Dekker 2011).

Nowadays, every AFM machine can operate under multimode. The basic four modes for imaging include contact, tapping, non-contact, and force-probe or AFM force spectroscopy mode. The mode can be selected in advance to define the probe movement during the scan. Apart from these modes, AFM can be extended into multiparametric, multifrequency, high-speed and topography and recognition imaging mode (TREC). Contact mode is static, whereas tapping and non-contact mode are dynamic. They are categorized based on the interaction force regime they work as illustrated graphically in Figure 9.2a. Repulsive force dominates in contact mode, while attractive force dominates in non-contact mode. Tapping mode has both repulsive and attraction forces; however, repulsive force dominates due to the positioning of the tip. The relationship between cantilever oscillation damping and tip-sample separation is linear if the interaction force is attractive or repulsive. The various modes of AFM for imaging are explained below.

9.2.1 CONTACT AND TAPPING MODE AFM

These are the widely used AFM modes for imaging purposes. In contact mode or static mode, cantilever is in direct contact with the sample and the setpoint is kept at the constant interaction force (Hodel and Hammond 2021). It is the historically first developed mode of AFM (Ruggeri et al. 2019). It depends on repulsive force (also known as short-range Coulomb force) or frictional force between tip atoms and sample atoms, which are localized, hence facilitating high spatial resolution AFM imaging (Ruggeri et al. 2019; Li et al. 2019). Due to the repulsive force, the cantilever deflects, which is measured by the optical lever method as explained previously. The deflection of the tip Δx is directly proportional to the interaction force F as per Hooke's law $F = k.\Delta x.$, where k is the spring constant of cantilever. Whenever laser

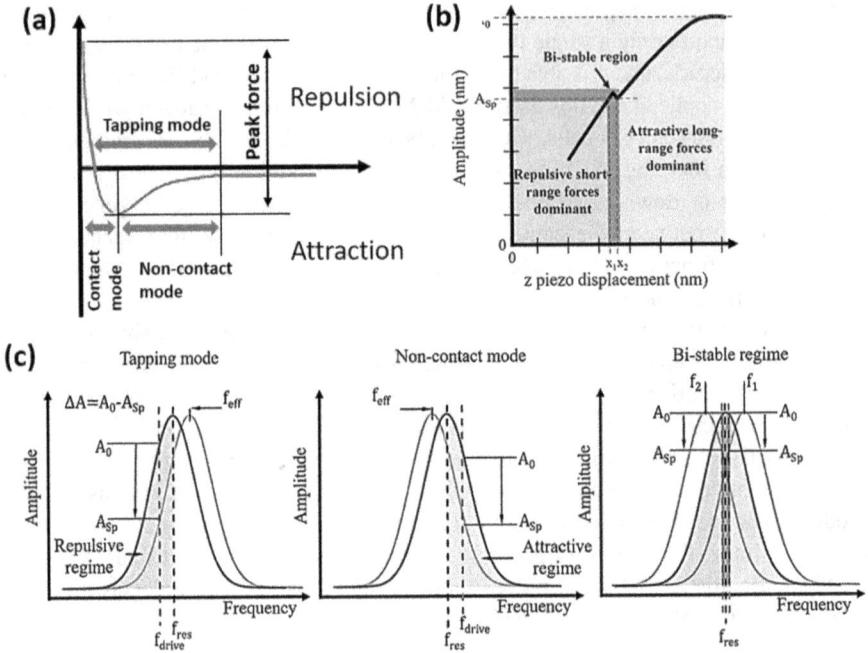

FIGURE 9.2 (a) The interaction force between AFM tip and sample as the tip-sample distance varies. Adapted from (Li et al. 2019). (b) Schematic plot of cantilever damping oscillation amplitude with respect to the tip-sample separation (or piezo displacement). If the cantilever oscillating frequency is the same as its resonance frequencies, then two displacement values exist; hence the curve is nonlinear, and it moves into a bi-stable regime, where attractive and repulsive force interaction exchanges. (c) schematic illustration of effective oscillating frequency of cantilever in tapping and non-contact mode imaging and bistable regime. As the cantilever moves towards or away from the sample, due to the interaction force, oscillating frequency changes. During tapping-mode, small tip-sample interaction causes repulsive force, leading to a higher value of effective oscillating frequency than resonance frequency (left). A large tip-sample distance causes attraction, thereby developing an effective oscillation frequency smaller than resonance frequency (middle). In a bistable regime, there exist two effective values of oscillating frequency and amplitude for the same distance. Hence detector keeps on toggling between these values (Reprinted with permission from Ruggeri et al. 2019.)

spot on the detector changes location, the feedback system provide information to the piezoelectric actuator by extending or retracting along the Z axis to correct the deviation from set point as the cantilever tip deflects (Ruggeri et al. 2019). Deflection does not affect the amplitude and frequency of oscillation.

Biological samples such as reconstructed membrane proteins in action, cytoskeleton, etc., can be visualized to obtain qualitative images using contact mode AFM. The issue with contact mode is that it induces mechanical scratch to the sample surface due to lateral tip movement introducing measurement artefacts, hence difficult to observe loosely adsorbed and soft biological samples. To overcome this issue, other modes of AFM have been evolved (Li et al. 2019; Ruggeri et al. 2019).

Generally, for AFM imaging, the cantilever with free amplitude of oscillation A_0 brings to sample surface until it reaches the set point amplitude A_{sp}. If the cantilever is excited exactly at its resonance frequencies f_{res}, the cantilever tip enters into a bi-stable regime, where the interchange between attractive and repulsive force happens, as shown in Figure 9.2b. It results in a nonlinear behaviour of signal as depicted. Hence, the cantilever is always excited by a driven frequency f_{drive}, whose value is near to one of its resonance frequencies, as plotted in Figure 9.2c. Together with the tip-sample distance, the effective resonance frequency f_{eff} changes. In contact mode, due to repulsive force, f_{eff} increases with the tip-sample distance. In a bi-stable regime, there exist two values of piezo displacement (x_1 and x_2) and effective resonance frequencies (f_1 and f_2) that corresponds to the same amplitude of oscillation. Hence, if the values are given to a feedback circuit, it toggles between these values resulting in uncertainty in determining the sample morphology. Hence, f_{drive} should be chosen slightly above the resonance curve (Ruggeri et al. 2019).

During tapping mode operation, the cantilever oscillates slightly below its resonance frequency, and the probe is only at periodic contact with the sample surface to avoid direct interaction with the sample surface. Here, the setpoint defines in terms of the resonance frequency. However, during scanning, the resonance frequency may get disturbed, so the setpoint is kept slightly below the resonance amplitude measured just above the surface (Hodel and Hammond 2021).

As mentioned, the oscillating behaviour of the cantilever, such as amplitude and resonance frequency changes, as a result of tip-sample contact. Figure 9.2c narrates that effective oscillation of cantilever is higher compared to resonance frequency during tapping mode. This might be due to the dominant repulsive interaction between tip and sample atoms, thereby modifying the oscillation. In addition, tip-sample interaction leads to phase shift in accordance with the dissipated energy during the interaction. A large phase shift is observed with high interaction force and dissipation energy. At the same time, low phase shift corresponds to low interaction force and dissipation energy. Hence, AFM works by measuring amplitude or estimating the changes in resonance frequency of cantilever or phase shift that connect the dynamics of the interaction force and distance. These parameters can be given as feedback parameters to obtain the topography of the sample (Ruggeri et al. 2019). With specific feedback parameters, there exists various tapping mode AFM. Amplitude modulation and frequency modulation-based tapping are common. Amplitude modulation AFM uses amplitude and its phase shift as its feedback parameters. Once the tip approaches the sample, the free oscillation amplitude damps, so the feedback loop adjust the tip-sample distance to keep the amplitude constant and estimated error is used to retrieve the topography of the sample (Ruggeri et al. 2019). Soft and loosely adsorbed biological samples such as mammalian suspended cells and DNA molecules can be imaged under amplitude tapping mode. Compared to resonance frequency feedback, the amplitude feedback loop is simpler. However, in comparison with amplitude feedback, the observed resonance frequency shift is more susceptible to the force acting on the cantilever tip near to the sample surface during oscillation in frequency tapping mode. Hence, resonance frequency tapping mode is capable to visualize fine structures of single biomolecules with higher spatial resolution such as DNA helix or antibodies' Y shaped structure (Li et al. 2019).

9.2.2   Non-contact Mode AFM

It is also a dynamic mode AFM. Here, the separation between tip and sample is comparatively high, so mechanical interaction is weak. Hence deformation and friction effects are removed (Ruggeri et al. 2019). Due to attractive interaction force, the cantilever's oscillation varies and results in an effective oscillation with its amplitude is lower than the resonance frequency, as illustrated in Figure 9.2c. The non-contact mode AFM is used to probe the atomic level visualization of chemical structures of molecules. In this mode, tips have to be specific (or functionalized). For example, metallic copper tip often functionalizes with a CO molecule, in which the carbon bonds with the metallic tip and oxygen protrudes outwards, as schematically shown in Figure 9.3a. In order to visualize the atomic-level structures, AFM uses a low-temperature scanning tunnelling microscopy and qPlus (based on quartz tuning fork) sensors, which bring down the AFM operating oscillations to fractions of an 'A^0'. Images are captured with constant tip height-mode, where it keeps the tip at a constant height above the sample during scanning while recording the frequency. This mode can visualize for soft molecules such as DNA structures atomically (Li et al. 2019).

9.2.3   AFM Force Spectroscopy

AFM force spectroscopy provides the mechanical properties of samples. In this mode, AFM scans the sample by performing a vertical approach-retract cycle. If the tip touches the top layer of water (sample is immersed in water) in ambient conditions, due to capillary force of attraction, the tip jumps to contact the sample. However, if the tip is fully in contact with the water, capillary force does not work. Hence it cannot jump to the sample surface. So, the tip indents the sample until the preset force exerts through the cantilever is achieved. Once it is achieved, the tip retracts from the sample. Firm adsorption of the tip on the sample is achieved due to the adhesion force between them. If the pulling force exerted on the cantilever is more than this adhesion force, the tip detaches from the sample. This process is graphically illustrated in Figure 9.3b (Li et al. 2019). In a typical AFM force spectroscopy, force ranging from 10^{-13} to 10^{-5}N and deflection of 0.1 nm to several μm can be measured (Ruggeri et al. 2019). In the case of biomolecules, this contact-detach of tip to the biological sample records the mechanical properties of the molecule by analyzing the obtained force curves. This force peak of curves corresponds to the intramolecular interaction as in the case of protein unfolding and DNA base-pair unzipping. As the AFM tip is unable to distinguish various biomolecules based on morphology, it is not possible to use this mode if the sample is not purified. Hence, the tip should be functionalized. Intermolecular dynamics of the cell surface can be studied by functionalizing the tip under force spectroscopy mode. The functionalized tip can ligand with a specific receptor (cognate receptors) on the cell surface, thereby forming a molecular pair. By pulling out the ligands from the receptor, this molecular pair ruptures, resulting in the characteristic force changes, which is recorded by the AFM. By studying the curves, receptors on the cell surface can be located (Li et al. 2019).

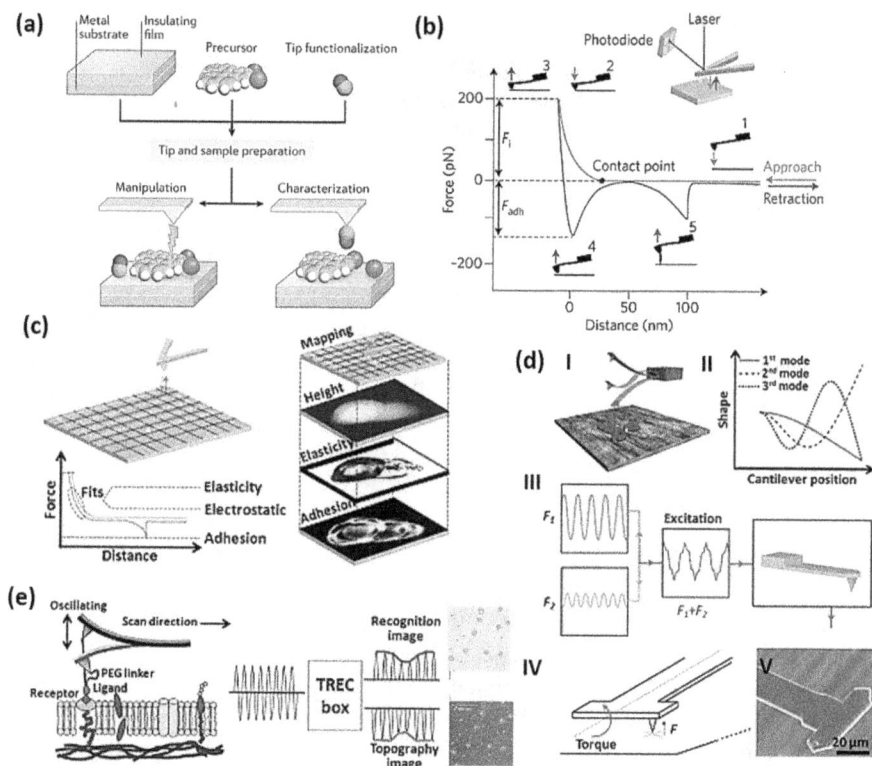

FIGURE 9.3 (a) Sample preparation and single-molecule imaging using non-contact mode AFM. Metal substrate was prepared by coating with a thin insulating layer such as NaCl. The analyte molecule and tip functionalization molecule are deposited from the gas phase. (Reprinted with permission from Pavlicek and Gross 2017.) (b) Force-distance curves recorded during the approach-retract procedure of the cantilever tip in AFM force spectroscopy. (1) The tip of the cantilever is away from the sample, (2) bought near, (3) during retraction, (4) tip movement at different distance and (5) at specific distance. (c) Multiparametric mode AFM to obtain both morphologies as well as mechanical information or physics of biological sample. Parameters are displayed in the colour map and correlate with the topography. (Reprinted with permission from Dufrêne et al. 2017.) (d) Multifrequency AFM for understanding the mechanics of biological samples; deflection of cantilever in multi-harmonic mode AFM (I), corresponding modal shapes of first three eigen modes of a rectangular tip (II), excitation of the cantilever using two external driving forces (III), and Torsional AFM uses T-shaped tip (IV), an SEM image of a liquid torsional harmonic cantilever tip (V). ((I), (II), (III) and (IV) are Reprinted with permission from Garcia and Herruzo 2012. (V) is Reprinted with permission from Dong, Husale, and Sahin 2009.) (e) TREC mode AFM to record the morphology and molecular recognition simultaneously (Reprinted with permission from Chtcheglova and Hinterdorfer 2018 and Alexander Reese and Xu 2019.)

9.2.4 MULTIPARAMETRIC AFM

Multiparametric AFM can be considered as an extension of tapping mode, which is used to obtain the multiple physical information of samples simultaneously. It works on the peak force tapping mode of AFM; the oscillating (or vibrating) frequency of the cantilever is much smaller than the resonance frequency of the cantilever. Therefore, applying tapping frequency is smaller than that of conventional tapping mode AFM. At each sampling event, the oscillating AFM tip indents the sample to record the force curves. The peak force of these curves corresponds to the force between the contact point and maximum extension of tip deflection, which is used as a feedback parameter. Analysis of these force curves gives information about the various physical properties (adhesion, elasticity, deformation, etc.) of the sample along with the topography of the sample, as illustrated in Figure 9.3c. As the exerting tapping force is small, this mode is suitable for imaging fragile biological samples such as proteins, viruses, membranes, etc. and to connect with its structure, physics and functions (Li et al. 2019).

9.2.5 MULTIFREQUENCY AFM

This mode works with the excitation of the cantilever at multiple frequencies. As the information acquired with conventional tapping mode AFM corresponds to the single frequency of excitation and detection, we miss the information that corresponds to other frequencies irreversibly. With multiple frequencies of excitation and detection, the complementary information about the sample can be recorded. It works with various techniques, such as multi-harmonic, bimodal and torsional harmonic AFM. In multi-harmonic AFM, it records the multiple harmonic amplitudes of the oscillating cantilever. Figure 9.3d-I, II show the recording of the first three flexural modes. Whereas, in bimodal harmonic AFM, the first two flexural modes were triggered by external driving forces, as shown in Figure 9.3d-III. The output signal of the first mode is used to obtain the topographical information of the sample, while the output signal of the second mode is used to obtain the details of properties. For torsional harmonic AFM, a T-shaped cantilever with a tip kept at an offset from the cantilever axis is required to generate the torque around the axis of the cantilever. Here, AFM records the higher harmonics of flexural (bending of the cantilever) and torsional (twisting of the cantilever) signal of the cantilever (Figure 9.3d-IV, V). Due to these features and complexities, multifrequency AFM requires specially designed cantilevers. Compared to other modes of AFM previously explained, multifrequency AFM possesses high spatial resolution. For example, it was able to visualize the chemical groups involved in the protein complexes and elastic modulus of heterogeneous materials in the 'A^0' scale (Li et al. 2019).

9.2.6 HIGH-SPEED AFM

High-speed AFM (HS-AFM) was evolved as a solution to improve the temporal resolution of AFM imaging. Conventional AFM imaging takes a few minutes to tens of minutes to visualize the typical biological sample. However, this duration is quite longer than the timescale at which physiological activities occur. Temporal resolution was achieved by improving the AFM design parts, namely, cantilever, cantilever deflection systems, and electronic circuits. For example, HS-AFM can be achieved by miniaturizing the cantilever design to improve the resonance frequency and providing a feedback control loop to suppress the mechanical vibration of the scanner. With HS-AFM, real-time dynamics of biomolecules such as clock proteins, rotary molecules, and myosin, where molecular activity happens at milliseconds timescale, were captured easily. In general, this mode of AFM is suitable to visualize stiff biomolecules or molecules adsorbed on a stiff surface. However, imaging of living mammalian cells is still challenging, as the imaging speed is very limited (10 frames per second) even after using a wide-area scanner. Therefore, it is required to use a high sensitivity deflection system with a tiny cantilever to improve the speed of imaging (Li et al. 2019).

9.2.7   TREC MODE AFM

Generally, the AFM imaging on the cell surface works by mapping the arrays of force curves corresponding to the receptors on the cell surface, which is a very time-consuming process. The simultaneous topography and recognition imaging mode AFM (TREC AFM) improved the efficiency of biomolecular recognition. It works as follows; the functionalized cantilever tip operates at a vibrating frequency approximately matching with its resonance frequency. While performing a raster scan, it either binds or does not bind with the receptor molecule. If it does not bind, then the tip vibrates as it does without functionalizing. Once it binds to the receptor surface, the following unbinding causes changes in the upper part of the vibrational amplitude signal, while there is no influence on the lower part of the signal. By equipping with special electronic circuitry (TREC box), recognition images, as well as topographic images, can be reconstructed simultaneously, as illustrated in Figure 9.3e. Even though it provides information on a direct understanding of the structure and molecular recognition on the cell surface, it does not record the force curves like other modes. Hence the quantitative information about the tip-molecular binding information is missing.

9.3   REQUIREMENTS FOR SINGLE-MOLECULE AFM

By functionalizing the tip with a single-molecule (e.g., CO (carbon monoxide)) AFM can be used to characterize and manipulate a single molecule, known as single-molecule AFM (smAFM). In addition, it should meet other requirements such as operating at frequency modulation at a constant height mode, at ultralow temperature, ultrahigh vacuum and using a qplus sensor (Fang and Hu 2021). The important idea in smAFM is atomic contrast, which originates from atomic forces as well as the mechanical response of atoms to these forces. Pauli repulsion due to the overlap between electron densities of imaged molecule and tip-apex is the major contributor of atomic contrast. It leads to a positive shift of oscillation frequency of the cantilever, resulting in constant height smAFM. The brighter region in smAFM corresponds to high electron density. Hence molecule structure (atoms, bonds, etc.) can be clearly identified. On the other hand, attractive interactions cause negative frequency shifts. Consequently, local attractive interaction results in formation of dark halo surrounding the bright spot. In the case of non-site uniform tip-sample interaction, it results in grey background images (Fang and Hu 2021).

9.3.1 FREQUENCY MODULATION AFM AT A CONSTANT HEIGHT

To improve the resolution down to the 'A^0' scale, the imaging mode of operation used is frequency modulation AFM at a constant height. The operating principle is the same as frequency modulated tapping mode AFM. Higher atomic resolution is achieved by keeping the tip oscillation amplitude down to ~1 A^0, which ensures the sensitivity to short-range forces by keeping the tip-sample distance as ~4 A^0. Within this small amplitude, the frequency shift is directly proportional to the gradient of the interaction force between tip and sample. Compared to basic non-contact frequency-modulated AFM, non-contact frequency-modulated AFM at constant height is improvised in operational stability as well as signal to noise ratio as it does not require any feedback circuit to control the tip height. However, this imaging is not suitable for non-planar molecules due to undesirable contact and complex image contrast (Fang and Hu 2021).

9.3.2 ULTRAHIGH VACUUM AND ULTRALOW TEMPERATURE

Ultrahigh vacuum, generally below 10^{-10} mbar, is chosen to be the operating pressure value for smAFM. It guarantees a clean environment for working by avoiding contamination of the tips and enhancing the resolution of sample images. Suppose if it induces an on-surface reaction via atomic manipulation, ultrahigh vacuum can avoid the undesirable reactions that happen between sample and the environment. Further, an ultrahigh vacuum can control the temperature precisely, which is important for temperature-sensitive sample reconstruction. The problem with the ultrahigh vacuum is that, it cannot be used if the sample is volatile. Another requirement for smAFM is the low temperature. Temperature down to 5 K can be achieved with liquid helium bath cryostat. It can freeze, isolate and stabilize the molecules that are under imaging as well as a molecule for functionalization. Further, vacuum improves at low temperatures, thereby ensuring a clean environment for imaging. In addition, it enhances the quality factor of the sensor, thereby advancing the mechanical stability (Fang and Hu 2021).

9.3.3 TIP FUNCTIONALIZATION

Another important factor is the tip; the commonly used tip material for smAFM is Cu and PtIr (Platinum/Iridium). Due to the high reactivity of metal, the tip may interact with the sample molecule, results in lateral or vertical manipulation of the molecule. Therefore, the tip is usually functionalized using another inert molecule that not only prevents tip-molecule chemical interaction, but also enhances the resolution by further reducing the tip radius to the atomic level (Fang and Hu 2021; Li et al. 2019). Molecules used for functionalization include CO, CuO, NO, Si, halogens (I, Br

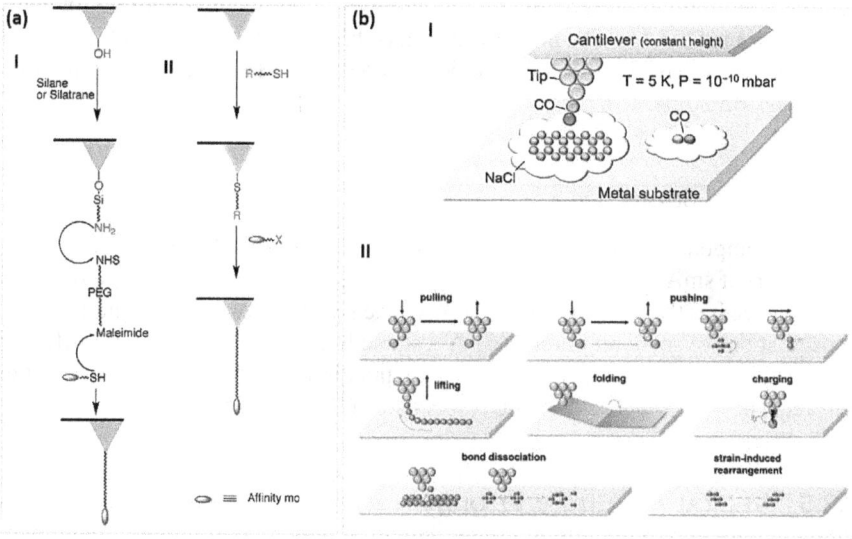

FIGURE 9.4 (a) Common approaches for functionalizing the AFM tip. (I) silanization of silicon on and silicon nitride tip, (II) gold-coated metallic tip. (Reprinted with permission from Alexander Reese and Xu 2019.) (b) Schematic illustration of CO molecule functionalized AFM tip for single-molecule imaging, (I) smAFM at the constant height at ultrahigh vacuum and ultralow temperature, (II) schematic illustration of different types of atomic manipulation. (Reproduced with permission from Fang and Hu 2021.)

and Cl) and noble gases (Kr and Xe) (Fang and Hu 2021). There are several schemes available to functionalize the tip with the above-listed molecules. Silanization of silica and silica nitride probes, and thiol functionalization of gold-coated tips, are the most common approaches (Figure 9.4a). As the commonly available tip material is silicon and silicon nitride, they are modified using organosilanes. 3-Aminopropyl triethoxysilane or APTES is a widely available organosilane and exposes a readily modified amine group. A bifunctional PEG (polyethylene glycol) linker can be coupled for further functionalization. As gold is an inert metal and its surface can be easily modified with amine group molecules, the gold coating is also preferred. A thin layer of gold (tens of nm) can be sputtered on the silicon probe tip followed by incubating with a bifunctional PEG linker to form a thiol bond (Alexander Reese and Xu 2019).

Diatomic molecule functionalized tip gives sharp features due to their easy tilting at the apex, thereby producing higher resolution images compared to single atomic molecule functionalized tip. Further, atomic contrast improves with the type of functionalized molecules, and stability improves with the specific molecule. Combining all the points, the CO molecule has been widely accepted as a functionalized molecule (Fang and Hu 2021). Figure 9.4b(I) shows the schematic illustration of CO molecule functionalization of tip using smAFM. With CO molecule, it adsorbs the metal surface tip with certain mechanical flexibility. However, due to Pauli's repulsion and Van der Waals force, the tip distorts and leads to a sharp bond feature. This amplifies the difference in bond length allows to discriminate the bond with its length differ-ences (down to 0.03 A^0). The main limitation is that CO tilting may result in artefacts that appear as bonds. This can be avoided by connecting lateral distortion and force (Fang and Hu 2021).

9.3.4 qPlus Sensor

Another component that is crucial for improving resolution is the qPlus sensor, which is the heart of smAFM. It is basically a quartz crystal that works on the piezoelectric effect, hence has the self-sensing capability. The signal to noise ratio is improved due to oscillations of small amplitude (<1 A^0), stable frequency (~ 30 kHz) and stiff crys-tal. Further, it can separate the conservative and dissipative interaction, which alters the frequency and feedback circuit, respectively (Fang and Hu 2021).

9.3.5 Vertical and Lateral Resolution

The smAFM imaging resolution depends on various factors, and it can be divided into two, vertical and lateral resolution. Thermal fluctuation of cantilever and noise from the deflection system are the limiting factors for vertical resolution. While lateral resolution arises from three factors; the radius of scanning tip, sensitivity and preci-sion of piezoelectric scanner in XY direction and instrumental resolution defined as the ratio of the size of the image to the number of pixels. The image resolution pos-sesses a typical height error of 0.5–1 A^0 mainly due to the stiffness of the object and roughness of the sample as well as noises contributed from several factors, including electrical noise, thermal vibration of the cantilever, environmental acoustic and elas-tic vibrations. Thermal vibration of the cantilever can down the height measurement of the sample at the A^0 scale and leads to distortion or artefacts in the resulted image. As the amplitude of thermal noise depends on absolute temperature and spring con-stant of the cantilever material, usage of stiff material at low temperature can reduce this noise. Electrical noise arose due to the grounding state (50–60 Hz) of the instru-ment and resulted in high gain feedback periodicity. It results in an error of 1–2 nm along the scanning axis. It can be avoided by administering a proper feedback circuit and connecting the sample stage to the ground to prevent electrostatic accumulation. Acoustic and mechanical vibration errors are due to external factors such as people walking, talking etc., that can be minimized using the acoustic hood. Further, verti-cal resolution can also be affected by tip-sample interaction, acquired map resolution and scanning speed (Ruggeri et al. 2019).

After scanning the sample, the topography is represented as a 3-D colour map with colour intensity representing the variation in height (Z). The resolution of this map depends on the pixels in the XY direction that stores specific intensity val-ues. By considering Shannon's sampling theorem and Nyquist sampling theorem, a minimum of two samples should be collected for each feature during sampling (or mapping). Also, features smaller than pixel size cannot be sampled, and it leads to losing information about small objects. As the pixel size increases, the sampling rate decreases, resulting in an insufficient sampling of features; hence the topography is not adequately represented. Hence, the choice of proper pixel size is needed in improving the lateral resolution. Apart from the sampling issue, increasing the pixel size results in large sample height deviations. If we reduce the pixel size close to zero, the lateral resolution is lower, due to the piezoelectric scanner's sensitivity and the fixed size of the sharp tip. The sharp tip can cause damage to the sample surface and results in convolution induced artefacts. The CO functionalized AFM tip can reduce this issue to a great extent (Ruggeri et al. 2019).

Single molecules can be manipulated using techniques such as pulling, pushing, lifting, folding, charging, bond-dissociation and strain-induced rearrangement as shown represented in Figure 9.4b(II) (Fang and Hu 2021). Repulsive interaction is realized in pushing, whereas attractive interaction between tip and sample is solely responsible for pulling, lifting, and folding. In the charging manipulation, charging state can be switched between electron attachment and detachment, whereas in bond-dissociation is due to the removal of atom(s) and succeeding intermolecular coupling or atomic rearrangement. If the atomic rearrangement is due to substrate-induced strain, then manipulation is known as strain-induced rearrangement.

9.4 AFM ASSISTED SINGLE-MOLECULE IMAGING

The AFM resolution is majorly affected by the sharpness of the tip. Fabrication of atomically sharp tip can achieve atomic and subatomic resolution (Gross et al. 2012). The sample dynamics also contribute to another factor in defining the resolution. The atomic level imaging of an organic molecule is possible at temperature 0K. This can help to measure the inter-atomic bond length (Gross et al. 2012). Nonetheless, at the room temperature, AFM can measure only up to nanoscale resolution. In addition to the tip geometry, the resolution of the imaging is also dependent on the environmental factors (Lyubchenko 2011). While performing imaging in the air medium, a water bridge is formed between the hydrated tip and sample. The capillary forces is large, as dozen of nano Newton (nN) is caused due to high surface tension (Moula et al. 2006). This causes the dragging of the sample over the surface during the scanning process leading to worsen the imaging. However, if the sample is dry and imaging is carried out at relative low humidity, the capillary effect is minimized. Also, if the imaging is carried out in a vacuum, it can further reduce the capillary effect. But imaging in vacuum is not very attractive method as it affects the biological samples. Furthermore, if the imaging is carried out in aqueous solutions, there is no chance of capillary effect and also absolute way to preserve the biological samples.

As seen in further sub-sections, the imaging of DNA with maximum resolution is carried out in aqueous solutions (Lyubchenko and Shlyakhtenko 1997; Mou et al. 1995). Such high-resolution imaging is often obtained using high-speed AFM (HS-AFM) with tip of it created by electron beam deposition followed by sharpening using a plasma-etching methodology (Suzuki et al. 2010; Ando, Uchihashi, and Fukuma 2008). A few examples illustrating the applications of HS-AFM instrument in the studies of DNA-protein complexes, protein assemblies with high resolution are discussed below.

9.4.1 STRUCTURAL DETERMINATION OF DNA

Traditionally, electron microscopy techniques were used for the DNA imaging. However, most studies report its limited structural resolution as the uncoated DNA lacks contrast against the sample grid (e.g., planar TEM) (Amzallag et al. 2006). Even the early AFM studies has inferior structural resolution of the DNA because of the distortion and movement of the DNA molecules (Hansma et al. 1992, 1993, 1998). The improvements in sample deposition methods and above discussed tapping modes of AFM has helped a lot to reveal the structure of DNA. The AFM imaging used to determine range of structural parameters such as periodicity, handedness, and minor/major groove angles. These parameters were found to be similar to the DNA structural characteristics studied by x-ray crystallography (Hansma 2003; Santos et al. 2013; Maaloum, Beker, and Muller 2011; Mou et al. 1995). Further, the improvements in the DNA immobilization methods have showed better reproducibility and resolution in AFM images (Hansma and Laney 1996). This led to the studies of various DNA conformations (Lyubchenko 2004). Recent advances in peak force tapping (PFT) mode and cantilever design refinement have led to the high-resolution periodicity measurements of the minor and major grooves of a single DNA molecule (Leung et al. 2012). This even made possible to visualize single phosphates in backbone of DNA (Ido et al. 2013). This was followed with the reproducible visualization of the uncoated secondary structure of DNA in aqueous environment. This assisted in observation of intramolecular variations of groove depths and direct measurements of twist (Pyne et al. 2021). Various unusual DNA conformations such as triplex DNA (Hansma et al. 1996; Tiner et al. 2001; Klinov et al. 2007; Cherny et al. 1998), G-quadruplex DNA (Marsh, Vesenka, and Henderson 1995; Bose et al. 2018; Neaves et al. 2009; Klejevskaja et al. 2016), and Z-DNA (Kominami, Kobayashi, and Yamada 2019) has also been visualized using AFM.

In many studies, AFM is also used to study the DNA damages caused by radiation via quantifying the linear, nicked, and supercoiled DNA proportions in a sample after the radiation exposure. Further this data is compared with the other techniques such as gel-electrophoresis (Jiang et al. 2007, 2009; Murakami, Hirokawa, and Hayata 2000; Pang et al. 2005; Jiang et al. 2010; Ke et al. 2008). DNA minicircles (< 500 bp) have shown to be useful tool to study the DNA conformation and structural changes in response to various biological phenomena. Such determination and analysis can be carried out using AFM at high-resolution, resulting in minimal complexities in the conformational landscape (Fogg et al. 2006; Klejevskaja et al. 2016; Pyne et al. 2021).

9.4.2 DYNAMICS OF PROTEIN-NUCLEIC ACIDS COMPLEXES

The contribution of AFM in the study of DNA-protein interaction leads to several fundamental roles in the regulation of DNA replication, repairs, and transcription. Disturbances in the arrangement of DNA-protein binding have strong possibility to start severe diseases such as cancer (Engin, Kreisberg, and Carter 2016). Thus, AFM helps to understand various cellular processes and its complexes. In this section, we will see improvements in AFM resolution, force control, and upgradation of softer cantilever which has helped to understand the interaction between DNA and protein. For imaging mode investigation, atomically flat mica surface is used for the attachment of DNA and protein. However, the negatively charged mica in the aqueous solution can interrupt the bonding between the similar charged biomolecules. Thus, either the mica surface needs to be pretreated or the divalent cations are required for the DNA attachment in the imaging medium. These solutions used for surface modifications can find quite a balance between mica surface and DNA-protein complexes. Indeed, a weak attachment can compromise the imaging, while the strong attachment of the biomolecules to the substrate may hinder in their functionality. Various researchers have studied the interactions between DNA-protein complexes. Few of the studies are cited below.

DNA restriction enzymes form synaptosomes to facilitate site-specific DNA cleavage. Several questions behind the mechanism of the formation of synaptosomes has been addressed by Gilmore et al. (2009). In his studies, the interaction of restriction enzyme (EcoRII), a site-specific DNA protein was visualized using HS-AFM (Gilmore et al. 2009). The time-lapse experiments showed that EcoRII slides and associates/dissociates, employing 2D and 3D diffusion while searching for the first site.

Clustered regularly interspaced short palindromic repeats/ protein-9 nuclease (CRISPR/ Cas9) is most widely used and mainly responsible for locating the target DNA and cleaving at the specific site (Jinek et al. 2014). The Cas9 is a multifunctional protein. It consists of 6 domains with the atomic structure provided by the crystallography (Nishimasu et al. 2014; Jinek et al. 2014). The mechanism of the protein suggests its highly dynamic features. This feature has been demonstrated by HS-AFM (Shibata et al. 2017). Figure 9.5a shows selected consecutive set of frames where initially Cas9 protein is in compact conformation, but not in perfect globular shape (Jinek et al. 2014; Nishimasu et al. 2014). As the time passes, single segments change their shape and contrast. This shows that the protein is undergoing intramolecular dynamics. The binding of Cas9 to RNA leads to the change in its geometry and dynamics. Figure 9.5b shows the set of images where protein has acquired a clear, bi-lobed geometry. This structure of protein as compared to a free protein remains stable for same time period. Figure 9.5c illustrates the graph of difference

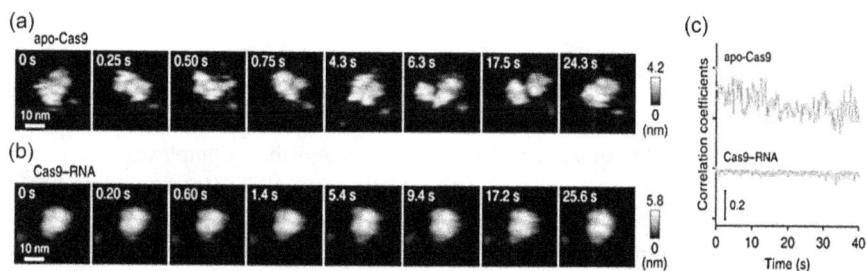

FIGURE 9.5 (a) HS-AFM images in sequence of apo-Cas9 and (b) Cas9–RNA on the surface of AP-mica. The change in color from black to white is in correlated with the height (scale bars -10 nm). (c) Correlation coefficients with respect to time between the sequential HS-AFM images of apo-Cas9 and Cas9-RNA. (Reprinted with permission Shibata et al. 2017.)

correlation coefficient where the subsequent images of AFM have been calculated. The fluctuations are larger for free Cas9 protein while the Cas9-RNA complex value remains constant. The dynamics of the DNA cleavage by the Cas9 protein was also further investigated. The study found the nuclease domain fluctuations in Cas9-RNA complexes corresponding to active and non-active enzyme states (Shibata et al. 2017). The overall study revealed the dynamic properties of the Cas9 and the role of the substrate. It also highlighted the successful characterizing such a large macromolecule system using HS-AFM.

Single-stranded DNA-binding protein (SSB) role was studied using AFM in the interaction of DNA fork with the RecG DNA helicase (Sun et al. 2015). In the DNA-SSB-RecG complex study, it was observed that, the RecG loading efficiency was enhanced due to the SSB protein onto the DNA fork. Moreover, the SSB-RecG interaction leads to enhanced attachment onto the DNA fork (Sun et al. 2015).

Recently, the tumor suppressing protein p53 interaction with DNA in the metal ion solution was studied using AFM (Yang Chen et al. 2017). The study showed the positive effect of the Mg^{2+} ions on DNA-p53 binding. In addition, p53 aggregation is promoted by the high concentrations of Mg^{2+} ions, thus leading to the formation of p53 and DNA self-assembly networks (Yang Chen et al. 2017).

In the centromere nucleosomes dynamics, a protein complex, kinetochore interacts specifically with chromosome centromere which allows the separation of sister chromosomes into daughter cells. The damage or removal of centromeres causes random segregation of chromosomes. The structural characteristics and mechanism of the centromere underlying the recognition process by kinetochore remained undetermined. In 2018, Stumme-Diers et al. used HS-AFM to study the dynamics of the CENP-A (Centromere protein A, also a histone H3 variant) mononucleosomes (Stumme-Diers et al. 2018). The AFM images show that CENP-A nucleosomes were qualitatively the same as that of H3 nucleosomes. Although with respect to quantitative analysis, it showed the wrapping efficiency ~1.7 turns for H3 nucleosomes, while just ~1.5 turns for CENP-A nucleosomes (Stumme-Diers et al. 2018).

9.4.3 Nanoscale Visualization of Protein-Based Interactions

This section briefly describes the outcomes of several research articles in which protein-based interactions were targeted using a unique nanoimaging technique acquired by the high-speed AFM (HS-AFM). The HS-AFM has numerous applications particularly in molecular biology. The AFM technique was developed as a time-lapse for direct imaging of fully hydrated biological molecules, while HS-AFM can visualize the dynamics of biomolecules and their complexes at a video-data acquisition rate (Fantner et al. 2006; Walters et al. 1998; Hansma et al. 2006). Nanometer-level spatial resolution is it's just another key characteristic.

Previously in the imaging studies of proteins using x-ray crystallography, electron microscopy, and nuclear magnetic resonance (NMR) have been successful and only producing static images of the protein interactions. For example, Myosin V plays an important role in motility processes and muscle contractions. Its major function is to retain all organelles and vesicles in the actin-rich periphery of the cell. Sakamoto et al. used single-molecule optical imaging and optical-trap probing techniques and

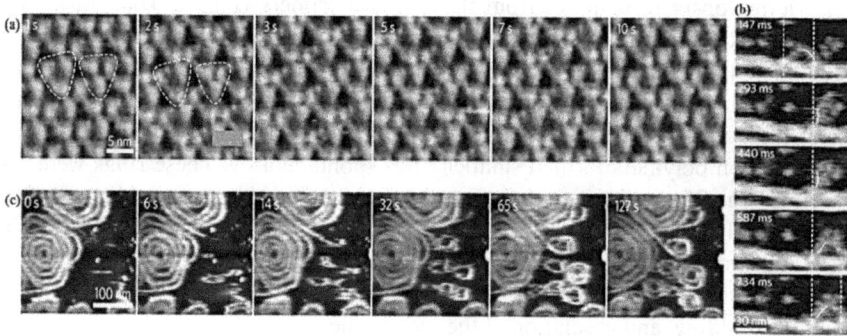

FIGURE 9.6   HS-AFM images of proteins in action. (a) Bacteriorhodopsin recorded under dark and illumination in its native purple membrane at 1 frame s⁻¹. Bacteriorhodopsin trimers is indicated with white triangles. 'Trefoils' is indicated with blue triangles consisting of three bacteriorhodopsin monomers, each belongs to an adjacent trimer. At the intervals of 2 seconds green light was illuminated and was switched off at 3 seconds. Outward dilation of bacteriorhodopsin trimers was observed on illumination, while bacteriorhodopsin monomers contact each other in trefoils. (Reprinted with permission Shibata et al. 2010.) (b) Myosin V walking unidirectional along with an actin filament, showing forward rotation of the leading lever-arm on trailing head detachment from actin. Dashed white lines indicate positions of actin-bound myosin heads. (Reprinted with permission Kodera et al. 2010.) (c) Formation of spiral filament by ESCRT-III protein Snf7 polymerization on a lipid membrane support. (Reprinted with permission Chiaruttini et al. 2015.)

showed Myosin V functionality. It has to 'walk' ~ 36 nm step along the actin filament and towards its positive end (2008). Whereas HS-AFM was able to record parallel structure and dynamics of myosin V molecules 'walking' along the actin filament (Figure 9.6b) (Kodera et al. 2010; Suzuki et al. 2010). This study also disclosed various myosin V dynamic properties, such as junction flexibility and also highlighted future perspectives in HS-AFM studies (Kodera and Ando 2014).

Recently, a unique insight into F_1-ATPase (Uchihashi et al. 2011), bacteriorhodopsin function, endosomal sorting complex that is necessary for transport (ESCRT) III (Chiaruttini et al. 2015), and nuclear pore complexes (Sakiyama et al. 2016) has been revealed by HS-AFM. In the F_1-ATPase rotary motor, the subunit γ rotates around the $(\alpha\beta)_3$ ring stator at the catalytic sites of the ATP hydrolysis located at subunit β (Uchihashi et al. 2011). The possibility of this rotation is due to the rotary propagation of three chemical state i.e., empty, ATP-bound and ADP-bound state. Hence the corresponding structural states over the subunit β. The γ-less $(\alpha\beta)_3$ rings visualization by the HS-AFM, has revealed that even without the γ subunit, the three states can propagate. Thus, β-β interplay across the α subunit causes this cooperativity. This rule out the previous theory of the γ-dictator model that γ-β interactions for the three subunits engenders cooperativity because of the γ subunit asymmetric structure.

For the light-driven proton pump bacteriorhodopsin, HS-AFM showed that, when light illuminates, each bacteriorhodopsin containing cytoplasmic E-F helix portion gives outward displacement by ~0.7 nm (Shibata et al. 2010). This helps the bacteriorhodopsin to contract from the adjacent trimers (Figure 9.6a). In another study, ESCRT-III (sucrose non-fermenting protein 7; snf7), is a key component in the budding and abscission of lipid membrane (Chiaruttini et al. 2015). Snf7 was placed on the supported planar lipid bilayer and using HS-AFM, it showed the concentric spiral filaments. Cantilever tip was used to disrupt the large spirals (Figure 9.6c). These broken polymers formed smaller rings spontaneously. These rings were preferred to be in 25 nm diameter of Snf7 and the spiral filaments remained unbent from their natural curvature. Hence, it was proposed that in cellular conditions, during the spiral spring growth, the energy will be accumulated, while the shrinking of the spiral diameter and inner spiral buckling, the energy will be released. This causes the buckling, bidding, and abscission of the membrane.

Nuclear pore complex (NPCs) helps to facilitate the exchange of molecules between the nucleus and cytoplasm in the eukaryotic cells. However, the selective barrier formed by the nucleoporins to facilitate the transport of molecules has been unclear. With the help of HS-AFM, it was possible to visualize the nucleoporins and their spatiotemporal dynamics inside the NPCs of *Xenopus laevis* oocytes at 100 ms timeline (Sakiyama et al. 2016). The observations resulted in the circumscribed cytoplasmic orifice with high flexibility, dynamic fluctuation of the nucleoporins, which rapidly elongates and retracts. This NPC channel transient entanglement demonstrates as a central plug.

Besides the above molecular studies, the HS-AFM has also been successfully used for the observation of dynamic process in prokaryotic (Yamashita et al. 2012; Fantner et al. 2010) as well as eukaryotic (Shibata et al. 2015) cells. However, HS-AFM has been dependent on the scanning sample stage. This excludes the large and heavy sample stage. Thus, making it difficult to combine with optical microscopy. The HS-AFM, tip-scan technique which is recently developed has remarkably expanded for the application studies in biological processes (Uchihashi et al. 2016). For example, monitoring live cells in petri plate, protein response to external forces, and proteins in suspended membranes can be attainable. The applications in cell biology requiring AFM combinations with other optical techniques will be made possible. There are possibilities of knowledge transfer to HS-SCIM for the monitoring of the live cell dynamics and isolation of intracellular organelles.

9.5 SINGLE-MOLECULE ADHESION FORCE KINETICS

It becomes necessary to classify the mechanical forces interaction at either the molecular or cellular level. This helps to determine the approach and methods to be used to obtain understanding on the sensitive of any mechanical systems. Cellular force interactions are measured in nN range, resulting in the application of force on the environment by the whole-cell. This has made our better understanding of cells with their environmental interactions and also an important information to study its behavior and morphology, affected by the forces. While at the molecular level interactions, the information of conformational changes occurred mainly due to the forces and the triggering of its mechanotransduction pathway is gained. The molecular force power is dependent on the notch receptor activation with the energy required about 4–12 pN (Mescola et al. 2019; Shinde et al. 2021). Thus, substantial variation in force magnitude is observed between molecular and cellular forces. In this section, we will mainly discuss AFM-assisted single molecule force spectroscopy technique in the field of protein-based interactions and molecular systems.

The cellular structural changes caused by the macromolecules under the influence of external mechanical force are probed directly with the help of single molecule force spectroscopy (SMFS) technique. The application of this technique was mainly concentrated on the studies of DNA (Smith, Cui, and Bustamante 1996) and RNA (Liphardt et al. 2001) (un) folding, leading to the successful research of DNA hairpins (Woodside et al. 2006) and complex RNA riboswitch aptamers (Greenleaf et al. 2008). Soon then the field of SMFS has been evolved vastly in the last two decades as a powerful approach to study the mechanical processes enabling single protein molecules to be (un) folded over a well-defined reaction co-ordinate. The distance between protein ends (where one end is tethered to tip/ surface and the other end interacts with either another molecule, tip or surface) is monitored by either protein unfolding as in single-molecule pulling experiments to monitor the extension of the polymer chain (force-extension mode) or under a constant force (force-clamp mode) (Schönfelder, De Sancho, and Perez-Jimenez 2016; Hughes and Dougan 2016). There are numerous cases such as cell adhesion, muscle contraction, and mechanotransduction, where the mechanical forces are exerted or resisted by the protein involved that can be in the range from 0.5 to 10 pN (protein domain motion), 1–100 pN (protein domain deformation) and 50–200 pN (protein domain unfolding) (Bao 2009). Thus, these are some favorable scenarios for the use of SMFS techniques, because of their ability to mimic the biological conditions, when investigating the protein folding, unfolding or more complex molecular systems.

The SMFS can be used to achieve a protein stretch in the low-force regime to control the entropic force. In proteins, non-covalent interactions though weaker but are very important than covalent bonds. Multiple non-covalent bonds make protein enough strong. Protein unfolding requires molecular structure modifications at a nanoscale resolution. This includes the breaking and reformation of numerous hydrogen bonds, Van der Waals interactions, and electrostatic interactions. To break these bonds, force in the range of 100–300 pN is required. When this is measured in an AFM experiment, the typical loading rate is 200–1000 nm s^{-1}. Covalent bonds require the strongest forces to break at about 1600 pN (Grandbois et al. 1999). Different length-scale forces correspond to different bonds and intermolecular interactions as shown in Figure 9.7a (Clausen-Schaumann et al. 2000). Figure 9.7b shows the AFM for SMFS where one end of the protein is bound to the substrate (e.g., gold) covered in a buffer droplet. The piezoelectric device is used to place on the gold substrate. A photo-detector is used to capture a reflected laser beam off the cantilever tip, so as to monitor the deflections. The force-extension mode is based on the tip retraction from the surface with a constant velocity (0). With increasing distance between tip and surface, the polyprotein (robust mechanical fingerprint for protein unfolding with a tandem array of single protein domain interlinked by polypeptide linkers) attached at the tip is elongated first (Figure 9.7c-iI). The extension of the polyprotein causes it to untangle further (Figure 9.7c-iII). This untangling results into releasing of amino acids, thus increasing in length of the polyprotein. At this point, there abrupt decrease in the forces acting on the cantilever due to loss

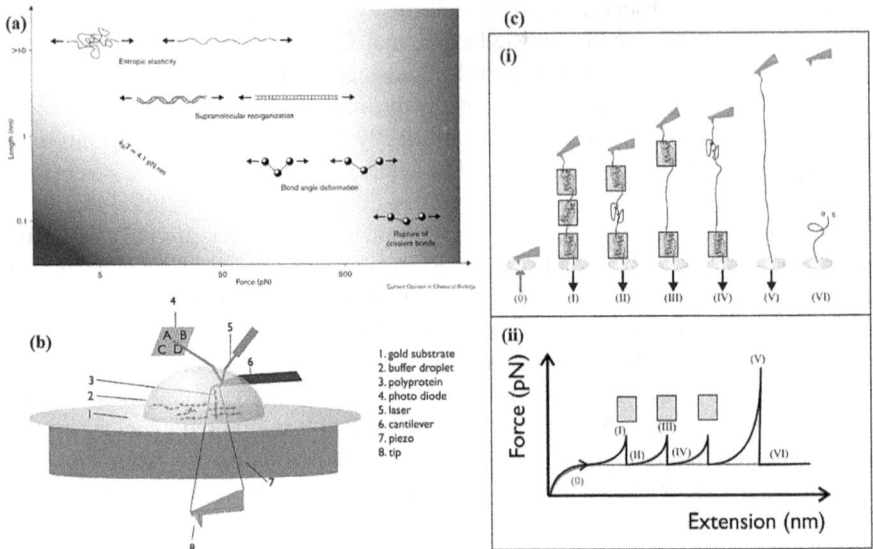

FIGURE 9.7 (a) Illustration of length scales and its corresponding forces associated with breaking of bond, for various intermolecular interactions. The forces that are available to SMFS is shown by a white region of the plot. shaded part are inaccessible regions, these corresponding regions are the thermal stability limit of molecular structures (low forces, left) and breaking of covalent bonds (high forces, right). (Reprinted with permission from Clausen-Schaumann et al. 2000.) (b) Illustration of AFM main components for SMFS. (Reprinted with permission from Hughes and Dougan 2016.) (c) (i) Illustration of the steps to obtain a force-extension trace; (ii) Illustration of the features associated with observed from (i). The substrate approach to the cantilever (0) is shown by the grey line, when the surface is in contact with the cantilever it deflects upwards causing initial negative force. The substrate retracts from the cantilever at constant velocity is shown by the black line. The increased distance of the tip from the surface is observed by an increase in restoring force. The protein resists the change in chain entropy. The protein bonds will be broken at a specific force (I). Resulting in the unfolded protein chain elongation with no resistance. As the force is applied through the next folded protein domain, there is further increase in restoring force on the cantilever (II). After the completion of unfolding of all domains (three in this example), the protein detaches either from its substrate or tip (V). (Reprinted with permission Hughes and Dougan 2016.).

in tension. Therefore, changes in cantilever deflection are observed. The unfolding of the protein domain will further continue until the fully elongated chain length (Figure 9.7c-iIII). A detachment peak corresponding to the detachment of protein from the substrate or tip is then observed (Figure 9.7c-iVI).

Many examples have been studied, where force acts as a denaturant in AFM-SMFS experiments. The unfolding activation barrier is lowered upon the application of the force across the protein domains. This increases the probability to unfold the protein. Some studies showed to recognize the role of mechanical unfolding forces in the case of protein degradation (Maillard et al. 2011; Aubin-Tam et al. 2011; Olivares et al. 2014). Detail mechanism of force generation in motor proteins has been explained

(Svoboda et al. 1993; Finer, Simmons, and Spudich 1994). They have also uncovered the importance of the plasticity of hydrogen bond networks in regulating the mechanochemistry of cell adhesion complexes (Chakrabarti, Hinczewski, and Thirumalai 2014), force generation (Svoboda et al. 1993; Finer, Simmons, and Spudich 1994), and cell signaling (Vogel 2006; Puchner et al. 2008; Yusko and Asbury 2014). These are some of the recent reviews, where authors discuss force-driven protein conformational changes, complex affinity, and modulating stability (Yun Chen, Radford, and Brockwell 2015).

The AFM technology is also used for the measurement of interacting forces between protein and DNA. Here, AFM tip tends to bind the DNA using polymer spacer, while the protein immobilization is carried out on the surface. The series of force-distance measurement is the result of multiple approach-retract cycles. Just before the rupture of the bond, the stretching profile of the polymer spacer can be identified as unbinding events. This procedure was used by Bartels et al., to investigate the ExpG, a transcriptional regulator found in *Sinorhizobium meliloti* (Gram-negative soil bacterium) (Bartels et al. 2003). Here the binding mechanism of exp gene cluster was studied between the ExpG and promoter fragment (consisting of expE1, expA1, and expG-expD1). Thus, the ExpG-specific binding to the promoter regions with AFM force spectroscopy experiments was confirmed to have an interaction force in the range of 50–165 pN (Bartels et al. 2003).

9.6 SUMMARY AND PERSPECTIVE

The AFM data combination with molecular dynamics simulations has great potential in single molecule studies. Atomic-level structural data is well provided by molecular dynamics simulations for larger systems. As these simulations provide set of structures, thus selection of appropriate structures is mainly a problem using computational modeling. The availability of high-resolution data is always a question. Thus, the need for alternative approaches is in demand. As explained before, HS-AFM can be used for the validation approach. Also, to keep in mind that, the in vivo environment is far different than that of the in-vitro conditions. Multiple organelles and vesicles are contained in the intracellular compartment. Thus, most of the biological systems operate via bounded with the organelles and cell membranes. However, AFM surface models with multiple characteristics to mimic the structures of plasma membrane and organelles are possible to design. As such, AFM is considered as an ideal and advanced topographic technique, it can help in characterizing the complex dynamics of molecular machinery as that of the in vivo conditions.

Although AFM explains bond-breaking mechanisms at molecular level, the bond formation dynamics has been hardly discussed. This is because of the limitations of technologies available that could monitor both bonds breaking at high force and formation of bond at low force range. Till now, AFM experiences thermal drift at lower force (<10 pN), thus it has been used only for the study of breaking bonds window at higher force. The recent achievements in AFM-SMFS have revealed the complex computational simulations which were predicted by protein folding theory. Upcoming insights have now allowed to explain the protein folding pathway with details by experiments. In a decade, time resolutions and AFM-SMFS setups have been evolved, which makes possible for new and exciting experiments to study the higher complexity in the protein folding process. The metallothionein (MT) used to measure protein repeats has shown the capacity for stable and long-term measurements in the low force regime. This has also complimented the AFM-SMFS experiments. For example, α-catenin protein binding under the sheer mechanical force was investigated by (Yao et al. 2014), where its unfolding was found in the equilibrium of titin during longtime measurements in low force regime (H. Chen et al. 2015). Also, a week-long study of protein dynamics for protein L was recently reported (Popa et al. 2016).

Some of SMFS experiments have already shown great advancement in the field of drug design. This is mainly based on altering the mechanical stability by binding the ligand to the substrate (Hu et al. 2014). In one of the articles, probing the HIV-1 receptor CD4 domain using AFM has reported that, Ibalizumab, a monoclonal antibody, has increased the mechanical stability of the CD4 domain (Perez-Jimenez et al. 2014). Such examples have been shown to test the SMFS experiments in pharmaceutical drug conformation change, folding pathway, and mechanical pathway of their target which will help to avoid the diseases related to protein misfolding, for example, HIV, to prevent their docking mechanism on the cell.

9.7 CONCLUSION

AFM is considered as one of the important technologies in the field of biological imaging. Among the conventional microscopes, AFM has vast versatility such as easy sample preparation and the possibility of dynamic imaging under physiological conditions. Structural and topological validation of DNA has only been possible with the help of established AFM imaging technology. It also further provides insights into DNA structural flexibility and changes affecting the enzymatic mechanisms. The interactions between DNA and protein have given a better understanding to characterize and address challenges in key biological processes. Also, the AFM has made it possible to understand the mechanical properties of the cellular environment leading to insight into important roles in physiology, development, and pathology. In this way, many other techniques have been complimentary with AFM involved in optimization and development in diverse biological, therapeutical and clinical landscape.

ACKNOWLEDGMENT

The authors greatly appreciate for the financial support from the DBT/Wellcome Trust India Alliance Fellowship under the grant number IA/E/16/1/503062. We also acknowledge all authors and publishers who provided the copyright permissions.

REFERENCES

Alexander Reese, R., and Bingqian Xu. 2019. "Single-molecule detection of proteins and toxins in food using atomic force microscopy." *Trends in Food Science and Technology* 87 (January 2018): 26–34. https://doi.org/10.1016/j.tifs.2019.03.031.

Amzallag, Arnaud, Cédric Vaillant, Mathews Jacob, Michael Unser, Jan Bednar, Jason D. Kahn, Jacques Dubochet, Andrzej Stasiak, and John H. Maddocks. 2006. "3D reconstruction and comparison of shapes of DNA minicircles observed by cryo-electron microscopy." *Nucleic Acids Research* 34(18): e125. https://doi.org/10.1093/NAR/GKL675.

Ando, Toshio, Takayuki Uchihashi, and Takeshi Fukuma. 2008. "High-speed atomic force microscopy for nano-visualization of dynamic biomolecular processes." *Progress in Surface Science* 83(7–9): 337–437. https://doi.org/10.1016/J.PROGSURF.2008.09.001.

Aubin-Tam, Marie Eve, Adrian O. Olivares, Robert T. Sauer, Tania A. Baker, and Matthew J. Lang. 2011. "Single-molecule protein unfolding and translocation by an ATP-fueled proteolytic machine." *Cell* 145(2): 257–67. https://doi.org/10.1016/J.CELL.2011.03.036.

Bao, G. 2009. "Protein mechanics: A new frontier in biomechanics." *Experimental Mechanics* 49 (1): 153–64. https://doi.org/10.1007/S11340-008-9154-0/TABLES/1.

Bartels, Frank Wilco, Birgit Baumgarth, Dario Anselmetti, Robert Ros, and Anke Becker. 2003. "Specific binding of the regulatory protein expg to promoter regions of the galactoglucan biosynthesis gene cluster of sinorhizobium meliloti - a combined molecular biology and force spectroscopy investigation." *Journal of Structural Biology* 143(2): 145–52. https://doi.org/10.1016/S1047-8477(03)00127-8.

Binnig, G., C. F. Quate, and Ch. Gerber. 1986. "Atomic force microscope." *Physical Review Letters* 56(9): 930–33. https://doi.org/10.1103/PHYSREVLETT.56.930.

Bose, Krishnashish, Christopher J. Lech, Brahim Heddi, and Anh Tuân Phan. 2018. "High-resolution AFM structure of DNA G-wires in aqueous solution." *Nature Communications* 9(1): 1–9. https://doi.org/10.1038/s41467-018-04016-y.

Chakrabarti, Shaon, Michael Hinczewski, and D. Thirumalai. 2014. "Plasticity of hydrogen bond networks regulates mechanochemistry of cell adhesion complexes." *Proceedings of the National Academy of Sciences of the United States of America* 111(25): 9048–53. https://doi.org/10.1073/PNAS.1405384111/SUPPL_FILE/PNAS.201405384SI.PDF.

Chen, Hu, Guohua Yuan, Ricksen S. Winardhi, Mingxi Yao, Ionel Popa, Julio M. Fernandez, and Jie Yan. 2015. "Dynamics of equilibrium folding and unfolding transitions of titin immunoglobulin domain under constant forces." *Journal of the American Chemical Society* 137(10): 3540–46. https://doi.org/10.1021/JA5119368/SUPPL_FILE/JA5119368_SI_001.PDF.

Chen, Yang, Tianyong Gao, Yanwei Wang, and Guangcan Yang. 2017. "Investigating the influence of magnesium ions on P53–DNA binding using atomic force microscopy." *International Journal of Molecular Sciences* 18(7): 1585. https://doi.org/10.3390/IJMS18071585.

Chen, Yun, Sheena E. Radford, and David J. Brockwell. 2015. "Force-induced remodelling of proteins and their complexes." *Current Opinion in Structural Biology* 30: 89–99. https://doi.org/10.1016/J.SBI.2015.02.001.

Cherny, Dimitry I., Alain Fourcade, Fedor Svinarchuk, Peter E. Nielsen, Claude Malvy, and Etienne Delain. 1998. "Analysis of various sequence-specific triplexes by electron and atomic force microscopies." *Biophysical Journal* 74(2): 1015–23. https://doi.org/10.1016/S0006-3495(98)74026-3.

Chiaruttini, Nicolas, Lorena Redondo-Morata, Adai Colom, Frédéric Humbert, Martin Lenz, Simon Scheuring, and Aurélien Roux. 2015. "Relaxation of loaded ESCRT-III spiral springs drives membrane deformation." *Cell* 163(4): 866–79. https://doi.org/10.1016/J.CELL.2015.10.017.

Chtcheglova, Lilia A., and Peter Hinterdorfer. 2018. "Simultaneous AFM topography and recognition imaging at the plasma membrane of mammalian cells." *Seminars in Cell and Developmental Biology* 73: 45–56. https://doi.org/10.1016/j.semcdb.2017.08.025.

Clausen-Schaumann, Hauke, Markus Seitz, Rupert Krautbauer, and Hermann E. Gaub. 2000. "Force spectroscopy with single bio-molecules." *Current Opinion in Chemical Biology* 4(5): 524–30. https://doi.org/10.1016/S1367-5931(00)00126-5.

Demir-Yilmaz, Irem, Pascal Guiraud, and Cécile Formosa-Dague. 2021. "The contribution of atomic force microscopy (AFM) in microalgae studies: A review." *Algal Research* 60. https://doi.org/10.1016/j.algal.2021.102506.

Dong, Mingdong, Sudhir Husale, and Ozgur Sahin. 2009. "Determination of protein structural flexibility by microsecond force spectroscopy." *Nature Nanotechnology* 4(8): 514–17. https://doi.org/10.1038/nnano.2009.156.

Dufrêne, Yves F., Toshio Ando, Ricardo Garcia, David Alsteens, David Martinez-Martin, Andreas Engel, Christoph Gerber, and Daniel J. Müller. 2017. "Imaging modes of atomic force microscopy for application in molecular and cell biology." *Nature Nanotechnology* 12(4): 295–307. https://doi.org/10.1038/nnano.2017.45.

Engin, H. Billur, Jason F. Kreisberg, and Hannah Carter. 2016. "Structure-based analysis reveals cancer missense mutations target protein interaction interfaces." *PLOS ONE* 11 (4): e0152929. https://doi.org/10.1371/JOURNAL.PONE.0152929.

Fang, Siyuan, and Yun Hang Hu. 2021. "Open the door to the atomic world by single-molecule atomic force microscopy." *Matter* 4(4): 1189–223. https://doi.org/10.1016/j.matt.2021.01.013.

Fantner, Georg E., Roberto J. Barbero, David S. Gray, and Angela M. Belcher. 2010. "Kinetics of antimicrobial peptide activity measured on individual bacterial cells using high-speed atomic force microscopy." *Nature Nanotechnology* 5(4): 280–85. https://doi.org/10.1038/nnano.2010.29.

Fantner, Georg E., Georg Schitter, Johannes H. Kindt, Tzvetan Ivanov, Katarina Ivanova, Rohan Patel, Niels Holten-Andersen, et al. 2006. "components for high speed atomic force microscopy." *Ultramicroscopy* 106(8–9): 881–87. https://doi.org/10.1016/J.ULTRAMIC.2006.01.015.

Finer, Jeffrey T., Robert M. Simmons, and James A. Spudich. 1994. "Single myosin molecule mechanics: Piconewton forces and nanometre steps." *Nature* 368(6467): 113–19. https://doi.org/10.1038/368113a0.

Fogg, Jonathan M., Natalia Kolmakova, Ian Rees, Sergei Magonov, Helen Hansma, John J. Perona, and E. Lynn Zechiedrich. 2006. "Exploring writhe in supercoiled minicircle DNA." *Journal of Physics: Condensed Matter* 18(14): S145. https://doi.org/10.1088/0953-8984/18/14/S01.

Garcia, Ricardo, and Elena T. Herruzo. 2012. "The emergence of multifrequency force microscopy." *Nature Nanotechnology* 7(4): 217–26. https://doi.org/10.1038/nnano.2012.38.

Gilmore, Jamie L., Yuki Suzuki, Gintautas Tamulaitis, Virginijus Siksnys, Kunio Takeyasu, and Yuri L. Lyubchenko. 2009. "Single-molecule dynamics of the DNA-EcoRII protein complexes revealed with high-speed atomic force microscopy." *Biochemistry* 48(44): 10492–98. https://doi.org/10.1021/BI9010368/SUPPL_FILE/BI9010368_SI_013.AVI.

Grandbois, Michel, Martin Beyer, Matthias Rief, Hauke Clausen-Schaumann, and Hermann E. Gaub. 1999. "How strong is a covalent bond." *Science* 283(5408): 1727–30. https://doi.org/10.1126/SCIENCE.283.5408.1727/ASSET/27062CCB-EEAB-4553-A348-A2F0C876AEDE/ASSETS/GRAPHIC/SE099730704A.JPEG.

Greenleaf, William J., Kirsten L. Frieda, Daniel A.N. Foster, Michael T. Woodside, and Steven M. Block. 2008. "Direct observation of hierarchical folding in single riboswitch aptamers." *Science* 319(5863): 630–33. https://doi.org/10.1126/SCIENCE.1151298/SUPPL_FILE/GREENLEAF.SOM.PDF.

Gross, Leo, Fabian Mohn, Nikolaj Moll, Bruno Schuler, Alejandro Criado, Enrique Guitián, Diego Peña, André Gourdon, and Gerhard Meyer. 2012. "Bond-order discrimination by atomic force microscopy." *Science* 337(6100): 1326–29. https://doi.org/10.1126/SCIENCE.1225621/SUPPL_FILE/1225621.GROSS.SM.PDF.

Hansma, Helen G. 2003. "Surface biology of DNA by atomic force microscopy." *Annual Review of Physical Chemistry* 52: 71–92. https://doi.org/10.1146/ANNUREV.PHYSCHEM.52.1.71.

Hansma, Helen G., and Daniel E. Laney. 1996. "DNA binding to mica correlates with cationic radius: Assay by atomic force microscopy." *Biophysical Journal* 70(4): 1933–39. https://doi.org/10.1016/S0006-3495(96)79757-6.

Hansma, Helen G., Magdalena Bezanilla, Frederic Zenhausern, Marc Adrian, and Robert L. Sinsheimer. 1993. "Atomic force microscopy of DNA in aqueous solutions." *Nucleic Acids Research* 21 (3): 505–12. https://doi.org/10.1093/NAR/21.3.505.

Hansma, Helen G., Irene Revenko, Kerry Kim, and Daniel E. Laney. 1996. "Atomic force microscopy of long and short double-stranded, single-stranded and triple-stranded nucleic acids." *Nucleic Acids Research* 24(4): 713–20. https://doi.org/10.1093/NAR/24.4.713.

Hansma, Paul K., Georg Schitter, Georg E. Fantner, and Craig Prater. 2006. "High-speed atomic force microscopy." *Science* 314(5799): 601–2. https://doi.org/10.1126/SCIENCE.1133497/ASSET/B4EA3234-AB7A-4493-8DAB-

Hansma, H. G., J. Vesenka, C. Siegerist, G. Kelderman, H. Morrett, R. L. Sinsheimer, V. Elings, C. Bustamante, and P. K. Hansma. 1992. "Reproducible imaging and dissection of plasmid DNA under liquid with the atomic force microscope." *Science* 256(5060): 1180–84. https://doi.org/10.1126/SCIENCE.256.5060.1180.

Hansma, H. G., A. L. Weisenhorn, S. A. C. Gould, R. L. Sinsheimer, H. E. Gaub, G. D. Stucky, C. M. Zaremba, and P. K. Hansma. 1998. "Progress in sequencing deoxyribonucleic acid with an atomic force microscope." *Journal of Vacuum Science & Technology B: Microelectronics and Nanometer Structures Processing, Measurement, and Phenomena* 9(2): 1282. https://doi.org/10.1116/1.585221.2EF3A392BC8A/ASSETS/GRAPHIC/601-1.GIF.

Hodel, Adrian W., and Katharine Hammond. 2021. "AFM Imaging of Pore Forming Proteins." In *Methods in Enzymology*, 1st ed., 649:149–88. Elsevier Inc. https://doi.org/10.1016/bs.mie.2021.01.002.

Hu, Xiaotang, Hongbin Li, Elias M. Puchner, Bo Huang, Hermann E. Gaub, and Wilhelm Just. 2014. "Force spectroscopy studies on protein–ligand interactions: A single protein mechanics perspective." *FEBS Letters* 588(19): 3613–20. https://doi.org/10.1016/J.FEBSLET.2014.04.009.

Hughes, Megan L., and Lorna Dougan. 2016. "The physics of pulling polyproteins: A review of single molecule force spectroscopy using the AFM to study protein unfolding." *Reports on Progress in Physics* 79(7): 076601. https://doi.org/10.1088/0034-4885/79/7/076601.

Ido, Shinichiro, Kenjiro Kimura, Noriaki Oyabu, Kei Kobayashi, Masaru Tsukada, Kazumi Matsushige, and Hirofumi Yamada. 2013. "Beyond the helix pitch: Direct visualization of native DNA in aqueous solution." *ACS Nano* 7(2): 1817–22. https://doi.org/10.1021/NN400071N/SUPPL_FILE/NN400071N_SI_001.PDF.

Jadavi, Samira, Paolo Bianchini, Ornella Cavalleri, Silvia Dante, Claudio Canale, and Alberto Diaspro. 2021. "Correlative nanoscopy: A multimodal approach to molecular resolution." *Microscopy Research and Technique* 84(10): 2472–82. https://doi.org/10.1002/jemt.23800.

Jiang, Yong, Changhong Ke, Piotr A. Mieczkowski, and Piotr E. Marszalek. 2007. "Detecting ultraviolet damage in single DNA molecules by atomic force microscopy." *Biophysical Journal* 93 (5): 1758–67. https://doi.org/10.1529/BIOPHYSJ.107.108209.

Jiang, Yong, Mahir Rabbi, Minkyu Kim, Changhong Ke, Whasil Lee, Robert L. Clark, Piotr A. Mieczkowski, and Piotr E. Marszalek. 2009. "UVA generates pyrimidine dimers

in DNA directly." *Biophysical Journal* 96(3): 1151–58. https://doi.org/10.1016/J.
BPJ.2008.10.030.

Jiang, Yong, Mahir Rabbi, Piotr A. Mieczkowski, and Piotr E. Marszalek. 2010. "Separating
DNA with different topologies by atomic force microscopy in comparison with gel
electrophoresis." *Journal of Physical Chemistry B* 114(37): 12162–65. https://doi.org/
10.1021/JP105603K/ASSET/IMAGES/LARGE/JP-2010-05603K_0003.JPEG.

Jinek, Martin, Fuguo Jiang, David W. Taylor, Samuel H. Sternberg, Emine Kaya, Enbo Ma,
Carolin Anders, et al. 2014. "Structures of Cas9 endonucleases reveal RNA-mediated
conformational activation." *Science* 343(6176). https://doi.org/10.1126/SCIENCE.
1247997/SUPPL_FILE/JINEK.SM.PDF.

Katan, Allard J., and Cees Dekker. 2011. "High-speed AFM reveals the dynamics of single
biomolecules at the nanometer scale." *Cell* 147(5): 979–82. https://doi.org/10.1016/j.
cell.2011.11.017.

Ke, Changhong, Yong Jiang, Piotr A. Mieczkowski, Garrett G. Muramoto, John P. Chute, and
Piotr E. Marszalek. 2008. "Nanoscale detection of ionizing radiation damage to DNA by
atomic force microscopy." *Small* 4(2): 288–94. https://doi.org/10.1002/smll.200700527.

Klejevskaja, Beata, Alice L.B. Pyne, Matthew Reynolds, Arun Shivalingam, Richard Thorogate,
Bart W. Hoogenboom, Liming Ying, and Ramon Vilar. 2016. "Studies of G-quadruplexes
formed within self-assembled DNA mini-circles." *Chemical Communications* 52(84):
12454–57. https://doi.org/10.1039/C6CC07110D.

Klinov, Dmitry, Benjamin Dwir, Eli Kapon, Natalia Borovok, Tatiana Molotsky, and Alexander
Kotlyar. 2007. "High-resolution atomic force microscopy of duplex and triplex DNA
molecules." *Nanotechnology* 18(22): 225102. https://doi.org/10.1088/0957-4484/18/
22/225102.

Kodera, Noriyuki, and Toshio Ando. 2014. "The path to visualization of walking Myosin V by
high-speed atomic force microscopy." *Biophysical Reviews* 6(3–4): 237–60. https://doi.
org/10.1007/S12551-014-0141-7/FIGURES/10.

Kodera, Noriyuki, Daisuke Yamamoto, Ryoki Ishikawa, and Toshio Ando. 2010. "Video imaging of walking myosin v by high-speed atomic force microscopy." *Nature* 468(7320):
72–6. https://doi.org/10.1038/nature09450.

Kominami, Hiroaki, Kei Kobayashi, and Hirofumi Yamada. 2019. "Molecular-scale visualization and surface charge density measurement of Z-DNA in aqueous solution." *Scientific
Reports* 9(1): 1–7. https://doi.org/10.1038/s41598-019-42394-5.

Leung, Carl, Aizhan Bestembayeva, Richard Thorogate, Jake Stinson, Alice Pyne, Christian
Marcovich, Jinling Yang, et al. 2012. "Atomic force microscopy with nanoscale cantilevers
resolves different structural conformations of the DNA double helix." *Nano Letters* 12(7):
3846–50. https://doi.org/10.1021/NL301857P/SUPPL_FILE/NL301857P_SI_001.PDF.

Li, Mi, Ning Xi, Yuechao Wang, and Lianqing Liu. 2019. "Atomic force microscopy in probing
tumor physics for nanomedicine." *IEEE Transactions on Nanotechnology* 18: 83–113.
https://doi.org/10.1109/TNANO.2018.2882383.

Liphardt, J., B. Onoa, S. B. Smith, I. Tinoco Jr., and C. Bustamante. 2001. "Reversible unfolding of single RNA molecules by mechanical force." *Science* 292(5517): 733–37. https://
doi.org/10.1126/SCIENCE.1058498/SUPPL_FILE/EQ4.GIF.

Lyubchenko, Yuri L. 2004. "DNA structure and dynamics." *Cell Biochemistry and Biophysics*
41(1): 75–98. https://doi.org/10.1385/CBB:41:1:075.

Lyubchenko, Yuri L. 2011. "Preparation of DNA and nucleoprotein samples for AFM imaging." *Micron* 42: 196–206.

Lyubchenko, Yuri L., and Luda S. Shlyakhtenko. 1997. "Visualization of supercoiled
DNA with atomic force microscopy in situ." *Proceedings of the National Academy
of Sciences of the United States of America* 94(2): 496–501. https://doi.org/10.1073/
PNAS.94.2.496/ASSET/7A444637-1172-43D1-9B69-6F04CC309DF8/ASSETS/
GRAPHIC/PQ0272327005.JPEG.

Maaloum, M., A. F. Beker, and P. Muller. 2011. "secondary structure of double-stranded DNA under stretching: Elucidation of the stretched form." *Physical Review E - Statistical, Nonlinear, and Soft Matter Physics* 83(3):031903. https://doi.org/10.1103/PHYSREVE.83. 031903/FIGURES/7/MEDIUM.

Maillard, Rodrigo A., Gheorghe Chistol, Maya Sen, Maurizio Righini, Jiongyi Tan, Christian M. Kaiser, Courtney Hodges, Andreas Martin, and Carlos Bustamante. 2011. "ClpX(P) generates mechanical force to unfold and translocate its protein substrates." *Cell* 145(3): 459–69. https://doi.org/10.1016/J.CELL.2011.04.010/ATTACHMENT/24E8D899-F330-4229-BCAF-31BF7820B2ED/MMC1.PDF.

Marsh, Thomas C., James Vesenka, and Eric Henderson. 1995. "A new DNA nanostructure, the G-wire, imaged by scanning probe microscopy." *Nucleic Acids Research* 23(4): 696–700. https://doi.org/10.1093/NAR/23.4.696.

Mescola, Andrea, Nathaly Marín-Medina, Gregorio Ragazzini, Maurizio Accolla, and Andrea Alessandrini. 2019. "Magainin-H2 effects on the permeabilization and mechanical properties of giant unilamellar vesicles." *Journal of Colloid and Interface Science* 553: 247–58. https://doi.org/10.1016/J.JCIS.2019.06.028.

Mou, Jianxun, Daniel M. Czajkowsky, Yiyi Zhang, and Zhifeng Shao. 1995. "High-resolution atomic-force microscopy of DNA: The pitch of the double helix." *FEBS Letters* 371(3): 279–82. https://doi.org/10.1016/0014-5793(95)00906-P.

Moula, Md Golam, Shushi Suzuki, Wang Jae Chun, Shigeki Otani, S. Ted Oyama, and Kiyotaka Asakura. 2006. "Surface structures of Ni_2P (0001) - scanning tunneling microscopy (STM) and low-energy electron diffraction (LEED) characterizations." *Surface and Interface Analysis* 38(12–13): 1611–14. https://doi.org/10.1002/SIA.2404.

Müller, Daniel J., and Yves F. Dufrêne. 2008. "Atomic force microscopy as a multifunctional molecular toolbox in nanobiotechnology." *Nature Nanotechnology* 3: 261–69. https://doi.org/10.1142/9789814287005_0028.

Murakami, Masahiro, Hideo Hirokawa, and Isamu Hayata. 2000. "Analysis of radiation damage of DNA by atomic force microscopy in comparison with agarose gel electrophoresis studies." *Journal of Biochemical and Biophysical Methods* 44(1–2): 31–40. https://doi.org/10.1016/S0165-022X(00)00049-X.

Neaves, Kelly J., Julian L. Huppert, Robert M. Henderson, and J. Michael Edwardson. 2009. "Direct visualization of G-quadruplexes in DNA using atomic force microscopy." *Nucleic Acids Research* 37(18): 6269–75. https://doi.org/10.1093/NAR/GKP679.

Nishimasu, Hiroshi, F. Ann Ran, Patrick D. Hsu, Silvana Konermann, Soraya I. Shehata, Naoshi Dohmae, Ryuichiro Ishitani, Feng Zhang, and Osamu Nureki. 2014. "Crystal structure of Cas9 in complex with guide RNA and target DNA." *Cell* 156(5): 935–49. https://doi.org/10.1016/J.CELL.2014.02.001.

Olivares, Adrian O., Andrew R. Nager, Ohad Iosefson, Robert T. Sauer, and Tania A. Baker. 2014. "Mechanochemical basis of protein degradation by a double-ring AAA+ machine." *Nature Structural & Molecular Biology* 21(10): 871–75. https://doi.org/10.1038/nsmb.2885.

Pang, Dalong, James E. Rodgers, Barry L. Berman, Sergey Chasovskikh, and Anatoly Dritschilo. 2005. "Spatial distribution of radiation-induced double-strand breaks in plasmid DNA as resolved by atomic force microscopy." *Radiation Research* 164(6): 755–65. https://doi.org/10.1667/RR3425.1.

Pavlicek, Niko, and Leo Gross. 2017. "Generation, manipulation and characterization of molecules by atomic force microscopy." *Nature Reviews Chemistry* 1(5). https://doi.org/10.1038/s41570-016-0005.

Perez-Jimenez, Raul, Alvaro Alonso-Caballero, Ronen Berkovich, David Franco, Ming Wei Chen, Patricia Richard, Carmen L. Badilla, and Julio M. Fernandez. 2014. "Probing the effect of force on HIV-1 receptor CD4." *ACS Nano* 8(10): 10313–20. https://doi.org/10.1021/NN503557W/SUPPL_FILE/NN503557W_SI_001.PDF.

Popa, Ionel, Jaime Andrés Rivas-Pardo, Edward C. Eckels, Daniel J. Echelman, Carmen L. Badilla, Jessica Valle-Orero, and Julio M. Fernández. 2016. "A halotag anchored ruler for week-long studies of protein dynamics." *Journal of the American Chemical Society* 138(33): 10546–53. https://doi.org/10.1021/JACS.6B05429/SUPPL_FILE/JA6B05429_ SI_002.TXT.

Puchner, Elias M., Alexander Alexandrovich, Lin Kho Ay, Ulf Hensen, Lars V. Schäfer, Birgit Brandmeier, Frauke Gräter, Helmut Grubmüller, Hermann E. Gaub, and Mathias Gautel. 2008. "Mechanoenzymatics of titin kinase." *Proceedings of the National Academy of Sciences of the United States of America* 105(36): 13385–90. https://doi. org/10.1073/PNAS.0805034105/SUPPL_FILE/SM1.MPG.

Pyne, Alice L.B., Agnes Noy, Kavit H.S. Main, Victor Velasco-Berrelleza, Michael M. Piperakis, Lesley A. Mitchenall, Fiorella M. Cugliandolo, et al. 2021. "Base-pair resolution analysis of the effect of supercoiling on dna flexibility and major groove recognition by triplex-forming oligonucleotides." *Nature Communications* 12(1): 1–12. https://doi. org/10.1038/s41467-021-21243-y.

Ruggeri, Fracesco Simone, Tomas Sneideris, Michele Vendruscolo, and Tuomas P. J. Knowles. 2019. "Atomic force microscopy for single molecule characterisation of protein aggregation." *Archives of Biochemistry and Biophysics* 664: 134–48.

Sakamoto, Takeshi, Martin R. Webb, Eva Forgacs, Howard D. White, and James R. Sellers. 2008. "Direct observation of the mechanochemical coupling in myosin va during processive movement." *Nature* 455(7209): 128–32. https://doi.org/10.1038/nature07188.

Sakiyama, Yusuke, Adam Mazur, Larisa E. Kapinos, and Roderick Y.H. Lim. 2016. "Spatiotemporal dynamics of the nuclear pore complex transport barrier resolved by high-speed atomic force microscopy." *Nature Nanotechnology* 11(8): 719–23. https:// doi.org/10.1038/nnano.2016.62.

Santos, Sergio, Victor Barcons, Hugo K. Christenson, Daniel J. Billingsley, William A. Bonass, Josep Font, and Neil H. Thomson. 2013. "Stability, resolution, and ultra-low wear amplitude modulation atomic force microscopy of DNA: Small amplitude small set-point imaging." *Applied Physics Letters* 103(6): 063702. https://doi.org/10.1063/1.4817906.

Schön, Peter. 2018. "Atomic force microscopy of RNA: State of the art and recent advancements." *Seminars in Cell and Developmental Biology* 73: 209–19. https://doi.org/ 10.1016/j.semcdb.2017.08.040.

Schönfelder, Jörg, David De Sancho, and Raul Perez-Jimenez. 2016. "The power of force: Insights into the protein folding process using single-molecule force spectroscopy." *Journal of Molecular Biology* 428(21): 4245–57. https://doi.org/10.1016/J.JMB.2016. 09.006.

Schroeder, J. I., R. Hedrich, and J. M. Fernandez. 1984. "Potassium-selective single channels in guard cell protoplasts of vicia faba." *Nature* 312: 361–62.

Shibata, Mikihiro, Hiroshi Nishimasu, Noriyuki Kodera, Seiichi Hirano, Toshio Ando, Takayuki Uchihashi, and Osamu Nureki. 2017. "Real-space and real-time dynamics of CRISPR-Cas9 visualized by high-speed atomic force microscopy." *Nature Communications* 8(1): 1–9. https://doi.org/10.1038/s41467-017-01466-8.

Shibata, Mikihiro, Takayuki Uchihashi, Toshio Ando, and Ryohei Yasuda. 2015. "Long-tip high-speed atomic force microscopy for nanometer-scale imaging in live cells." *Scientific Reports* 5(1): 1–7. https://doi.org/10.1038/srep08724.

Shibata, Mikihiro, Hayato Yamashita, Takayuki Uchihashi, Hideki Kandori, and Toshio Ando. 2010. "High-speed atomic force microscopy shows dynamic molecular processes in photoactivated bacteriorhodopsin." *Nature Nanotechnology* 5(3): 208–12. https://doi. org/10.1038/nnano.2010.7.

Shinde, Ashwini, Kavitha Illath, Pallavi Gupta, Pallavi Shinde, Ki-Taek Lim, Moeto Nagai, and Tuhin Subhra Santra. 2021. "A review of single-cell adhesion force kinetics and applications." *Cells* 10(3): 577. https://doi.org/10.3390/cells10030577.

Smith, Steven B., Yujia Cui, and Carlos Bustamante. 1996. "Overstretching B-DNA: The elastic response of individual double-stranded and single-stranded DNA molecules." *Science* 271(5250): 795–99. https://doi.org/10.1126/SCIENCE.271.5250.795.

Stumme-Diers, Micah P., Siddhartha Banerjee, Mohtadin Hashemi, Zhiqiang Sun, and Yuri L. Lyubchenko. 2018. "Nanoscale dynamics of centromere nucleosomes and the critical roles of CENP-A." *Nucleic Acids Research* 46(1): 94–103. https://doi.org/10.1093/NAR/GKX933.

Sun, Zhiqiang, Hui Yin Tan, Piero R. Bianco, and Yuri L. Lyubchenko. 2015. "Remodeling of RecG helicase at the DNA replication fork by SSB protein." *Scientific Reports* 5(1): 1–7. https://doi.org/10.1038/srep09625.

Suzuki, Yuki, Yuji Higuchi, Kohji Hizume, Masatoshi Yokokawa, Shige H. Yoshimura, Kenichi Yoshikawa, and Kunio Takeyasu. 2010. "Molecular dynamics of DNA and nucleosomes in solution studied by fast-scanning atomic force microscopy." *Ultramicroscopy* 110(6): 682–88. https://doi.org/10.1016/J.ULTRAMIC.2010.02.032.

Svoboda, Karel, Christoph F. Schmidt, Bruce J. Schnapp, and Steven M. Block. 1993. "Direct observation of kinesin stepping by optical trapping interferometry." *Nature* 365(6448): 721–27. https://doi.org/10.1038/365721a0.

Tiner, William J., Vladimir N. Potaman, Richard R. Sinden, and Yuri L. Lyubchenko. 2001. "The structure of intramolecular triplex DNA: Atomic force microscopy study." *Journal of Molecular Biology* 314(3): 353–57. https://doi.org/10.1006/JMBI.2001.5174.

Uchihashi, Takayuki, Ryota Iino, Toshio Ando, and Hiroyuki Noji. 2011. "High-speed atomic force microscopy reveals rotary catalysis of rotorless F 1-ATPase." *Science* 333(6043): 755–58. https://doi.org/10.1126/SCIENCE.1205510/SUPPL_FILE/UCHIHASHI.SOM.PDF.

Uchihashi, Takayuki, Hiroki Watanabe, Shingo Fukuda, Mikihiro Shibata, and Toshi Ando. 2016. "Functional extension of high-speed AFM for wider biological applications." *Ultramicroscopy* 160: 182–96.

Veigel, Claudia, and Christoph F. Schmidt. 2011. "Moving into the cell: Single-molecule studies of molecular motors in complex environments." *Nature Reviews Molecular Cell Biology* 12(3): 163–76. https://doi.org/10.1038/nrm3062.

Vogel, Viola. 2006. "Mechanotransduction involving multimodular proteins: Converting force into biochemical signals." *Proceedings of the National Academy of Sciences of the United States of America* 35: 459–88. https://doi.org/10.1146/ANNUREV.BIOPHYS.35.040405.102013.

Walters, D. A., J. P. Cleveland, N. H. Thomson, P. K. Hansma, M. A. Wendman, G. Gurley, and V. Elings. 1998. "Short cantilevers for atomic force microscopy." *Review of Scientific Instruments* 67(10): 3583. https://doi.org/10.1063/1.1147177.

Woodside, Michael T., Peter C. Anthony, William M. Behnke-Parks, Kevan Larizadeh, Daniel Herschlag, and Steven M. Block. 2006. "Direct measurement of the full, sequence-dependent folding landscape of a nucleic acid." *Science* 314(5801): 1001–4. https://doi.org/10.1126/SCIENCE.1133601/SUPPL_FILE/WOODSIDE.SOM.PDF.

Yamashita, Hayato, Azuma Taoka, Takayuki Uchihashi, Tomoya Asano, Toshio Ando, and Yoshihiro Fukumori. 2012. "Single-molecule imaging on living bacterial cell surface by high-speed AFM." *Journal of Molecular Biology* 422(2): 300–9. https://doi.org/10.1016/J.JMB.2012.05.018.

Yao, Mingxi, Wu Qiu, Ruchuan Liu, Artem K. Efremov, Peiwen Cong, Rima Seddiki, Manon Payre, et al. 2014. "Force-dependent conformational switch of α-catenin controls vinculin binding." *Nature Communications* 5(1): 1–12. https://doi.org/10.1038/ncomms5525.

Yusko, Erik C., and Charles L. Asbury. 2014. "Force is a signal that cells cannot ignore." *Molecular Biology of the Cell* 25(23): 3717–25. https://doi.org/10.1091/MBC.E13-12-0707/ASSET/IMAGES/LARGE/3717FIG2.JPEG.

10 SERS Analysis for Single-Molecule Detection of Disease Biomarkers

M. Muthu Meenakshi
Vel Tech Rangarajan Dr. Sagunthala R&D
Institute of Science and Technology

Gowri Annasamy
Indian Institute of Information Technology

P. Kaavya and M. Hema Brindha
SRM Institute of Science and Technology

Yih Bing Chu
UCSI University

N. Ashwin Kumar
SRM Institute of Science and Technology

CONTENTS

10.1 INTRODUCTION TO SINGLE-MOLECULE SURFACE-ENHANCED RAMAN SPECTROSCOPY (SM-SERS)

Surface-Enhanced Raman Spectroscopy (SERS) is receiving attention recently as an essential ultrasensitive spectroscopic tool for the trace-level detection of chemicals and biomolecules such as proteins, nucleic acids, and even cells [1]. This

DOI: 10.1201/9781003409472-10

non-destructive technique offers label-free detection of molecules by investigating their respective intrinsic Raman spectral fingerprint. The sensitivity of the SERS technique for quantification and biophysical characterization of biomolecules is dependent on the size, shape, and distribution of the nanostructures on the substrate resulting in nano-roughened interparticle junctions (approximately less than 10 nm) called plasmonic hot spots. Generally, SERS sensing involves electromagnetic enhancement (EM) and chemical enhancement (CE) in which EM arises from the coherent oscillation of electrons around the metal surface and CE arises from the charge transfer mechanism through interaction between the adsorbate and metal [2]. The plasmonic hot spots generate electromagnetic enhancement in which the local field's magnitude could be explored to detect the presence of a single molecule when trapped between the nanogaps [3].

The appearance of time-dependent signal fluctuations at the hotspots is correlated to identifying the presence of a single molecule. Detecting a single molecule could be crucial and involves critical analysis of signal fluctuations and reproducibility to confirm the presence of the single molecule with precision [4]. The collective response at a particular time due to the interaction between the analyte and plasmonic nanostructure is considered for detecting single molecules. It is determined by the shape and arrangement of plasmonic nanostructures for instance dimers, trimers, and clusters of anisotropic nanoparticles. There is a possibility of non-specific interaction due to which signal may be observed. To increase the specificity, various methods were reported including bi-analyte detection in which the signal from the single or few molecules is differentiated thus resolving the inhomogeneous broadening of single-molecule spectral peaks. Blinking in SERS arises as intrinsic fluctuations from single-molecule electromagnetic enhancement which is reflected as changes in spectral signal and total intensity [5,6]. In the case of non-resonant molecules, the blinking activity is unclear where the necessity of more confined hotspots is expected to achieve higher electromagnetic enhancement. Due to weak signal fluctuation and charge transfer complex formation, the blinking effect is smaller for non-resonant molecules. Thus, the SERS substrates and the generated hot spots are vital to investigating the detection of single molecules [7,8]. There are various techniques reported in the literature for the fabrication of SERS substrates including beam lithography, metal-induced etching, sputtering, and nanoparticle immobilization [9]. Plasmonic nanostructures could be immobilized on the solid substrates through the generation of functional group moieties including silanes, amines, thiols, carboxylates, epoxides, or aldehydes. Subsequently, the biomolecules are immobilized onto the substrates directly through spacers or a bifunctional cross-linking conjugation scheme with $-SH$, $-NH2$, $-COOH$ functional groups for controlled positioning [10].

10.2 FABRICATION OF SM-SERS SUBSTRATES

Nanoslits, dimers, nanopores, holes, nanoneedles, and other anisotropic structures are covered in this section. Single biomolecule sensing relies heavily on nanostructures. Many studies have reported on SERS substrates for single biomolecule sensing including glucose, ions, lipids, live cells of bacteria, plants, yeast, HeLa cell lines, and intracellular chemical imaging [11]. In a study, electrodeposited gold substrates were investigated for single-molecule protein detection. The interaction of the protein with a crosslinker resulted in the formation of thiol groups thereby promoting better adhesion with the gold-based SERS substrates. The molecular movement in and around the hotspots was acquired as multiple SERS spectra and analyzed using statistical methods [12]. Single-cell SERS analysis was carried out using silver (Ag) colloids deposited in bacterial cells by the *in-situ* method. Due to the positive charge of silver nanoparticles, it is attached to the cells easily through electrostatic attraction, forming hotspots and could easily discriminate between bacterial strains such as *E. coli* and *S. epidermis* [13]. PDMS films were coated with gold concave nanocubes (Au CNCs) that were self-assembled at the hexane-water interface with alkanethiol in the organic phase. Au CNCs are prepared by seed-mediated synthesis using cetyltrimethylammonium chloride, these structures are packed in a monolayer in such a way that nanoholes were created between the face of two gold CNCs (Figure 10.1). The edges and nanoholes of gold CNCs act as hotspots for protein binding, which was validated for SERS detection of biomarkers insulin and cytochrome complex [14].

The hybrid graphene oxide (GO) gold nanoarray structures were fabricated using laser interference lithography and electrochemical deposition technique for single-molecule sensing of dopamine. The nanoarrays were modified with a dye-conjugated aptamer to specifically detect dopamine release from a single neuronal stem cell. During stem cell differentiation, dopamine is released as a significant factor for neurological functions. The hybrid nanoarray of gold nanostructures with uniform gap sizes will provide Raman enhancement with less signal variation, GO allows chemical enhancement through π-π interactions, which facilitates efficient attachment of dye-conjugated aptamer [15]. The decrease in SERS signal intensity of Raman dye was investigated which is directly correlated with the presence of dopamine and achieved dopamine detection limits in the range of 10^{-4}–10^{-9} M [16]. Single-molecule sensing of Rhodamine 6G (R6G) was studied using silver nanoparticles (Ag) deposited SiNW which was fabricated using the wet chemical method in a controlled manner resulting in the gap between nanoparticles in the order of several nanometers (Figure 10.2). Further, the spectral intensity bands of R6G and Ag-SiNW were observed to fluctuate over time followed by the appearance and decay of strong intensity bands in the case of larger molecules. As the number of molecules decreased, there was the appearance of weak intensity bands and finally temporal

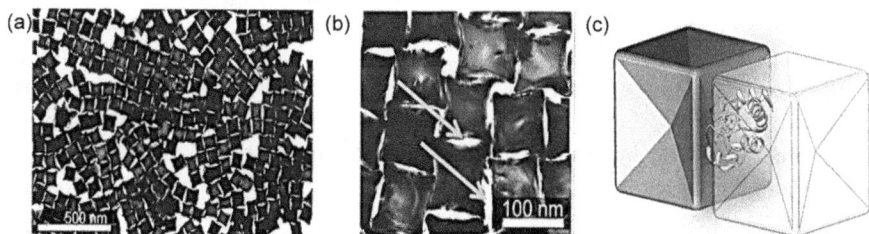

FIGURE 10.1 (a) -(b) TEM image of gold concave nanotube array with nanoholes showed in different scale bar of 500 and 100 nm (c) Nanohole formed between nanotube array for single protein detection. (Adapted with permission from Ref. [14].)

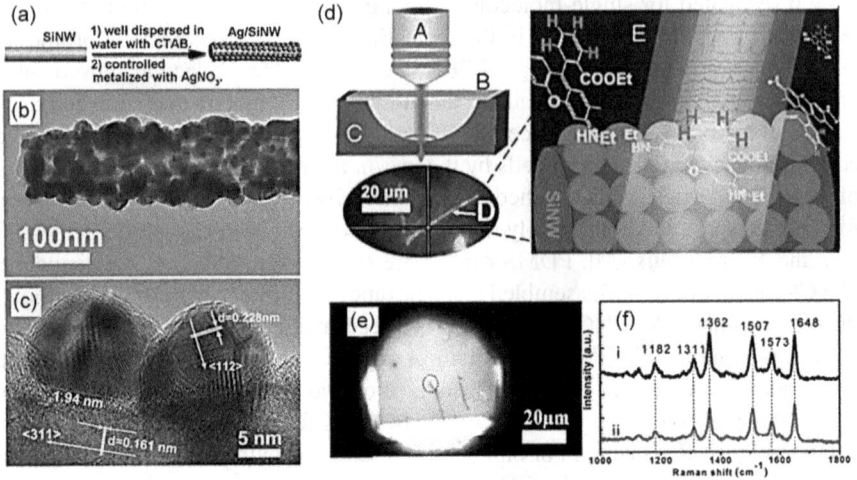

FIGURE 10.2 (a) Illustration depicting the synthesis of the SiNW SERS sensor by depositing AgNPs on SiNWs under controlled conditions. (b) The AgNP–SiNW TEM image with the scale bar of 100nm. (c) HRTEM image of the AgNP-SiNW interface with the scale bar of 5nm. (d) A single SiNW serves as the sample container for single-molecule detection. Within figure D, the following label A represents the Lens of the Raman spectrometer; label B represents cover glass; label C represents the glass substrate; label D represents the Image of the AgNP–SiNW under the Raman microscope; and label E represents the schematic representation of the R6G molecules in the detecting volume. (e) Optical representation of AgNP–SiNWs. (f) SERS spectra of R6G (5 108 M) were taken from (i) newly synthesized AgNP–SiNWs and (ii) AgNP–SiNWs that had been stored for one week. Both spectra were captured using the same detection point. (Adapted with permission from Ref. [17].)

blinking without decay within this period which confirmed the presence of a single R6G molecule seen using a Raman spectrometer [17]. Single-molecule sensing of Hemoglobin (Hb) was carried out using citrate-reduced silver (Ag) colloids. The silver colloids are incubated with adult Hb with a ratio of 1:3 (Hb: Ag) and a very low concentration of 10^{-11} M. The incubated Ag/Hb solution was deposited over a polymer-coated Si wafer for SERS analysis. The dimers and triplets formed on the Si wafer were analyzed in SEM and chosen as hot sites for SERS analysis. The presence of single-molecule Hb appeared as intense peaks with fluctuations over time [18].

Single-molecule sensing of DNA strands was investigated using gold (Au) bowtie nanostructures with a DNA origami pattern using a bottom-up assembly approach with a precise gap of 5 nm. Au nano prisms (80nm) were functionalized with thiolated ssDNA and bowtie structures were formed by hybridization of thiolated strands and capture strands. Also, the probing stand was kept at the nanogaps formed in the bowtie structure. DNA conjugated Raman probe (Cy5, Cy3, and alkyne group) allowed to bind with capture strand on the origami structure between two Au nano prisms for SM detection with the SERS enhancement factor of 10^9 [19]. Au nano bowtie structures formed by two triangular nano prisms, the nanogap formed between the prisms allows detection of DNA, as it transports through the nanopore and the SERS signal intensity variations were analyzed for nucleotide composition followed by sequencing of DNA. Single-molecule detection of protein, nucleotide, bacteria, and viruses was made possible with the fabrication of plasmonic nanopores with pore dimensions of less than 10 nm. These solid-state nanopore structures are formed by lithography and milling techniques. The principle behind sensing is the ability to spatially locate the target molecules in nanopores [20]. An oblique angle gold nanoneedle array was fabricated by an ion-beam sputtering technique on a Si substrate with a diameter of 300 nm and axis length of 700 nm (Figure 10.3). These Au nanoneedle arrays were functionalized with Human Angiotensin-converting-enzyme 2 to capture the SARS-CoV-2 virus through affinity binding with the S protein. The structure and geometry of gold nanoneedle arrays create virus traps that help in capturing the virus. The nanoarray edges induce a lightning rod effect and serve as hotspots for single virus SERS detection with a limit of detection of 80 copies per ml and an enhancement factor of 10^9 [21].

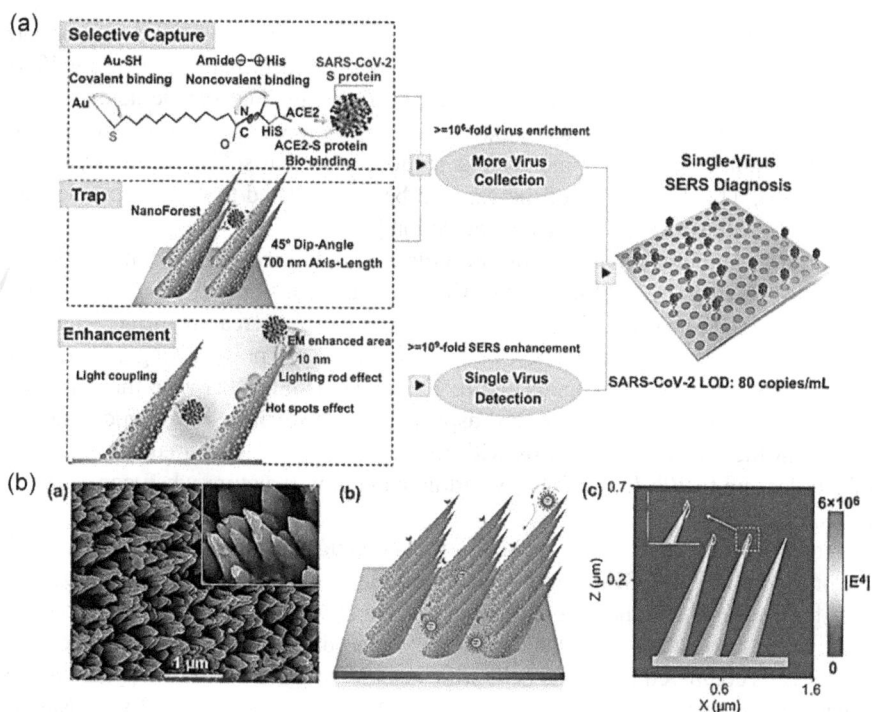

FIGURE 10.3 (A) Schematic for identifying a single SARS CoV-2 virus by selectively capturing and trapping it using a multi-SERS enhancement technique using oblique angle-deposited Gold nanoneedle arrays. (B) Characterization of the fabricated subtract was done using SEM. (a) Represents the SEM image of the Au nanoneedle array with a scale bar of 1μm. (b) A diagram of a "virus-traps" nanoforest made of a tilting array of gold nanoneedles was shown (c)Intensity distribution at the edges of the Au nanoneedle array. (Adapted with permission from Ref. [21] under the CC BY 4.0 license.)

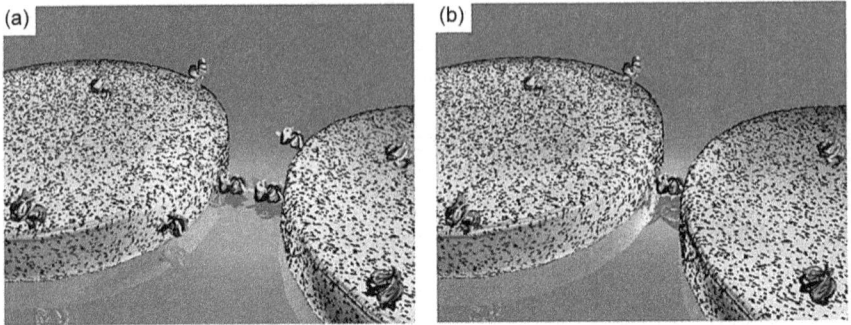

FIGURE 10.4 (a)-(b) Schematic representation of BSA (Bovine Serum Albumin, BSA) binding in the gap of a gold dimers array with consistent nanogaps for single-molecule protein (Bovine Serum Albumin, BSA) detection. (Adapted with permission from Ref. [22] under Copyright © 2009, American Chemical Society.)

E-beam lithography and reactive ion etching techniques were used to fabricate gold dimers with gaps as small as 10 nm between adjacent particles (Figure 10.4). These dimers were functionalized with MUA (Mercaptoundecanoic acid) to bind with the BSA molecule. By manipulating the gap between the dimers based on the size of a biomolecule, single-molecule detection, as well as sensitivity enhancement is achieved. The dimensions of the gap can be manipulated to fit the biomolecules and thereby offering single-molecule detection [22].

Elevated Au bowtie nanoantenna structures were fabricated by a combination of processes electron beam lithography (EBL), metal deposition, liftoff, and reactive ion etching (RIE) for the single-molecule sensing of p-mercaptoaniline. The gold is controllably deposited over the Cr adhesion layer on Si posts having a 20 nm gap size through the EBL technique (Figure 10.5). The distance between the Au nanoantenna is 8 ± 1 nm, these structural aspects provide intrinsic plasmonic coupling effects in the nanocavities or tips with the increase in SERS signal enhancement of 2 orders up to 10^{11} found for elevated nanoantenna structure when compared to non-elevated [23].

Another study reported the gold nanorods conjugated to anti-epidermal growth factor receptor (anti-EGFR) antibodies for single oral cancer cell detection. As the rod shape gives the highest electromagnetic field (10^{14}) due to curvatures, known as the lightning rod effect. Ab conjugated Au nanorods are specifically aligned with cancer cells compared to normal cells which can be seen from microscopy and SERS spectroscopy [24]. Ag nanoparticles embedded in silica shells tagged with Raman reporter and fluorescent molecule reported for detection single cancer biomarker. Further, it was conjugated with protein Annexin V to target and image apoptosis in tumor cells. In another study, Ag-embedded silica shells were conjugated with antibodies to target proteins expressed on tumor cells for imaging [25]. A low-cost biocompatible SERS substrate developed using silver nanoparticles and chitosan biopolymer for single-molecule sensing of adenine. Chitosan-coated anisotropic silver nanoparticles (Ag NPs) were prepared by seed-mediated growth method using

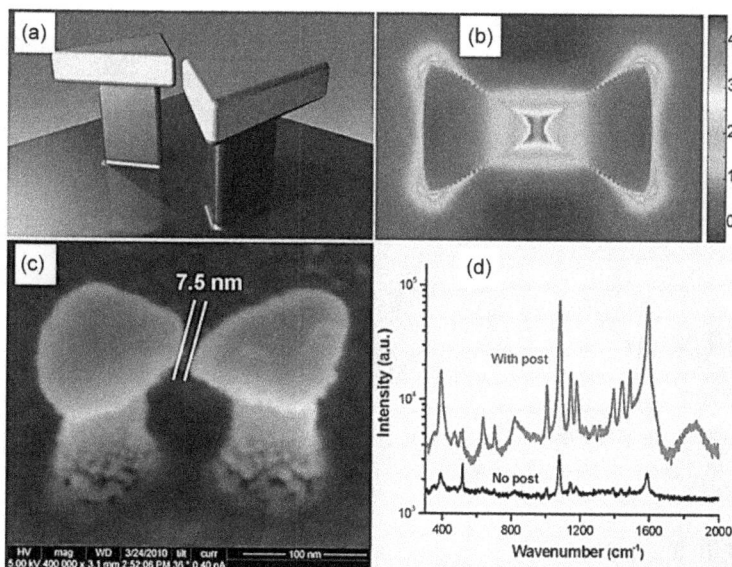

FIGURE 10.5 (a) Schematic of gold bowties on Si posts etched into a Si wafer with a magenta-colored Cr adhesion layer. (b) The spatial distribution of E field intensity is determined by FDTD simulations with 100-nm-side bowtie equilateral triangles, 10-nm gaps, 40-nm-diameter, 200-nm-tall Si posts, and 10-nm apex widths. The logarithmic color bar on the graph shows intensity. (c) SEM Image representing the Elevated Au nano bowtie nanoantenna structures with and without Si post for detection of p-mercaptoaniline with the nanogap of 7.5 nm. (d) The graph compares p-SERS mercaptoaniline's spectra from raised and non-elevated bowtie array substrates. (Adapted with permission from Ref. [23] under Copyright © 2010 American Chemical Society.)

reducing agents and drop cast onto the silica slide substrate for SERS detection of adenine as shown in Figure 10.6a and b. Ag NPs enveloped in the chitosan film acted as molecule traps as well as efficient SERS substrate without any aggregating agent and linker molecule. Here, clusters of Ag NPs such as dimers, and trimers helped in forming the hotspot for single-molecule detection of adenine with the detection limit of 12×10^{-12}M (Figure 10.6e) [26]. The film morphologies were studied using both AFM and SEM imaging techniques (Figures 10.6c and d). Images obtained with an atomic force microscope (AFM) and a scanning electron microscope (SEM) reveal the presence of lone Ag NPs and very small clusters on the surface of the solid substrate. Ag NPs are coated with chitosan shells to prevent the NPs from forming huge aggregates. However, after solvent evaporation, the penetration of the chitosan shell on a solid substrate permitted the construction of 2–3 anisotropic nanoparticles. Due to their small size and potential for containing heat spots, these clusters of films could serve as excellent SERS substrates. Even the SERS characterization was done for the adenine molecule regarding the orientation relative to the silver surface (Figures 10.6e and f).

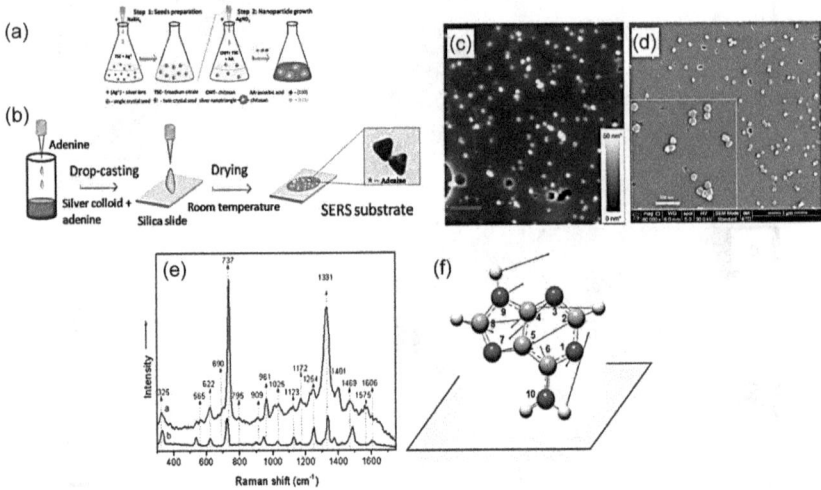

FIGURE 10.6 (a) Diagrammatic representation of the two-step chitosan-coated anisotropic Ag NPs growth. (b) The preparation of the adenine mixed Ag NPs is drop-casted to silica subtracted to form the solid film shown in the diagram. (c) AFM image of Ag clusters deposited SERS active film. (d) SEM image of Ag NPs. (e) Solid film adenine SERS spectrum using Raman spectroscopy. (f) The adenine molecule's orientation on the silver surface film is depicted in this diagram. (Adapted with permission from Ref. [26].)

Silver dimers for the detection of cancer biomarker miRNA-21 were fabricated by the in-situ hotspot assembly (Figure 10.7). Initially, the glass slide was functionalized with APTES, followed by treatment with 4,4′-biphenyldithiol (DBDT) to create hotspots of Ag dimers with a gap of 1 nm. The DBDT molecule acted as both linker and the Raman reporter. The capture strand functionalized on the PDMS chip specifically binds with the mi-RNA 21 expressed from the single cancer cell, further it was hybridized with a trigger sequence which reacts with the probe sequence 1 followed by probe sequence 2. Cascade of these hybridization reactions formed the hotspots by assembling a large number of Ag dimers thereby enhancing the Raman signal [27].

A fiber optic nano biosensor was developed for in vivo single-cell imaging and analysis. Fabrication of fiber at nanometer dimensions with biosensing elements enables sensing at the intracellular level. The sensor was fabricated by the heat and pull method using a fiber-pulling device. SERS nanoprobes were made with Raman active dye 2,4-dintrobenzoic acid (DNBA) adsorbed silver nano colloids and anti-EGF to image the expression of the epidermal growth factor in the targeted cells. In another way, antibody functionalized fiber tip is used to target and detect the intracellular activity in cancer cells. Also, apoptotic activity in live single cells was measured using caspases activity, which is an early predictor of cell death. Tetrapeptide-based optical nanoprobes (Leucine–glutamic acid–histidine–aspartic acid–7-amino-4-methyl coumarin (LEHDAMC), covalently immobilized to the nanotips are employed for sensing specific caspases [7,9] to monitor cell death with high sensitivity and specificity [28].

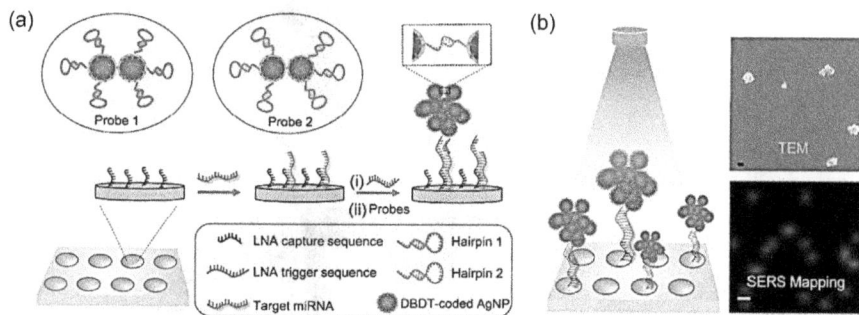

FIGURE 10.7 (a) *In-situ* hotspot assembly technique for miRNA detection using DBDT encoded AgNP dimer probes and protocol was illustrated in the diagrammatic form. (b) The detection schematic representation of the cancer biomarker miRNA-21 by the assembly of silver nanodimers. The side images represent the TEM image of Ag nanodimers and SERS mapping of cancer cells. (Adapted with permission from Ref. [27] under Copyright © 2017 American Chemical Society.)

Nanoslit devices are fabricated using a combination of techniques such as electron beam/UV lithography and wet etching. Plasmonic nano slits formed after Au sputtering with a thickness of 100–200 nm, the final dimensions of the slit with wide 10 nm, length of 1 μm, and depth of 800–900 nm. The device is comprised of a nanoslit, a hollow, and two Bragg mirror gratings embedded in the membrane (Figure 10.8). Surface Plasmon Polaritons (SPPs) can be reflected into the slit-cavity by the plasmonic Bragg-mirror gratings, further strengthening the optical field. The plasmonic gap mode only generates a hot spot region with a significant and broad electromagnetic field amplification for SERS inside the slit. For nucleotide sequencing and identification, a higher concentration of four nucleotide bases was analyzed using nanoslit SERS and spectral information was gathered to form a library. Single-molecule sensitivity is assessed using the BiASERS method, in which independent combinations of two analytes are introduced into the hotspots of nanoslit and analyzed using SERS, the resulting spectrum that can be temporal fluctuations or blinking will be attributed to the presence of only one analyte inside the nanoslit. Two analytes of adenine[14]N-A and[15]N-A, have different Raman spectroscopic characteristics analyzed for single-molecule sensitivity at 10^{-7} M [29].

Generally, smart molecular traps consist of self-assembled gold/silica-coated substrates fabricated by evaporation of gold film using 3-mercaptopropyl tri methoxy silane. Further, the surface was modified with 6-amino-1-hexanediol (AHT) to form a self-assembled monolayer of the gold film with a positive charge. Monothiolated DNA conjugated AuNPs were assembled to produce an array through electrostatic interaction. By treatment of DTT (Dithiothreitol), the AuNPs were attached to the underlying gold film. Molecular traps were created through the adsorption of a thermoresponsive polymer thiolated poly (N-isopropyl acrylamide) over the surface of AuNPs and gold film (Figure 10.9). The SERS activity of molecular traps was investigated for analyte R6G. The substrate was exposed to the analyte at a higher

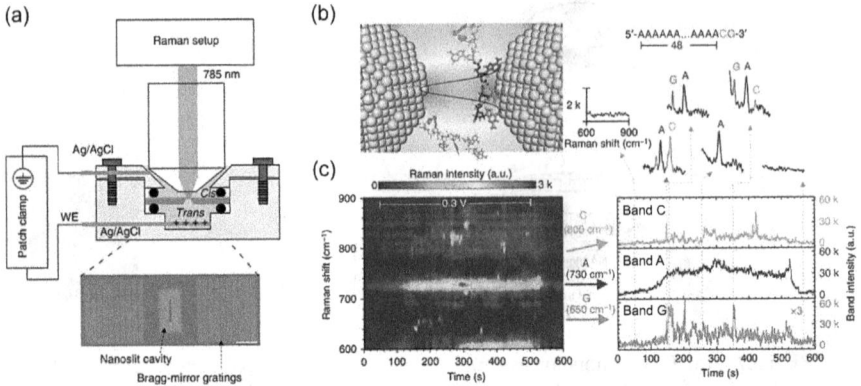

FIGURE 10.8 (a) Schematic of the nanoslit SERS arrangement. Axon patch 200B amplifies transmembrane voltages and ionic currents between Ag/AgCl electrodes. Inset shows a top-view SEM image of the nanoslit structure, which has an inverted prism nanoslit cavity and 1 μm Bragg-mirror gratings. (b) Schematic representation of the nanoslit device for single nucleobase sensing. (c) Nanoslit SERS contour map of 50-mers ss-DNA oligonucleotides (1×10^{-8} M in 10 mM KNO3). The variously colored dashed-line frames relate to the Raman bands B and G, B and A, and B and C. The band intensity of the nucleobases highlighted in C changes over time. (Adapted with permission from Ref. [29] under the CC BY 4.0 license.)

temperature of 50°C higher than the low critical solution temperature (34.5°C) of the polymer. At this temperature, the polymer will shrink and the analyte solution enters into molecular traps cooling down to a lower temperature (4°C) will expand the polymer and traps the analyte in the hotspots. In this way, the substrate can be reused after SERS measurements by repeating the same procedure. Single-molecule sensitivity of the substrate was assessed using the binalyte method in which a mixture of R6G and crystal violet at a concentration of 0.5 μM was used to ensure single-molecule presence inside the molecular traps. SERS measurements at different spots resulted in the presence of only crystal violet (44.6%) or R6G (15.4%). The substrate exhibited good SERS enhancement up to 10^9 enhancement factors with high single-molecule sensitivity and reusability [30].

Thus, this section summarized (Table 10.1) the principle of single-molecule SERS detection and various physical and chemical methods of fabricating anisotropic metallic nanostructures for developing SERS substrates with single-molecule sensitivity. The various noble metallic nanostructures for the detection of single proteins, nucleotides, and disease biomarkers are discussed. Further sections cover the development of SM-SERS-based application which helps in the early identification of disease and deficiency of metabolites with single-molecule sensitivity.

10.3   SM SERS FOR SENSING DISEASE BIOMARKERS

Detection and diagnosis of cancer are currently being identified with various invasive and non-invasive technologies [32]. X-ray imaging, magnetic resonance imaging, and

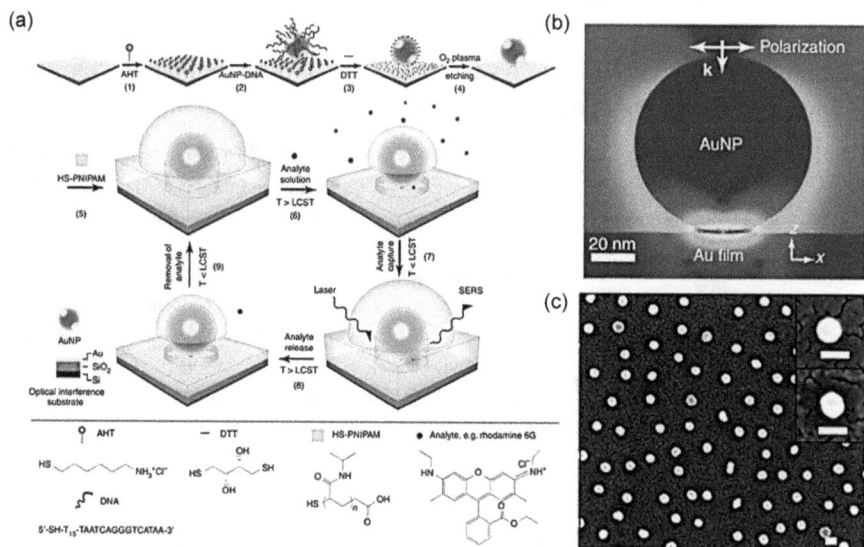

FIGURE 10.9 (a) Plasmonic smart molecular traps based on thermoresponsive polymer for single-molecule detection are demonstrated step by step using a schematic representation of an optical interference substrate made from a silicon wafer and a monolayer of AHT. (b) Simulated SERS enhancement factor distribution of Au nanoparticles over the Au film. (c) SEM image of gold (Au) nanoparticles array (80 nm) onto the Au film. (Adapted with permission from Ref. [30] under the CC BY 4.0 license.)

positron emission tomography (PET) are non-invasive imaging technologies used to study the effects of therapy. Optical imaging techniques such as fluorescence microscopes, confocal microscopes, and single-photon and two-photon microscopic techniques are used to reveal molecular information about cancer [33]. However, the diagnosis using these methods limits the quantitative molecular information of cells. Quantitative information is obtained using standard technologies such as ELISA, optical techniques including fluorescence systems, chemiluminescence, bioluminescence, and microfluidic devices with nanotechnology to provide an improved solution [34–36]. In this line of interest, SERS is advantageous and enables faster identification of biochemical reactions, and is sensitive to a single molecule with unique specificity (Figure 10.10). Designing a rapid detection platform for effective diagnosis requires nano-based tags to help to improve the sensitivity of the system [37]. Several combinations of nanoparticles made of metal, composites, and core-shell along with functionalized biomarkers enhance the SERS detection capability.

10.3.1 CANCER DIAGNOSIS

Cancer has become the most prevalent and life-threatening disease in the world. According to cancer statistics from the World Health Organization (WHO), more than 10 million people die from cancer in 2020, and 19.3 million new cancer cases

TABLE 10.1

A Summary of the Different Ways to Make SERS Substrates that are Sensitive to Single-Molecule Detection.

S. No	Substrate	SERS Biosensor	Application	Reference
1.	A PDMS film fabricated with a bidimensional array of gold CNCs.	Gold concave nanocubes (CNCs) for protein detection	SERS fingerprinting is used to identify the protein once the protein is trapped in a nanohole between the constructed gold nanocubes.	[14]
2.	Fabricated by Controlled deposition of silver nanoparticles onto a silicon nanowire (AgNP–SiNW).	Silver nanoparticles (Ag NPs) for single-molecule detection	One-dimensional nanowire (NW) optical sensors incorporating a single molecule can be integrated as a potent instrument for cell monitoring and disease diagnostics.	[17]
3.	SERS- Silicon substrates with arrays of 2D Gold nanoneedles fabricated by oblique angle ion-beam sputtering.	Gold nanoneedles-based COVID-19 SERS biosensor	Nanostructures with an ACE2-functionalized gold "viral traps" nanostructure quickly detect coronavirus S-protein expression and so detect SARS-CoV-2 in contaminated water.	[21]
4.	Gold dimer arrays were fabricated using reactive ion etching and electron beam lithography.	Ultra-sensitive protein detection can be achieved using LSP gold nanostructures.	Differentiation and analysis of the range of protein molecules that can fit through the nanogap between dimers can be made using this technique. This has even more specificity because of the size exclusion.	[22]
5.	To generate a gap between the gold layer and the Si post, Cr was etched into a Si wafer.	Gold bowtie nanoantenna array for environmental remedies	Real groundwater samples were analyzed using a nanofabricated bowtie array substrate and portable Raman spectrometer for the identification and analysis of environmental pollutants like perchlorate and TNT.	[23,31]
6.	Anisotropic silver nanoparticle clusters embedded in a chitosan biopolymer thin film	Chitosan-coated silver nanoparticle for molecular detection	Plasmonic substrates give enough field enhancement for single-molecule adenine SERS detection by directly sensing the Raman signal on the film surface.	[26]

(continued)

TABLE 10.1 (*Continued*)
A Summary of the Different Ways to Make SERS Substrates that are Sensitive to Single-Molecule Detection.

S. No	Substrate	SERS Biosensor	Application	Reference
7.	The assembly of an *in-situ* hot spot using DBDT-encoded Ag NP dimer probes	Ag NP dimer probes encoded with 4, 4'-biphenyldithiol (DBDT) for miRNA detection	The expression of miRNA-21 in single cancer and normal cells can be precisely quantified utilizing SM-SERS single-molecule technology.	[27]
8.	Plasmonic nanoslit was made in a silicon chip using micromachining techniques and sputtered with Au NPs.	Gold nanoparticles over the nanoslit for DNA sensing	DNA analytes are trapped inside the plasmonic nanoslit and nucleotides identification using SERS.	[29]
9.	Spherical AuNPs on silicon optical interference substrates coated with gold and silica.	Gold nanoparticles (Au NPs) coated with HS-PNIPAM	Electrostatic self-assembly can be used to create a high AuNP density and minimize interparticle gap to improve SM-SERS detection.	[30]

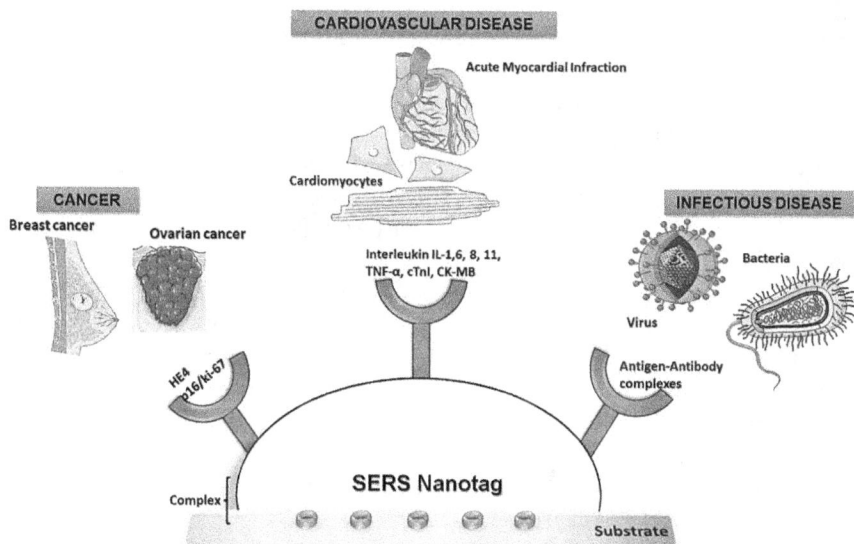

FIGURE 10.10 A graphical illustration of the SERS-based disease diagnosis. Representation is utilized from smart.servier.com.

are being diagnosed. Statistics indicate that 1 in 5 people have a lifetime risk of cancer [38]. There is still a significant challenge in the early detection of cancer. The detection of biomarkers generated by cancer cells is crucial for predicting the patient's prognosis and treatment response. The nanoprobe-based SERS detection of cancer biomarkers provides an effective and rapid approach for early-stage cancer diagnosis [39,40]. Ovarian cancer is the prevalent type of cancer that predominantly affects women. The four biomarker panels are Cancer antigen (CA-125), Human epididymis protein (HE4), Carcinoembryonic Antigen (CEA), and vascular cell adhesion molecule 1 (VCAM-1) exhibited the highest diagnostic potential, with 86% sensitivity for early-stage and 93% sensitivity for late-stage ovarian cancer. In neoplastic ovarian tissue, the biomarker HE4 is overexpressed and shows a promising diagnostic tool for ovarian cancer [41,42]. A method for the highly specific detection of HE4 has rarely been reported. The Korean research laboratory led by Taejoon Kang worked on the attomole-level detection of HE4 by SERS immunoassay employing Nanoplate coated with a single crystalline layer of Gold (Au) [43]. A thermal furnace technology was utilized in the synthesis of single-crystalline gold nanoplates (Au NPls) to create an ideal bioactive surface. The (Cys)3-protein G was bound with the capture antibody over the Au NPls, providing an efficient immunological substrate for HE4. The detection antibody was immobilized with Au NPs that had been surface modified with malachite green isothiocyanate (MGTIC) to function as an immunoprobe for HE4. The surface-modified Au NPs with the MGTIC significantly improve Raman scattering, which works as a localized surface plasmon resonance (LSPR) for detecting HE4 [44]. The complex created due to the interaction between the HE4 substrate antigen and the detector antibody was detected using a high-resolution dispersive Raman microscope (Figure 10.11a). Ultra-fat and ultra-clean Au NPls act as an ideal bioactive surface with bioreceptor orientation, which were characterized using AFM (Figures 10.11b and c). The bioreceptor on Au NPls platforms was tested to detect various concentrations of biomolecules to identify their attomole level using SERS. The presence of SERS hotspots in the interparticle connections of Au NPs and Au NPls complexes is dependent on the concentration of HE4 in the sample. Figure 10.11e shows the SERS spectra produced by an excitation 633 nm laser beam showing the band intensity at $1617\,cm^{-1}$ against varying molar concentrations of HE4. The intensity of the band increases as the HE4 molarity increases. In comparison to other approaches, such as immunosensors, optical nanosensors, and paper-based biosensors, the sensitivity of HE4 utilizing SERS shows high specificity at a $10^{-17}\,M$ concentration level. The nanosized fat Au NPls substrate inhibits non-specific binding, resulting in excellent sensitivity and selectivity for cancer biomarker detection using SERS. SERS-based immunoassay approaches not only enable an accurate diagnosis of ovarian cancer but also have the potential to detect a variety of different biomarkers. However, single-cell-based cancer detection is more reliable than bioanalytical cancer detection.

SERS-encoded particles for single-cell characterization will increase the chances of an early diagnosis [45,46]. Cervical cancer is another prevalent kind of cancer in women, with over 500,000 cases detected annually [47]. It is crucial to evaluate the stage of cervical cancer, whether it is benign or malignant. Existing approaches for examining cervical cancer stages include cytology, histology, immunohistochemistry,

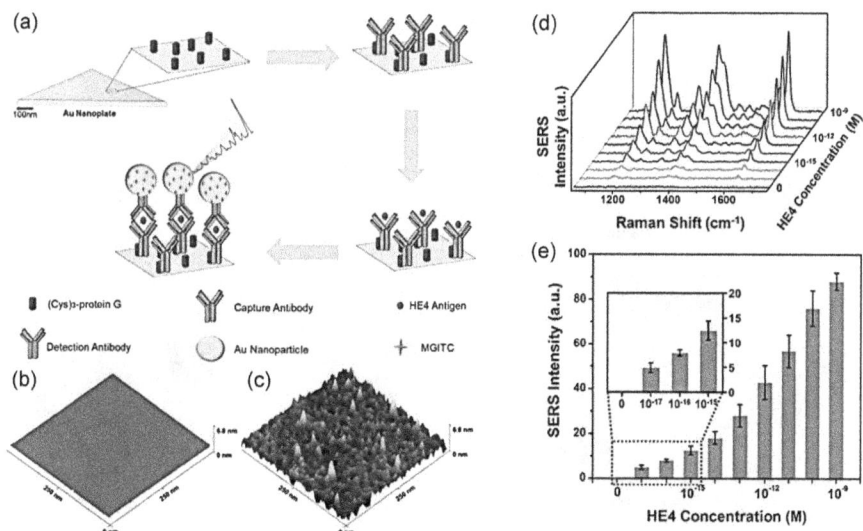

FIGURE 10.11 (a) Diagram depicting the detection of HE4 using the Au NP1 SERS immunoassay. (b) and (c) AFM images of naked Au NP1 and capture of anti-body immobilized Au NP1 with HE4. (d) SERS spectra produced from Au NPs with MGTIC and Au NP1 complexes following HE4 detection ($0–10^{-9}$ M). (e) The intensity of the 1617 cm^{-1} band plotted against the concentration of HE4. The enlarged plot is shown in the inset. The data indicate the mean plus/minus standard deviation of 10 measurements. (Adapted with permission from Ref. [43].)

and molecular-based analysis. These techniques require tissue samples that are obtained through a biopsy or surgery causing damage to the cervix's region. The non-invasive method for identifying cancer in its early stages employs exfoliated cells from the cervical cancer site. The team of Kaustabh Kumar Maiti reported that the SERS-nanotags platform enabled the detection of the dual biomarkers p16/ki-67 from cervical exfoliated normal and abnormal cells [48]. P16 is a highly sensitive and specific biomarker that is minimally expressed in the normal state but overexpressed in glandular neoplasia of the cancerous cervix. K-i67 is a nuclear antigen that prevents mitotic chromosomal aggregation. K-i67 is a cellular proliferative marker for cancer diagnosis because it is inactive during the G0 phase of the cell cycle. p16/ki67 co-occurrence is recognized as the gold standard for detecting human cervical cancer [49]. The simultaneous detection of p16 and ki-67 utilizing Nano-tagged SERS functions as a dual biomarker for the diagnosis of cervical cancer. A nanoflower-shaped SERS-tag was made by merging a hybrid gold nanostar with silver tips (AuNS@SQ-RM@Ag) to maximize fingerprint enhancement from the reporter molecule (Figure 10.12). Then, a combination of monoclonal antibodies targeting p16/K-i67 was used to functionalize it. The coexistence of p16/Ki-67 is not detected in the Immunohistochemistry method. Therefore, tagging antibodies coated on the surface of Gold nanostar identifies the grading of cervical cancer through Raman mapping (Figure 10.12a). The Raman microscope and charge-coupled device

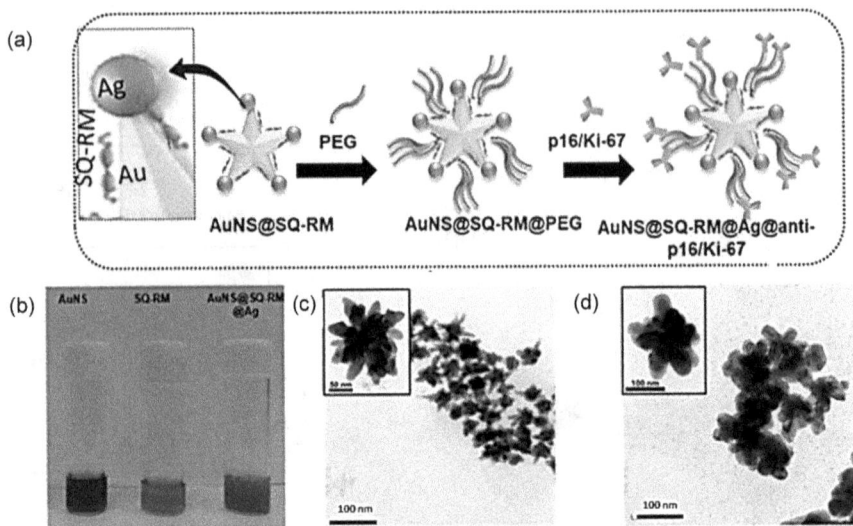

FIGURE 10.12 (a) Synthetic Scheme for the AuNS@SQ-RM@Ag@PEG@anti-p16/Ki-67 Compound (b) Photographic representation of AuNS, SQ-RM, and AuNS@SQ-RM@Ag. (c) AuNS HR-TEM analysis with a scale bar of 100 nm (Inset image scale bar is 50 nm). (d) HR-TEM study of AuNS@SQ-RM@Ag with a 100 nm scale bar (inset scale bar: 100 nm). (Adapted with permission from the Ref. [48] under Copyright © 2021 American Chemical Society.)

(CCD) detector unit were utilized for SERS experiments. The samples were stimulated using a 633 nm laser wit h a power of 10 mW, and the spectra were obtained in the range of 400–1800 cm^{-1}. Raman spectra differentiate between normal cells and abnormal cells by revealing the cellular molecular alterations that occur in individual cells. The detection of Raman spectra in conjunction with a confocal Raman microscope equipped with a CCD detector emerges as a promising method for cancer diagnosis in real-time. This approach to cancer clinical diagnosis is a simple and time-consuming procedure compared to immunocytochemistry approaches.

Magnetoplasmonics (MP) is a new growing branch of study for the identification of SERS-based biomarkers in bio-magnetoplasmonics (BMP). The SERS intensities that are based on MP demonstrate a high level of specificity and sensitivity in the identification of biomarkers [50]. The magneto-plasmonic biosensor system was developed by Hu and his coworkers to detect CRP protein biomarkers using SERS [51]. The falling of reporter molecules for SERS detection limits the capability with low sample concentration. Aggregation followed by magnetic capture is the method used to maximize detection while preventing receptor protein molecule fall. The probe is made of 4-ATP functionalized with Gold nanoparticles (Au NPs) and Silver coated magnetic nanoparticles (Ag MNPs) to isolate them for SERS detection (Figures 10.13a and b). Raman peak captured at 1140 cm^{-1} showed a linear relationship between the concentration and Raman intensity with nanoparticles. The maximum limit of the detection range of Raman peak CRP is 10 FM (1.14 pg/mL) compared to other techniques (Figure 10.13d). The same is tested with the human

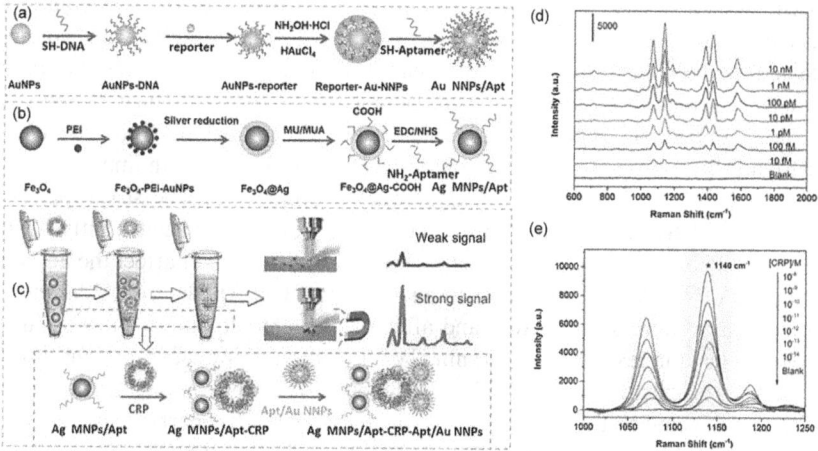

FIGURE 10.13 (a) Schematic of Au NNPs SERS tag synthesis and (b) Magnetic substrate using Ag MNPs. (c) The sandwich construction of Ag MNPs-CRP-Au NNPs for SERS protein detection. (d) SERS spectra of different CRP concentrations in PBS buffer. (e) Graph plot of Raman peak intensity at 1140cm^{-1} changing as CRP content increased between and 1250cm^{-1}. (Adapted with permission from Ref. [51].)

serum samples illustrating the detection efficacy was around 93%. The specificity of the CRP is high compared to other protein molecules present in the human samples such as IgG, HSA, Myo, THR, Lys, and CytC. Micro-RNA (miRNA), molecular biomarkers provide information on diagnosis, prognosis, tumor response, and planning of cancer treatment. The mi-RNA is involved in the regulation of cancer cells and their anti-mi-RNAs are used as a treatment modality. Detection of these biomarkers facilitates personalized medicine. The combination of the Raman signal along with catalytic hairpin assembly (CHA) amplifies the low concentration due to the rehybridization of hairpin probes than the conventional SERS method. Two sensitive miRNA-21 and miRNA-155 in breast, cancer serum were detected in the substrate made of Au-Si coated with 4-mercaptobenzoic acid (4-MBA) with silver nanoparticles to form core-shell nanoparticles. The highly efficient technique combined with CHA-SERS enables detection limits in the range of femtomolar concentration was achieved. Similarly, Mao and his coworkers detected miRNA-21 and miRNA-196a-5p for the diagnosis of lung cancer cell markers in urine. Raman molecules 4-mercaptobenzoic acid to detect the miRNA detection limit as low as 3.3 and 2.18 picomolar respectively [52]. Kiatnida and his team detected miRNA-29a using highly sensitive Raman signal 4-MBA linked to a DNA detection probe coated onto gold nanorods. The complex formation with gold nanorods and capture of DNA conjugated with magnetic nanoparticles resulted in high SERS signals. The complex formation between the gold nanorods and magnetic nanoparticles can detect 10 picomolar concentrations of miRNA-29a [53].

SERS is a method of increasing the Raman signal of a molecule by using nanometer-sized metal substrates such as metal colloids. SERS has received a lot of attention as a method for analyzing biofluids at the point of care, particularly for cancer diagnosis [54]. Preliminary observations in a murine model, for example, revealed that the Raman spectra of urine samples had different characteristics in rats with breast cancer. SERS spectra of urine samples from 53 breast cancer patients and 22 healthy controls. The chemisorption of anionic species (including purine metabolites) to the silver surface and activation of the SERS effect are both improved when cations, such as Ca^{2+} or Mg^{2+}, are added to the silver colloid [55]. The SERS spectra of uric acid 10^{-4}M are brilliant in the presence of Ca^{2+}, but they are significantly weaker in the absence of Ca^{2+}. The addition of Ca^{2+} up to 10^{-4}M did not affect the aggregation of hya-AgNPs. The UV-Vis absorption spectra of hya-AgNPs were examined to evaluate if they were stable before and after subsequent activation with Ca^{2+} up to 10^{-4} M. A Plasmon resonance at 408 nm UV-Vis spectroscopy measurement showed the effects of uric acid on hya-AgNPs. SERS detection is not confined to biomarkers associated with cancer. Biomarkers for medical disorders like heart disease and communication disease can also be identified and assessed using the SERS.

10.3.2 CARDIOVASCULAR DISEASE

Cardiovascular diseases (CVD), such as myocardial infarction, ischemic stroke, and chronic inflammatory disease are diagnosed using a variety of biochemical tests and imaging studies. Clinicians will use blood panels to test cholesterol and plasma ceramide levels or chest X-rays, echocardiograms, electrocardiograms, and electrocardiograms to obtain additional information for CVD diagnosis. Performing in-depth investigation for the diagnosis of CVD demands advanced equipment and trained medical experts [56]. However, conventional diagnostic methods provide less information about the disease mechanism and early detection of CVDs. Patients who are prone to ischemic stroke require cardiac biomarkers for diagnosis, determining their prognosis, and stratifying their risk. Biomarkers for cardiovascular disease can be categorized according to the pathological process, which includes inflammation (that includes interleukin IL-1,6, 8, 11, Tumor Necrosis Factor-α (TNF-α), C reactive protein (CRP), G-CSF (granulocyte colony-stimulating factor), and oxidative stress (such isoprostanes), and also metabolic stress (such as low and high–lipoproteins). These biomarkers for CVD are to detect heart failure with heart accuracy and reliability [57]. SERS is highly efficient for identifying and quantifying biomarkers of CVD.

Activable molecular reporters are becoming more popular in bioanalytical research and bioimaging [58]. It contains low background signals and can be tweaked to respond to a wide range of stimuli. Leucomethylene blue (LMB), Benzoyl LMB, and Leucomalachite green (LMG) are some of the known Raman activable molecules that do not produce SERS signals due to their lack of aromatic conjugation [59–61]. Based on these Raman inactive reporters, Hui Wei and his team investigated an enzymatically active reduction bioassay utilizing SERS reporters for CRP detection. CRP is a small protein that is a member of the calcium-dependent ligand-binding plasma protein family. The greater level of CRP in the blood plasma implies an increase in the inflammatory response [62]. CRP has emerged as a promising biomarker for determining the risk of cardiovascular disease. CRP was detected utilizing Au NPs that were functionalized with a CRP antibody and horseradish per-oxidase (HRP) using Leucomalachite green (LMG). LMG is not a strong Raman scattering molecule on the Au NPs substrate in the presence of CRP antibodies. Due to the HRP-induced peroxidase reaction, LMG is reduced and transformed to malachite green (MG). The intensity of the MG SERS signal changes depending on the different concentrations of CRP. The SERS bioassay for the detection of CRP possesses both positive sensitivity and selectivity due to the utilization of activable compounds [63].

Acute myocardial infarction (AMI) is regarded as a serious cardiovascular dis-ease that contributes to increased rates of morbidity and mortality worldwide [64]. AMI is difficult to diagnose by ECG because most patients' signals have less signifi-cant alterations. Cardiac Troponins (I and T) and creatine kinase myoglobin (CK-MB) are two important diagnostic biomarkers for detecting AMI. Cardiac troponin T (cTnT) and troponin I (cTnI) are the regulatory proteins that are closely related to the heart muscle. CTnT and cTnI govern the myocardium of the heart through the calcium-mediated interaction between actin and myosin [65]. CK-MB is one of the isoenzymes found in cardiac muscle tissue. The increase in CK-MB in the blood is a result of cell injury in the heart muscle [66]. The early AMI diagnosis using the immunoassay method was limited due to less accuracy in detecting the low concentration of these biomarkers [67]. SERS-based AMI detection is currently gaining a lot of interest due to its speed and accuracy [68]. Fabiao Yu and his co-workers developed a sandwich immunoassay based on SERS and a nanoprobe for identifying AMI biomarkers. The gold-patterned chip was used to make a substrate for the SERS-based immunoassay (Figure 10.14a). The immersion approach was used to deposit gold nanoparticles (Au NPs) onto the surface of the silicon wafer to fabricate the gold-patterned chip. The gold chip was immobilized using CTnI and CK-MB antibodies (Figure 10.14b). The core-shell SERS nanoprobe was prepared using gold-silver nanoparticles (Au@Ag NPs) and malachite green isothiocyanate (MGITC) as the Raman reporter (Figure 10.14c). The addition of the sample with antigen biomarker with Au@Ag core-shell SERS probe on top of the subtract forms a polyclonal-antibody conjugation, which results in the formation of sandwich immu-nocomplexes. The quantitative study was performed with a Raman microscopy sys-tem by employing an excitation laser with a wavelength of 632.8 nm and a power of 17 mW. Quantification of both biomarkers was accomplished by using the MGITC Raman reporter, which centered the spectrum peak at 1615 cm^{-1}. The increase in the spectra intensity strength corresponds to the increase in the sample's varying concentration (Figure 10.14e). The limits of detection (LODs) for the SERS-based immunoassay were used to estimate the sensitivity of the AMI biomarkers detec-tion. SERS-based assay techniques for AMI biomarkers detection showed increased sensitivity of cTnI and CK-MB at 8.9 and 9.7 pg mL^{-1} respectively. The LODs of the SERS-assay technique are significantly higher than those of immunoassay and ELISA techniques [69].

The detection of cTnI is less sensitive than the detection of CK-MB in the first hour after the AMI. The difficulty with the CK-MB biomarker is that it began to an elevated level after three hours of AMI occurred. Although these biomarkers are considered effective for the diagnosis of AMI, more specific and sensitive biomarkers

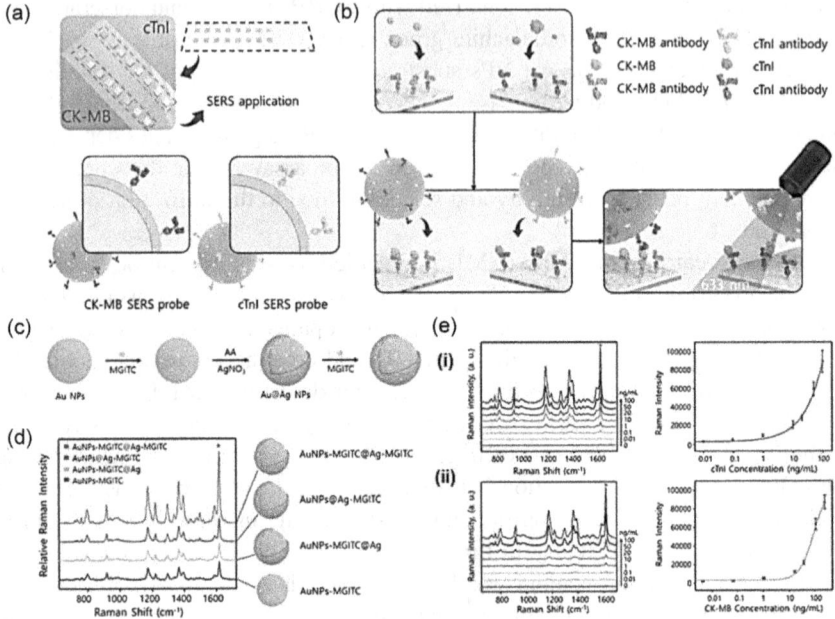

FIGURE 10.14 SERS-based sandwich immunoassays for the quantitative determination of cTnI and CK-MB are illustrated. (a) Dual biomarker detection can be achieved with the use of a gold-patterned chip with SERS probes. (b) Antibodies are conjugated to the gold-patterned chip for the capture of target antigens (cTnI and CK-MB). Laser-based detection of immuno-complexes utilizing 632.8 nm laser Raman spectroscopy. (c) Process of fabricating Au@Ag core-shell nanoparticles tagged with MGITC was shown using diagrammatic representation. (d) The Raman spectra of pure with gold-silver core-shell nanoparticles with MGITC had a significantly higher Raman signal intensity at 1615 cm^{-1} compared to the other nanoparticle types under equal detection conditions. (e) Spectra of SERS for increasing concentrations of (i) cTnI and (ii) CK-MB. SERS signal intensity at 1615 cm^{-1} was shown as a function of the logarithm of cTnI and CK-MB concentrations. (Adapted with permission from Ref. [69].)

are needed for the early detection of AMI injury [70]. Coronary artery disease (CAD) is the warning symptom of AMI. CAD is a heart disease caused by impaired coronary artery blood vessel perfusion caused by total obstruction. This condition causes chest pain and a heart attack, resulting in irreparable damage to the cardiac muscles [71]. A heart-type fatty acid-binding protein is required for metabolic activity within cardiomyocytes (H-FABP) [72]. When the heart muscles are injured by ischemia, H-FABP levels in the bloodstream rise rapidly. The level of H-FABP in the blood serum represents the severity of heart muscle damage [73]. Currently, electrochemistry, electrochemiluminescence, and electrochemical impedance spectroscopy are employed to detect ischemia to predict AMI injury. However, the detection efficiency of conventional approaches is less specific and sensitive.

Ming Guan and his team from China developed an early diagnostic method for AMI by performing simultaneous biomarker identification using a SERS-based

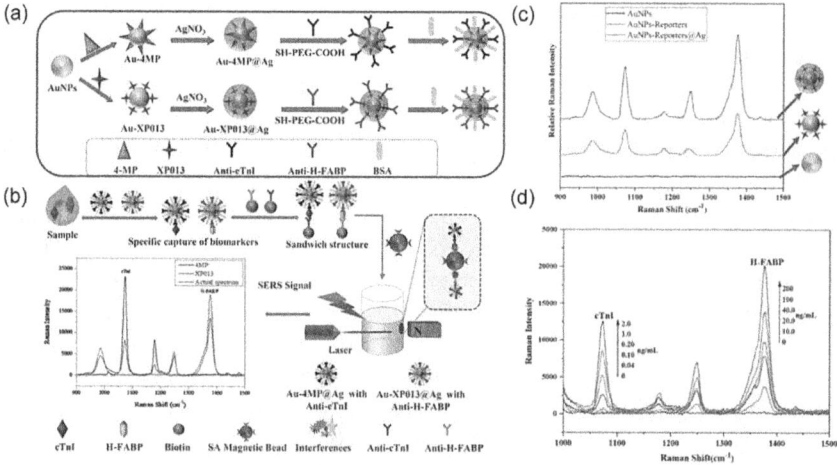

FIGURE 10.15 Schematic illustration of the (a) SERS nanoprobe preparation procedure and (b) SERS-based sandwich immunoassay for simultaneous detection of cTnI and H-FABP. (c) The graph compares enhancement effects for various nanoparticle types, Raman spectra for pure Au NPs, Au-reporters R_1 and R_2, then Au and Ag as a core-shell nanoprobe (Au-4MB/XP013@Ag). (d) H-FABP and cTnI SERS spectra at various concentrations. (Adapted with permission from Ref. [74].)

magnetic immunoassay [74]. SERS nanotag core-shell assembly used the gap between Au and Ag nanoparticles to form a hot spot for detecting AMI biomarkers cTnI and H-FABP (Figure 10.15). Synthesis was performed on the SERS nanotags Au-R_1@Ag and Au-R_2@Ag, where R_1 and R_2 refer to the Raman reporters 4-mercaptobenzonitrile (4-MP) and XP013, respectively (Figures 15a and c). The anti-CTnI and anti-H-FABP antibodies were used to conjugate these R_1 and R_2 SERS core-shell nanotags (Au-4MB/XP013@Ag). These nanotags were captured using biotin and streptavidin (SA) magnetic beads, which act as a capture probe paired with antibodies. Different sample concentrations containing two biomarkers tagged with a SERS nanotag are combined with a nanoprobe for capture to generate a double anti-body sandwich complex. These complexes were then enriched with an external magnetic field to segregate them for the identification of target-specific biomarkers (Figure 10.15b). The isolated complexes are then examined using a Raman spectrometer. Raman signals from the isolated compound created separate characteristic spectrum peaks for two biomarkers. Due to the R_1 and R_2 reporters, the corresponding cTnI and H-FABP biomarkers, and Raman spectra peaks are observed at 1073.5 and 1344.3 cm^{-1} (Figure 10.15d). The concentration of the two biomarkers present in the sample can be determined based on the spectrum's characterization. The limit of detection (LOD) for the AMI biomarkers cTnI and H-FABP was determined as 0.6396 ng/mL and 0.0044 ng/mL, respectively. This method provides simultaneous quantification of cTnI and H-FABP biomarker with high sensitivity and accuracy. SERS-based magnetic enrichment detection can also aid in the simultaneous detection of multiple biomarkers for the precise diagnosis of cardiovascular disease.

10.3.3 INFECTIOUS DISEASE

Infectious disease is caused by live pathogens such as bacteria, viruses, and parasites and is transmissible through injection, airborne, or waterborne ways [75]. Diseases caused by infectious agents continue to be a major contributor to the global health burden [76]. Even though there is a tremendous amount of knowledge about infectious disease treatment and diagnosis, it is still responsible for 15% of global deaths [77]. Infectious disease detection needs to be rapid and accurate to prevent the spreading of the disease [78]. When it comes to the diagnosis of infectious diseases, the gold standard approaches for molecular and immunoassay-based detections include Real-time polymerase chain reaction (RT-PCR), Enzyme-linked immunoassay (ELISA), and Electrochemiluminescence (ECL) [79]. The diagnosis using these traditional methods is difficult due to the need for sophisticated facilities and highly skilled technicians. Point-of-care (PoC) diagnostic techniques have received great attention for the early diagnosis of infectious diseases [80,81].

Testing for microorganisms in drinking water and food, raw or processed, is essential to preventing microbial contamination outbreaks. There are a large variety of water and foodborne pathogens that exist in low quantities and are difficult to detect [82]. Bacteria like Escherichia coli (E-coli) and coliforms are common in humans and other warm-blooded animals. The identification of these microorganisms can be used as a marker for detecting the level of contaminants in water and food [83]. Detection based on fluorescence is commonly employed for quantitative analysis of bacterial or viral identification. Fluorescence detection is hampered by background noise, which reduces sensitivity and limits the capacity to multiplex detection [84]. Recent studies have shown that the SERS-based detection method is fast and accurate in the identification of specific genes or proteins. Lingxin Chen and Institute of translational medicine researchers created a SERS-active nanotag for ultra-sensitive bacterial detection [85]. SERS-based detection is used not only for bacterial identification but also for antibiotic susceptibility testing (AST). The SERS active nanotag for the detection of E. coli bacteria was built of gold nanorods (AuNRs) coated with Ag NPs to form Au@AgNRs (Figure 10.16). Compared to Au nanocrystals (Au NCs) and Au nanoparticles (Au NPs), Ag nanocrystals (Ag NCs), AuNRs are better in their capacity to facilitate localized plasmon resonances in the near-ultraviolet spectral region. As shown in Figure 10.16a, Au@AgNRs are functionalized with the Raman reporter Rhodamine 6G (R6G) and magnetic bead SA to form an N1 (Au@AgNR@R6G@SA) nanotag. The low-abundance protein detector biotinylated antibodies are then linked to the N1 nanotag to make N1-Ab. The complex is formed by anchoring the SERS nanotag N1-Ab with the E. coli bacterial sample in mice blood. The Raman spectra band intensity was measured at $1517\,cm^{-1}$ with a Raman Microscope equipped with a laser wavelength of 785 nm and a power of 20 mW (Figure 10.16e). The changes in Raman intensity in the spectra directly reflect the effects and growth of bacteria (Figure 10.16f). Due to its rapid AST analysis provided by the SERS platform, effective treatment can be initiated promptly. To test the efficacy of the SERS N1-Ab nanotag, the experiment was conducted with PBS and tap water with varying numbers of E. coil colony-forming units per milliliter (CFU/mL). It demonstrated great repeatability and specificity for a wide range of CFU concentrations. The LOD

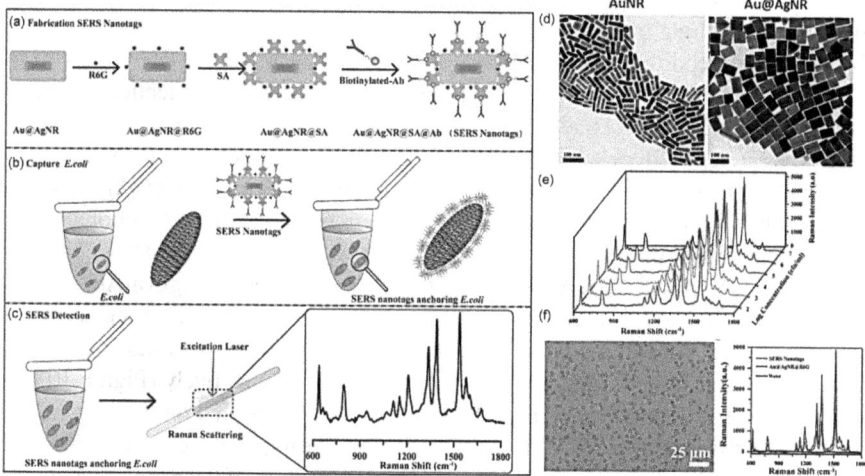

FIGURE 10.16 SERS nanotag functionalized N1-Ab preparation and attachment to E. coli is depicted in the following flowchart (a–c). (d) TEM images of AuNR and Au@AgNR to understand the morphology of the synthesized SERS-nanotag (scale bar of 100 nm). (e) SERS spectra of *E. coli* complex at various concentrations ranging from 10^7 to 10^2 CFU/mL. (f) SERS nanotag optical microscopy picture treated with mouse blood including *E. coli* at 10^4 CFU/mL and the measurement displayed in the graph. (Adapted with permission from Ref. [85].)

for the synthesized N1-Ab SERS nanotag for the detection of *E. coil* is up to 10^2 CFU/mL.

Diseases that can be transmitted from animals to humans are referred to as zoonotic diseases [86]. It encompasses a wide range of contagious diseases that can impact both humans and animals with a unique set of symptoms. The most prevalent form of transmission is direct contact, followed by inhaling or swallowing contaminated items and touching an animal-infected surface [87]. Specifically, viral zoonotic diseases today offer a significant threat to the general public's health. It is estimated that viruses are responsible for causing 75% of newly emergent zoonotic infections [88]. Pseudorabies virus (PrV), a member of the herpes viral family, causes severe mortality in pigs. As a result of this virus disease, people in the porcine animal husbandry sector suffer tremendously, and the death of the pigs results in significant financial losses [89]. The development of a more effective PrV detection technology is required. PrV has eleven envelope glycoproteins with different functions from gB to gN [90]. The attachment of PrV to the host cell is gB's role, while gE is responsible for the replication of the virus. The current vaccination for the prevention of wild-type PrV is Bartha-K61. The vaccine is gE-deleted and is intended to protect pigs from pseudorabies. When PrV is not detected in its early stages, it can create mutations in genes, which can lead to an epidemic of pseudorabies even in patients who have been vaccinated. Traditional laboratory techniques for detecting PrV include loop-mediated isothermal amplification (LAMP), Recombinase Polymerase Amplification, and polymerase chain reaction

(PCR). Detecting PrV necessitates the use of specialist tools, specialized materials, and time-consuming methods. There are only a few benchtop methods for detecting PrV have been reported. The lateral flow strip (LFS) and the SERS-lateral flow assay (LFS) have recently been reported as methods for the diagnosis of infectious diseases. Yong Tang and his team conducted a study on diagnosing wile-type and gE-deleted vaccine treatment utilizing SERS-LFS. SERS-nanotag synthesis of bimetallic (Bm) Au-Ag nanoparticles in core-shell form with Ag NP to produce AuAg@Ag NPs. The 4-amino thiophenol (4-ATP) is used as the bifunctional molecule to functionalize Au NPs and generate Bm1 (AuAg$^{4\text{-ATP}}$@Ag NPs) in the bimetallic formation between Au and Ag (Figure 10.17a). Anti-PrV gE monoclonal antibodies (gE-mAb) that are coupled with Bm1 to serve as Bm1-Ab nanotags. The SERS-LFS absorbent strip was spotted with gb-mAb and goat anti-mouse IgG separately as two lines were designated as the Test line (T line) and Control line (C line), respectively (Figure 10.17c). The wild-type PrV samples are mixed with Bm1-Ab, then placed on the strip, which moves towards the T or C line to form complexes. The sample that contains wild-type PrV and Bm1-Ab are captured at the T and C lines as a result of the expression of both protein gB and protein gE. Wherein, the samples that did not contain wild-type PrV and contained Bm1-Ab only captured it at the C line. The SERS spectra wild-type positive complexes are observed at 1585.7 cm^{-1} using a portable Raman spectrometer with a 785 nm excitation and a power of 100 mW (Figure 10.17d). The performance of SERS-LFS with wild-type PrV was tested utilizing a variety of concentrations ranging from 5 to 650 ng/mL. It was observed that the Mb1-Ab's color intensity decreased when its concentration dropped below 81 ng/mL. As a result, the LOD level

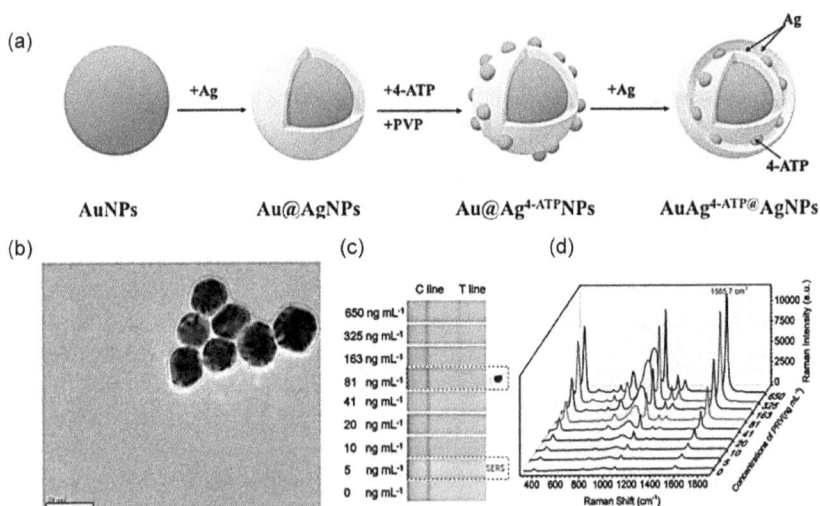

FIGURE 10.17 (a) Diagrammatic representation for the synthesis of the AuAg$^{4\text{-ATP}}$@AgNP molecule. (b) Transmission electron micrographs of Bm1 with a 50 nm scale bar. (c) Images of the Bm1-Ab at the strip taken in visible light at different PrV concentrations. (d) The SERS spectra at the strip T line were acquired by employing Bm1-Ab with an excitation wavelength of 785 nm (350 mW, 1s). (Adapted with permission from Ref. [91].)

of wild-type PrV can be detected using SERS-LFS at a detection rate of 5 ng/mL in 15 minutes. The SERS-LFS assays using Bm1-Ab were investigated for potential clinical applications and were shown to be comparable with gE-specific PCR. It demonstrates that SERS-LFS with Bm1-Ab is a novel method for detecting the wild-type PrV and gE-deleted vaccine.

10.4 SUMMARY AND OUTLOOK

A two-step method for medical diagnostics utilizing SERS-based single-molecule detection has recently received much attention. SERS with microarray and qPCR analysis is frequently used for the detection of single molecules with minimal procedures, time-consuming, and clinical application as a point-of-care (PoC) diagnostic. SERS has made significant strides in resolving some of the system's shortcomings to maintain its high level of sensitivity in the future. SERS applications are divided into two categories: SM-SERS and non-SM-SERS detection. An investigation on Surface-enhanced spectroscopy using SM-SERS. The second approach makes use of SM-SERS to delve deeper into the unique properties of single molecules that are otherwise inaccessible through traditional ensemble research. Bi-analyte SERS (BiASERS) approach gained considerable attention in the section on diagnostics. BiASERS is essentially a contrast technique for observing the statistics of single-molecule events associated with one dye against the background of signals generated by the other. The complication with the BiASERS is that there are multiple counter signals as a result of massive oscillations caused by different molecules during the research. A long-held objective in the research has been to enable SERS to detect non-resonant molecules at the sub-micron level with oscillation in the detection. Non-resonant molecules are more difficult to obtain as SM-SERS detection parameters than resonant molecules. Importantly, any contaminant is likely to display a SERS signal that is equivalent to or even greater than that of the non-resonant target analyte. This problem must be addressed in the future development of non-resonant SM-SERS. SM-SERS is being used to study the SERS effect in a few more fascinating ways. Graphene as a substrate has been used in studies of SERS, perhaps down to the SM level. SERS research is actively looking into nonlinear optical effects and electron transport processes at the sub-micron level. The conductivity and thermal properties of sub-micron molecules are now being studied with SM-SERS. A possible future investigation may be the electrochemistry of single molecules, as described in the preceding section. However, it is anticipated that the list will expand considerably in the next years, creating multiple subfields of research. In retrospect, the expected future path of the SM-SERS technique detection is as surprising and unexpected as one may imagine.

REFERENCES

1. Lu Y., Lin L., Ye J. Human metabolite detection by surface-enhanced Raman spectroscopy. *Mater Today Bio.* 2022;13:100205.
2. Chen R., Jensen L. Interpreting the chemical mechanism in SERS using a Raman bond model. *J Chem Phys.* 2020;152(2):024126.

3. Wang Y., Irudayaraj J. Surface-enhanced Raman spectroscopy at single-molecule scale and its implications in biology. *Philos Trans R Soc Lond B Biol Sci.* 2013;368(1611): 20120026.

4. Ayas S., Cinar G., Ozkan A.D., Soran Z., Ekiz O., Kocaay D., et al. Label-free nanometer-resolution imaging of biological architectures through surface enhanced Raman scattering. *Sci Rep.* 2013;3:2624.

5. Emory S.R., Jensen R.A., Wenda T., Han M., Nie S. Re-examining the origins of spectral blinking in single-molecule and single-nanoparticle SERS. *Faraday Discuss.* 2006;132:249–59; discussion 309–19.

6. Wrzosek B., Kitahama Y., Ozaki Y. SERS blinking on anisotropic nanoparticles. *J Phys Chem C.* 2020;124(37):20328–39.

7. Qian X.M., Nie S.M. Single-molecule and single-nanoparticle SERS: From fundamental mechanisms to biomedical applications. *Chem Soc Rev.* 2008;37(5):912–20.

8. Yamamoto Y.S., Ishikawa M., Ozaki Y., Itoh T. Fundamental studies on enhancement and blinking mechanism of surface-enhanced Raman scattering (SERS) and basic applications of SERS biological sensing. *Front Phys.* 2013;9(1):31–46.

9. Luo S.C., Sivashanmugan K., Liao J.D., Yao C.K., Peng H.C. Nanofabricated SERS-active substrates for single-molecule to virus detection in vitro: A review. *Biosens Bioelectron.* 2014;61:232–40.

10. Chan T.Y., Liu T.Y., Wang K.S., Tsai K.T., Chen Z.X., Chang Y.C., et al. SERS detection of biomolecules by highly sensitive and reproducible raman-enhancing nanoparticle array. *Nanoscale Res Lett.* 2017;12(1):344.

11. Radziuk D., Moehwald H. Prospects for plasmonic hot spots in single molecule SERS towards the chemical imaging of live cells. *Phys Chem Chem Phys.* 2015;17(33):21072–93.

12. Almehmadi L.M., Curley S.M., Tokranova N.A., Tenenbaum S.A., Lednev I.K. Surface enhanced Raman spectroscopy for single molecule protein detection. *Sci Rep.* 2019;9(1):12356.

13. Zhou H., Yang D., Ivleva N.P., Mircescu N.E., Niessner R., Haisch C. SERS detection of bacteria in water by in situ coating with Ag nanoparticles. *Anal Chem.* 2014;86(3): 1525–33.

14. Matteini P., de Angelis M., Ulivi L., Centi S., Pini R. Concave gold nanocube assemblies as nanotraps for surface-enhanced Raman scattering-based detection of proteins. *Nanoscale.* 2015;7(8):3474–80.

15. Pan X., Li L., Lin H., Tan J., Wang H., Liao M., et al. A graphene oxide-gold nanostar hybrid based-paper biosensor for label-free SERS detection of serum bilirubin for diagnosis of jaundice. *Biosens Bioelectron.* 2019;145: 111713.

16. Choi J.H., Kim T.H., El-Said W.A., Lee J.H., Yang L., Conley B., et al. In situ detection of neurotransmitters from stem cell-derived neural interface at the single-cell level via graphene-hybrid SERS nanobiosensing. *Nano Lett.* 2020;20(10):7670–9.

17. Wang H., Han X., Ou X., Lee C.S., Zhang X., Lee S.T. Silicon nanowire based single-molecule SERS sensor. *Nanoscale.* 2013;5(17):8172–6.

18. Xu H., Bjerneld E.J., Käll M., Börjesson L. Spectroscopy of single hemoglobin molecules by surface enhanced Raman scattering. *Phys Rev Lett.* 1999;83(21):4357–60.

19. Zhan P., Wen T., Wang Z.G., He Y., Shi J., Wang T., et al. DNA origami directed assembly of gold bowtie nanoantennas for single-molecule surface-enhanced Raman scattering. *Angew Chem Int Ed Engl.* 2018;57(11):2846–50.

20. Spitzberg J.D., Zrehen A., van Kooten X.F., Meller A. Plasmonic-nanopore biosensors for superior single-molecule detection. *Adv Mater.* 2019;31(23). doi: 10.1002/adma. 201900422.

21. Yang Y., Peng Y., Lin C., Long L., Hu J., He J., et al. Human ACE2-functionalized gold "Virus-Trap" nanostructures for accurate capture of SARS-CoV-2 and single-virus SERS detection. *Nanomicro Lett.* 2021;13:109.

22. Acimovic S.S., Kreuzer M.P., Gonzalez M.U., Quidant R. Plasmon near-field coupling in metal dimers as a step toward single-molecule sensing. *ACS Nano.* 2009;3(5):1231–7.
23. Hatab N.A., Hsueh C.H., Gaddis A.L., Retterer S.T., Li J.H., Eres G., et al. Free-standing optical gold bowtie nanoantenna with variable gap size for enhanced Raman spectroscopy. *Nano Lett.* 2010;10(12):4952–5.
24. Huang X., El-Sayed I.H., Qian W., El-Sayed M.A. Cancer cells assemble and align gold nanorods conjugated to antibodies to produce highly enhanced, sharp, and polarized surface Raman spectra: a potential cancer diagnostic marker. *Nano Lett.* 2007;7(6):1591–7.
25. Yu K.N., Lee S.M., Han J.Y., Park H., Woo M.A., Noh M.S., et al. Multiplex targeting, tracking, and imaging of apoptosis by fluorescent surface enhanced Raman spectroscopic dots. *Bioconjug Chem.* 2007;18(4):1155–62.
26. Potara M., Baia M., Farcau C., Astilean S. Chitosan-coated anisotropic silver nanoparticles as a SERS substrate for single-molecule detection. *Nanotechnology.* 2012;23(5): 055501.
27. Liu H., Li Q., Li M., Ma S., Liu D. In situ hot-spot assembly as a general strategy for probing single biomolecules. *Anal Chem.* 2017;89(9):4776–80.
28. Vo-Dinh T., Kasili° Wabuyele M. Nanoprobes and nanobiosensors for monitoring and imaging individual living cells. *Nanomedicine.* 2006;2(1):22–30.
29. Chen C., Li Y., Kerman S., Neutens P., Willems K., Cornelissen S., et al. High spatial resolution nanoslit SERS for single-molecule nucleobase sensing. *Nat Commun.* 2018; 9(1):1733.
30. Zheng Y., Soeriyadi A.H., RP.osa L., Ng S.H., Bach U., Justin Gooding J. Reversible gating of smart plasmonic molecular traps using thermoresponsive polymers for single-molecule detection. *Nat Commun.* 2015;6:8797.
31. Das G.M., Manago S., Mangini M., De Luca A.C. Biosensing using SERS active gold nanostructures. *Nanomaterials (Basel).* 2021;11(10):2679.
32. Chen X., Gole J., Gore A., He Q., Lu M., Min J., et al. Non-invasive early detection of cancer four years before conventional diagnosis using a blood test. *Nat Commun.* 2020; 11(1):3475.
33. Chen X., Zheng B., Liu H. Optical and digital microscopic imaging techniques and applications in pathology. *Anal Cell Pathol (Amst).* 2011;34(1–2):5–18.
34. Sakamoto S., Putalun W., Vimolmangkang S., Phoolcharoen W., Shoyama Y., Tanaka H., et al. Enzyme-linked immunosorbent assay for the quantitative/qualitative analysis of plant secondary metabolites. *J Nat Med.* 2018;72(1):32–42.
35. Nasseri B., Soleimani N., Rabiee N., Kalbasi A., Karimi M., Hamblin M.R. Point-of-care microfluidic devices for pathogen detection. *Biosens Bioelectron.* 2018;117:112–28.
36. Mazetyte-Stasinskiene R., Köhler J.M. Sensor micro and nanoparticles for microfluidic application. *Appl Sci.* 2020;10(23):8353.
37. Litti L., Trivini S., Ferraro D., Reguera J. 3D printed microfluidic device for magnetic trapping and SERS quantitative evaluation of environmental and biomedical analytes. *ACS Appl Mater Interfaces.* 2021;13(29):34752–61.
38. Sung H., Ferlay J., Siegel R.L., Laversanne M., Soerjomataram I., Jemal A., et al. Global cancer statistics 2020: GLOBOCAN estimates of incidence and mortality worldwide for 36 cancers in 185 countries. *CA Cancer J Clin.* 2021;71(3):209–49.
39. Li J., Skeete Z., Shan S., Yan S., Kurzatkowska K., Zhao W., et al. Surface enhanced Raman scattering detection of cancer biomarkers with bifunctional nanocomposite probes. *Anal Chem.* 2015;87(21):10698–702.
40. Cheng H.-W., Luo J., Zhong C.-J. SERS nanoprobes for bio-application. *Front Chem Sci Eng.* 2015;9(4):428–41.
41. Hellstrom I., Heagerty P.J., Swisher E.M., Liu P., Jaffar J., Agnew K., et al. Detection of the HE4 protein in urine as a biomarker for ovarian neoplasms. *Cancer Lett.* 2010; 296(1):43–8.

42. Fritz-Rdzanek A., Grzybowski W., Beta J., Durczynski A., Jakimiuk A. HE4 protein and SMRP: Potential novel biomarkers in ovarian cancer detection. *Oncol Lett.* 2012;4(3):385–9.

43. Eom G., Hwang A., Kim H., Moon J., Kang H., Jung J., et al. Ultrasensitive detection of ovarian cancer biomarker using Au nanoplate SERS immunoassay. *BioChip J.* 2021;15(4):348–55.

44. Schlucker S. Surface-enhanced Raman spectroscopy: Concepts and chemical applications. *Angew Chem Int Ed Engl.* 2014;53(19):4756–95.

45. Guerrini L., Alvarez-Puebla R.A. Surface-enhanced Raman spectroscopy in cancer diagnosis, prognosis and monitoring. *Cancers (Basel).* 2019;11(6):748.

46. Guerrini L., Pazos-Perez N., Garcia-Rico E., Alvarez-Puebla R. Cancer characterization and diagnosis with SERS-encoded particles. *Cancer Nanotechnol.* 2017;8(5):1–24.

47. Kamemoto L.E., Misra A.K., Sharma S.K., Goodman M.T., Luk H., Dykes A.C., et al. Near-infrared micro-Raman spectroscopy for in vitro detection of cervical cancer. *Appl Spectrosc.* 2010;64(3):255–61.

48. Karunakaran V., Saritha V.N., Ramya A.N., Murali V.P., Raghu K.G., Sujatha K., et al. Elucidating Raman image-guided differential recognition of clinically confirmed grades of cervical exfoliated cells by dual biomarker-appended SERS-tag. *Anal Chem.* 2021;93(32):11140–50.

49. Shi Q., Xu L., Yang R., Meng Y., Qiu L. Ki-67 and P16 proteins in cervical cancer and precancerous lesions of young women and the diagnostic value for cervical cancer and precancerous lesions. *Oncol Lett.* 2019;18(2):1351–5.

50. Rizal C. Bio-magnetoplasmonics, emerging biomedical technologies and beyond. *J Nanomed Res.* 2016;3(3). doi: 10.15406/jnmr.2016.03.00059.

51. Hu Z., Zhou X., Duan J., Wu X., Wu J., Zhang P., et al. Aptamer-based novel Ag-coated magnetic recognition and SERS nanotags with interior nanogap biosensor for ultrasensitive detection of protein biomarker. *Sens Actuators B Chem.* 2021;334:129640.

52. Cao X., Sun Y., Mao Y., Ran M., Liu Y., Lu D., et al. Rapid and sensitive detection of dual lung cancer-associated miRNA biomarkers by a novel SERS-LFA strip coupling with catalytic hairpin assembly signal amplification. *J Mater Chem C.* 2021;9(10):3661–71.

53. Treerattrakoon K., Roeksrungruang P., Dharakul T., Japrung D., Faulds K., Graham D., et al. Detection of a miRNA biomarker for cancer diagnosis using SERS tags and magnetic separation. *Anal Methods.* 2022;14(20):1938–45.

54. Perez-Jimenez A.I., Lyu D., Lu Z., Liu G., Ren B. Surface-enhanced Raman spectroscopy: benefits, trade-offs and future developments. *Chem Sci.* 2020;11(18):4563–77.

55. Moisoiu V., Socaciu A., Stefan A., Iancu S., Boros I., Alecsa C., et al. Breast cancer diagnosis by surface-enhanced Raman scattering (SERS) of urine. *Appl Sci.* 2019; 9(4):806.

56. Mabbott S., Fernandes S.C., Schechinger M., Cote G.L., Faulds K., Mace C.R., et al. Detection of cardiovascular disease associated miR-29a using paper-based microfluidics and surface-enhanced Raman scattering. *Analyst.* 2020;145(3):983–91.

57. Dhingra R., Vasan R.S. Biomarkers in cardiovascular disease: Statistical assessment and section on key novel heart failure biomarkers. *Trends Cardiovasc Med.* 2017;27(2):123–33.

58. Chan J., Dodani S.C., Chang C.J. Reaction-based small-molecule fluorescent probes for chemoselective bioimaging. *Nat Chem.* 2012;4(12):973–84.

59. Zhang X., Li G., Liu J., Su Z. Bio-inspired nanoenzyme synthesis and its application in a portable immunoassay for food allergy proteins. *J Agric Food Chem.* 2021; 69(49):14751–60.

60. Arabi M., Ostovan A., Zhang Z., Wang Y., Mei R., Fu L., et al. Label-free SERS detection of Raman-Inactive protein biomarkers by Raman reporter indicator: Toward ultrasensitivity and universality. *Biosens Bioelectron.* 2021;174:112825.

61. Eremina O.E., Eremin D.B., Czaja A., Zavaleta C. Selecting surface-enhanced Raman spectroscopy flavors for multiplexed imaging applications: Beyond the experiment. *J Phys Chem Lett.* 2021;12(23):5564–70.

62. Badimon L., Pena E., Arderiu G., Padro T., Slevin M., Vilahur G., et al. C-reactive protein in atherothrombosis and angiogenesis. *Front Immunol.* 2018;9:430.

63. Guo W., Hu Y., Wei H. Enzymatically activated reduction-caged SERS reporters for versatile bioassays. *Analyst.* 2017;142(13):2322–6.

64. Go A.S., Mozaffarian D., Roger V.L., Benjamin E.J., Berry J.D., Blaha M.J., et al. Heart disease and stroke statistics--2014 update: A report from the American Heart Association. *Circulation.* 2014;129(3):e28–e292.

65. Sharma S., Jackson P.G., Makan J. Cardiac troponins. *J Clin Pathol.* 2004;57(10):1025–6.

66. Ahmed K.A., Al-Attab W.M. Prognostic performance of combined use of high-sensitivity troponin T and creatine kinase MB isoenzyme in high cardiovascular risk patients with end-stage renal disease. *Kidney Res Clin Pract.* 2017;36(4):358–67.

67. Chen Y.X., Chen M.W., Lin J.Y., Lai W.Q., Huang W., Chen H.Y., et al. Label-free optical detection of acute myocardial infarction based on blood plasma surface-enhanced raman spectroscopy. *J Appl Spectrosc.* 2016;83(5):798–804.

68. Moore T.J., Moody A.S., Payne T.D., Sarabia G.M., Daniel A.R., Sharma B. In Vitro and In Vivo SERS biosensing for disease diagnosis. *Biosensors (Basel).* 2018;8(2):46.

69. Cheng Z., Wang R., Xing Y., Zhao L., Choo J., Yu F. SERS-based immunoassay using gold-patterned array chips for rapid and sensitive detection of dual cardiac biomarkers. *Analyst.* 2019;144(22):6533–40.

70. Aydin S., Ugur K., Aydin S., Sahin I., Yardim M. Biomarkers in acute myocardial infarction: Current perspectives. *Vasc Health Risk Manag.* 2019;15:1–10.

71. Ye X.D., He Y., Wang S., Wong G.T., Irwin M.G., Xia Z. Heart-type fatty acid binding protein (H-FABP) as a biomarker for acute myocardial injury and long-term post-ischemic prognosis. *Acta Pharmacol Sin.* 2018;39(7):1155–63.

72. Chmurzynska A. The multigene family of fatty acid-binding proteins (FABPs): function, structure, and polymorphism. *J Appl Genet.* 2006;47(1):39–48.

73. Jirak P., Fejzic D., Paar V., Wernli B., Pistilli R., Rohm I., et al. Influences of Ivabradine treatment on serum levels of cardiac biomarkers sST2, GDF-15, suPAR and H-FABP in patients with chronic heart failure. *Acta Pharmacol Sin.* 2018;39(7):1189–96.

74. Hu C., Ma L., Guan M., Mi F., Peng F., Guo C., et al. SERS-based magnetic immunoassay for simultaneous detection of cTnI and H-FABP using core-shell nanotags. *Anal Methods.* 2020;12(45):5442–9.

75. Johnson P.T., de Roode J.C., Fenton A. Why infectious disease research needs community ecology. *Science.* 2015;349(6252):1259504.

76. Li Z., Leustean L., Inci F., Zheng M., Demirci U., Wang S. Plasmonic-based platforms for diagnosis of infectious diseases at the point-of-care. *Biotechnol Adv.* 2019;37(8):107440.

77. De Cock K.M., Simone P.M., Davison V., Slutsker L. The new global health. *Emerg Infect Dis.* 2013;19(8):1192–7.

78. Tadesse L.F., Safir F., Ho C.S., Hasbach X., Khuri-Yakub B.P., Jeffrey S.S., et al. Toward rapid infectious disease diagnosis with advances in surface-enhanced Raman spectroscopy. *J Chem Phys.* 2020;152(24):240902.

79. Greatorex J., Ellington M.J., Koser C.U., Rolfe K.J., Curran M.D. New methods for identifying infectious diseases. *Br Med Bull.* 2014;112(1):27–35.

80. Wang C., Liu M., Wang Z., Li S., Deng Y., He N. Point-of-care diagnostics for infectious diseases: From methods to devices. *Nano Today.* 2021;37:101092.

81. Gowri A., Ashwin Kumar N., Suresh Anand B.S. Recent advances in nanomaterials-based biosensors for point of care (PoC) diagnosis of Covid-19- A minireview. *Trends Analyt Chem.* 2021;137:116205.

82. Molina F., Lopez-Acedo E., Tabla R., Roa I., Gomez A., Rebollo J.E. Improved detection of Escherichia coli and coliform bacteria by multiplex PCR. *BMC Biotechnol.* 2015; 15:48.

83. Zhou X., Hu Z., Yang D., Xie S., Jiang Z., Niessner R., et al. Bacteria detection: From powerful SERS to its advanced compatible techniques. *Adv Sci (Weigh).* 2020;7(23): 2001739.

84. Yoon S.A., Park S.Y., Cha Y., Gopala L., Lee M.H. Strategies of detecting bacteria using fluorescence-based dyes. *Front Chem.* 2021;9:743923.

85. Bi L., Wang X., Cao X., Liu L., Bai C., Zheng Q., et al. SERS-active Au@Ag core-shell nanorod (Au@AgNR) tags for ultrasensitive bacteria detection and antibiotic-susceptibility testing. *Talanta.* 2020;220:121397.

86. Hubalek Z. Emerging human infectious diseases: Anthroponoses, zoonoses, and prognoses. *Emerg Infect Dis.* 2003;9(3):403–4.

87. Bueno-Mari R., Almeida A.P., Navarro J.C. Editorial: Emerging zoonoses: Eco-epidemiology, involved mechanisms, and public health implications. *Front Public Health.* 2015;3:157.

88. Chomel B.B., Belotto A., Meslin F.X. Wildlife, exotic pets, and emerging zoonoses. *Emerg Infect Dis.* 2007;13(1):6–11.

89. En F.X., Wei X., Jian L., Qin C. Loop-mediated isothermal amplification establishment for detection of pseudorabies virus. *J Virol Methods.* 2008;151(1):35–9.

90. Klupp B.G., Baumeister J., Dietz P., Granzow H., Mettenleiter T.C. Pseudorabies virus glycoprotein gK is a virion structural component involved in virus release but is not required for entry. *J Virol.* 1998;72(3):1949–58.

91. Shen H., Xie K., Huang L., Wang L., Ye J., Xiao M., et al. A novel SERS-based lateral flow assay for differential diagnosis of wild-type pseudorabies virus and gE-deleted vaccine. *Sens Actuators B Chem.* 2019;282:152–7.

11 Single Molecule Imaging Using State-of-the-Art Microscopy Techniques

Arun Bhupathi
Vishnu Dental College

M. Hema Brindha
SRM Institute of Science and Technology

Ganapathy Krishnamurthi
Indian Institute of Technology

N. Ashwin Kumar
SRM Institute of Science and Technology

CONTENTS

11.1   INTRODUCTION

In life science, single-cell analysis and the interaction of biomolecules is the fundamental experimental methods for personalized medicine. A population of cells is naturally diverse, exhibiting a wide range of physical, chemical, and biological features, even though each cell is genetically unique [1]. These heterogeneities of a cell can be utilized to determine its response to a drug or pathogen in a specific environment [2]. A cell-based analysis often leads to misinterpretation owing to the difficulty in distinguishing which cell activity is responsible for the change in signaling pathways

DOI: 10.1201/9781003409472-11

[3]. Therefore, a cell-by-cell examination of the complete population is necessary to determine the molecular contribution of each cell [4,5]. Single-molecule biophysics research, which focuses on individual molecules within a biological system, is a new frontier in biological exploration. It is possible to examine cellular responses and precise biomolecular pathways by detecting and examining the biomolecules inside cells [6].

Biophysics is an interdisciplinary field of physics and biology that blends mathematical analysis with scientific models. It has applications not just in the realm of engineering but also in the domain of medical practice [7]. Biophysics is prevalent in imaging technologies, in which it aids in the understanding of the structure, dynamics, interactions, and function of biological systems [8]. Biophysics strives to explain biological function by distinguishing between the various sizes ranging from 1 nm to 1 μm of biomolecules, such as long DNA molecules, small fatty acids, sugars, macro, and microproteins, and starches [9]. The development of instruments that can determine the structure and dynamics of biomolecules is receiving more attention in the field of biophysics [10].

Biophysics imaging technologies cover a wide variety of techniques, including microscopy, spectroscopy, electrophysiology, single-molecule approaches, and molecular signalling [11]. The need for relying on the biophysics imaging instrument in research provides very sensitive detection and is a non-invasive analysis tool for biomolecule examination. It is an emerging method for investigating the heterogeneous cell state of a single biomolecule activity. It gives specific information about the properties of a single molecule without ensemble averaging [12]. The approach for detecting a single molecule emphasizes the optical method for identification and evaluation [13]. The optical technique provides accurate detection of biomolecules and the ability to label molecules of interest for subsequent analysis [14]. Recent technological breakthroughs in optical method-based detection and manipulation by light-emitting fluorophores improve the signal-to-noise ratio of the measurement [15].

A fluorophore, also known as a fluorescent molecule or fluorochrome, is a material that emits light when excited by photons of specific energy [16]. Fluorophores are aromatic molecular groups used as tracer agents or dyes for staining, markers, and tags in the molecule of interest for quantitative and qualitative detection using spectroscopic and imaging technologies [17]. Small organic compounds, fluorescent proteins, and antibodies are employed to create fluorophores for sequentially identifying biomolecules on specimens to provide a high-resolution image. Similarly, photo-switching or blinking behavior was exhibited by conjugated fluorescent polymers, quantum dots, and individual photosynthetic reaction centers [18,19]. The first single molecule experiment in live cell imaging of biomolecules was performed by tagging with the green fluorescent protein (GFP) [20,21]. With the rapid development of several fluorophores, imaging techniques can identify the localization of biomolecules within a few nanometers.

In biomedical science, optical microscopy techniques are widely preferred for imaging cellular and organelle activities. Light microscopes, wide-field microscopes, and fluorescence microscopes are examples of conventional optical microscopes [22]. Since the 1950s, various experiments have made it possible to examine biomolecules within cells using single-molecule techniques [23]. In 1956, the first images of DNA, proteins, and collagen were acquired by Transmission Electron Microscopy (TEM). Smaller biomolecules have fewer carbon atoms with a lower atomic number, making TEM imaging difficult [24]. In 1961, Boris Rotman was the first person to measure single molecules by fluorescence microscopy in enzyme-catalyzed reactions that were labeled with a fluorescent dye [25]. In 1976, Thomas Hirschfeld's research team detected single globulin protein biomolecules using hundreds of fluorescent chemical dye molecules [26]. Barack and Webb scientists in 1982 explored the diffusion effectiveness of multiple fluorescent tags for tracking the lipoprotein in the fibroblast cell [27]. In 1989, Moerner and his team detected single molecules optically by frequency modulation absorption spectroscopy [28]. Later Dickson and his team reported a study in 1997 detailing the reversible photo-switching of fluorescent protein molecules [29]. When excited with light as the excitation source, the image obtained by fluorescence from a conventional optical microscope appears blurry and photobleached on the detector side. The primary reason is an inadequate excitation source that illuminates the entire focusing volume, resulting in out-of-focus regions above and below the image plane. This renders conventional microscopy unsuitable for detecting single molecules in three-dimensional images [30].

Recent imaging technology advancements have enabled the detection of single molecules with increased sensitivity and specificity [31]. Single-molecule fluorescence microscopy is a subset of conventional fluorescence microscopy with distinct properties. Single-molecule fluorescence microscopy can differentiate and distinguish single fluorophore molecules from background signals. The study of dynamic molecular interactions in living cell plasma membranes can be greatly aided by the use of single-molecule imaging. Various ways can be used to decrease background signals by constructing a spatially constrained structure that allows molecules of interest to be detected. All the single-molecule such as have a similar fundamental concept of temporal separation of fluorophore emission and then reconstruction of the image after multiple localization events with sub-diffraction-limited features. The chapter offers a thorough overview of the various imaging techniques (Figure 11.1) based on fluorescence that make it possible to identify single molecules, along with a rundown of the technical particulars and practical applications of each technique.

11.2 TECHNIQUES THAT ENABLE SINGLE-MOLECULE FLUORESCENCE-BASED DETECTION

11.2.1 CONFOCAL MICROSCOPY

Confocal microscopy, also known as confocal laser scanning microscopy (CLSM) or laser confocal scanning microscopy (LCSM), is an optical imaging technique that provides exact location and identification at distinct depth levels in thick specimens. CLSM provides numerous advantages for imaging live or fixed cells [32]. Confocal microscopy overcomes the limitation of simultaneous visualization of in-focus and out-of-focus in wide-field microscopy by employing the laser as the source of energy to excite the fluorophore along with other parameters of instrumentation [33]. Laser

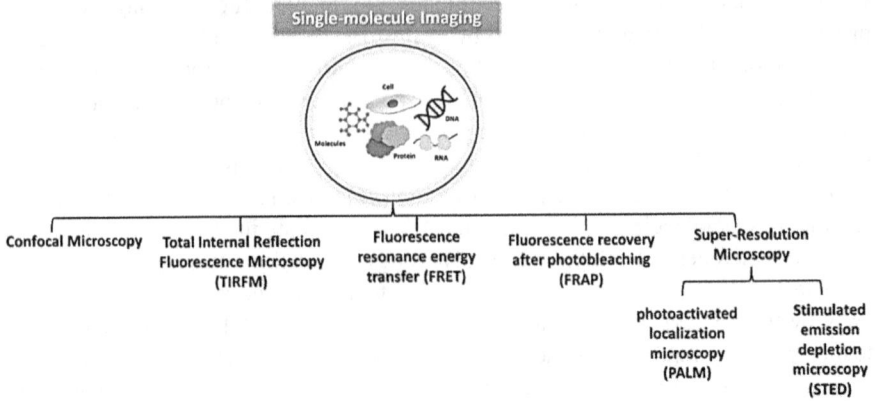

FIGURE 11.1   The overall hierarchical representation of single-molecule imaging.

beam offers focused imaging rather than illuminating the entire specimen with xenon arc lamps. Out-of-focus lights are efficiently removed by positioning the pinhole at the source and image planes before reaching the detector [34]. The presence of two pinholes one at the light source projection area and the other before the detector enables the focused laser beam to reach the specimen and detector [35]. The rest of the scattered emission light below the focal plane is blocked out. CLSM provides localized detection by the elimination of background noise [36].

CLSM consists of pinholes, objective lens, rapid scanning mirror, wavelength selection filter, detector, and laser illumination are the essential components shown in Figure 11.2. Different sources of lasers used in CLSM include Argon-ion laser, krypton-ion laser, air-cooled laser using argon-krypton mixture laser, helium-neon (He-Ne) laser, helium cadmium (He-Cd) laser, diode laser, diode-pumped neodymium-yttrium aluminum garnet (Nd: YAG) laser, titanium doped sapphire (Ti-sapphire) laser [37]. Among them, air-cooled argon-ion lasers are widely preferred as the light source for CLSM because it contains two usable wavelength emission ranges 488 and 514 nm, which is equal to 75% of the total laser power. Krypton-ion lasers are found to have fewer wavelength outputs compared to the argon-ion. The raster scanning by focusing laser beam over the specimens facilitates the acquisition of three-dimensional structures by stacking multiple focal plane images [38]. The stacking of the specimen images is done using the suitable software commonly known as microscopy deconvolution software (z-stack) [33,39].

The basic comparison of widefield and confocal was performed earlier by Nathan and his colleagues [40], to show a substantial amount of signal difference using fluorescent dyes (Figure 11.3). To achieve high-sensitivity detection, LSCM for biological sample analysis uses fluorescence as an imaging mode. Target-oriented visualization is essential in biological samples for identifying the structural and functional activities of live cells and tissues [41]. Fluorescent probes are intended to bind with biomolecules such as protein-nucleic acid with its aromatic chemical interaction [42]. The fluorescent probes are localized in the specific biomolecule, such as the cytoskeleton,

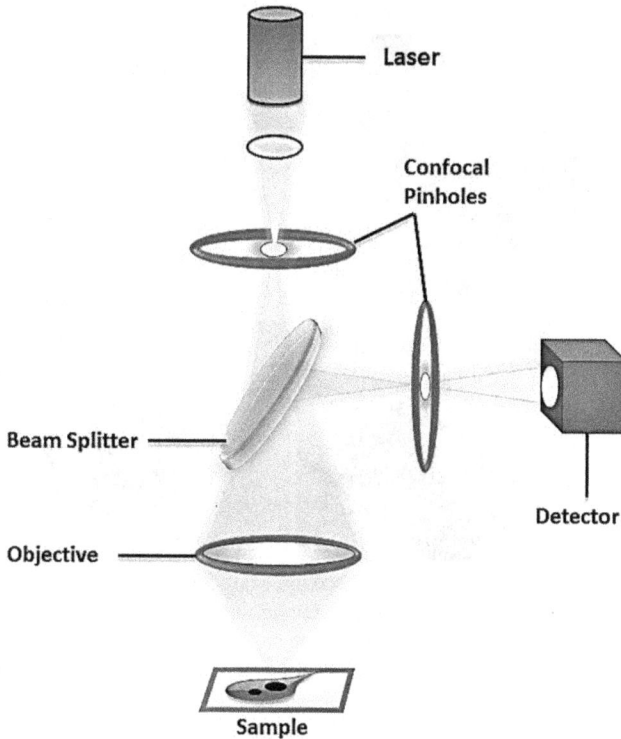

FIGURE 11.2   A schematic representation of the key parts present in confocal microscopy.

mitochondria, and other subcellular organelles. These probes are also employed in monitoring cellular viability, endocytosis, exocytosis, protein trafficking, signal transduction, and enzyme activities [43]. The classic fluorescent probe for confocal microscope successfully utilized for many years includes fluorescein (495 nm), iso-thiocyanate, lissamine rhodamine (575 nm), Texas red [44], phenanthridine derivative ethidium bromide [45], propidium iodide (617 nm). The incident laser light is directly absorbed by fluorescent dye-stained molecules and emitted at a certain wavelength. In this line of interest, the hippocampus of the mouse brain has thick sections treated with anti-bodies like glial fibrillary acidic protein (GFAP) gives red emission [46]. GFAP highlights the protein found in the glial cell. Furthermore, these sections were stained by counter dye known as Hoechst 33342, which emits a blue color that high-lights the nuclei. However, a large number of glares are seen mouse hippocampus section using wide-field microscopy (Figures 11.3a and c). viewed but, visualizing the same samples under a confocal microscope provides excellent structural details of the hippocampus. Similarly, they also tested using the thick specimen to test the penetration ability of the confocal microscope vs wide-field. The rat smooth muscle fibers were stained with Alexa Fluro 568 and Oregon green 488 to highlight the actin and glycoproteins, respectively which are counterstained with DRAQ5 for nuclei. The thick specimen also produced blurred images in wide-field with a lack of details

Microscopic Image

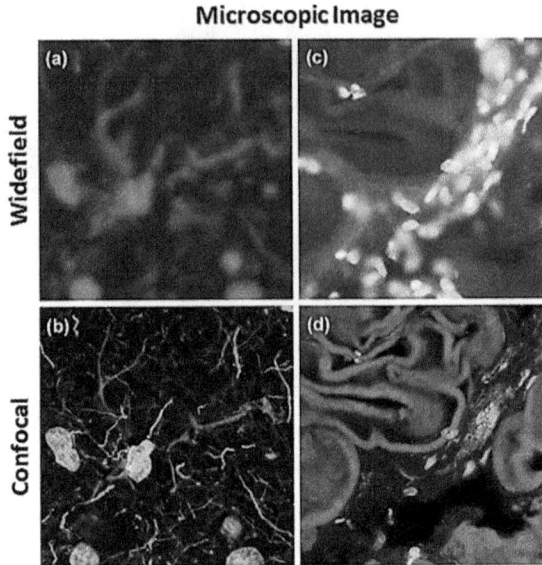

FIGURE 11.3   The image depicts a comparison between the wide-field microscope and the confocal microscope. (a) and (b) The image of the mouse brain's hippocampus. (c) and (d) The image of a thick slice of rat smooth muscle. (The images were adapted with permission from the reference [40].)

while confocal microscopy reveals the topography of the rat's smooth muscles (Figure 11.3b and d). When compared to other types of optical microscopy, confocal microscopy can produce images of biomolecules that are superior in quality, both in terms of resolution and accuracy. Furthermore, the confocal microscope's high spatial image quality provides a better understanding of the biomolecule's activities.

The LSCM technique can be utilized to examine the single-molecule interactions that take place between DNA and proteins. Segers and his co-workers studied the protein-DNA complexes involved in Nucleotide Excision Repair (NER) using LSCM imaging [47]. NER is a critical strategy involving the mammalian DNA repair process by removing damaged DNA caused by UV radiation. The NER process requires the ordered assembly of several proteins at the damaged DNA site and requires complex intermediate structures [48]. The reaction mechanism behind the intermediate structures remains unclear due to the lack of optical fluorophore methods to represent distinct DNA-protein complexes. The author employed human DNA Damage Recognition and Repair Factor (XPA protein) tagged with an enhanced green fluorescent protein (EGFP) to investigate the DNA repair mechanism in XPA-deficient human fibroblasts. DNA substrates with a 5' Cy3.5 label were designed to represent NER intermediates. EGFP is a unique marker gene product that is easily visible using imaging techniques. It can be activated by the 488 nm laser line and is best detected at 510 nm [49].

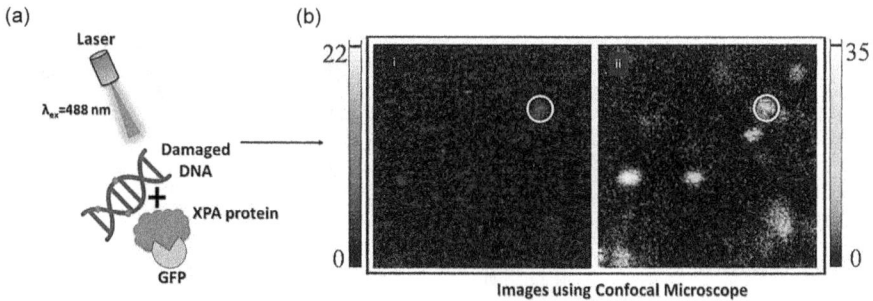

FIGURE 11.4 (a) The Diagram of GFP-tagged protein-DNA complexes is to study the DNA repair mechanism using LSCM. (b) Excitation and emission are detected simultaneously in confocal images of EGFP–XPA and Cy3.5 DNA complexes in a 1% agarose matrix. (i) Cy3.5 emission, 577–632 nm, and (ii) EGFP emission, 500–550 nm. (The images were modified and adapted with permission from the reference [47] with Copyright © 2002 Oxford University Press.)

Cyanine dyes are used to mark proteins, antibodies, peptides, nucleic acid probes, and other biomolecules for use in fluorescence imaging and detection techniques. The emission wavelengths of cyanine excitation using a laser source are generally around 532 and 635 nm [50]. The binding of an EGFP-labeled XPA protein to a Cy3.-5-labeled DNA substrate was quantified using simultaneous excitation and emission detection of both fluorophores (Figure 11.4a). The co-localization of Cy3.5 and EGFP signals inside a single diffraction-restricted region showed XPA-DNA complexes. Measurements were made on samples in a 1% agarose matrix under circumstances that are consistent with protein activity and allow for the study of reactions under equilibrium conditions. The EGFP sample was detected in Cy3.5 with the wavelength 577–632 indicated in red seen in Figure 11.4a. From Figure 11.4b, the bright spot indicated the emission profile of the immobilized EGFP-XPA molecules detection in the 500–550 nm region. The low intensity of this image, as well as the lack of localized spots, indicate that there was no cross-talk from EGFP emission into the Cy3.5 detecting channel. DNA was freely diffusing but protein-bound DNA remained stationary, allowing them to be distinguished and yield quantitative data on DNA binding. The NER was examined on a single molecule level under varied conditions using LSCM. It offers novel insights into the dynamic analysis of molecular processes in biological reactions.

A confocal imaging technique is a valuable tool for quantifying the functions of biomolecules, which is essential for knowing the activity of biological system responses. Several in-vitro studies are carried out to describe the mechanism and reaction of the cells, which is often detected by using various assay techniques. Understanding cell biology can be significantly improved by knowing about the behavior and interactions of biomolecules in their native environment [51]. One such advanced technique used for imaging the live animal is confocal intra-vital microscopy (CIVM) [52]. In this context, Pedro and his colleagues imaged the liver cells and microvasculature of live animals by adapting the low-

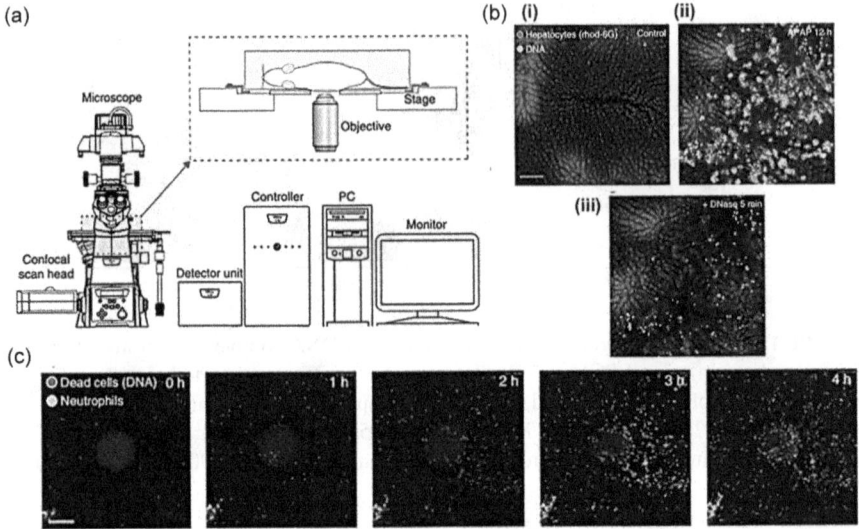

FIGURE 11.5 A confocal intravital microscope was used to capture images of liver cells in a rat. (a) Cost-effective inverted laser-scanning microscope for in-vivo imaging. (b) Confocal microscopy image of rat liver hepatocytes I Normal liver hepatocytes without necrosis stained with rhodamine 6G (ii) Necrotic hepatocytes stained with fluorophore sytox after 12 hours (iii) Necrotic hepatocytes following the injection of DNase. (c) Neutrophil migration to the necrotic site was seen throughout 0–4 hours at the necrotic site. (The images were reproduced with permission from the reference [53].)

cost inverting laser-scanning microscope for in-vivo imaging [53]. The visualization of the liver cell by doing laparotomy on the mice and placing the liver in the static position on the stage. Before this procedure, the injured liver used acetaminophen (APAP) to understand the liver cell's failure mechanism in its native environment. The use of antibodies for the in-vivo cell staining includes Anti-f4/80 for monocytes, Ly-6G dye for neutrophil staining, CD31 staining the endothelial cells, and DAPI for staining the DNA to identify the damage. shows the normal liver stained with rhodamine 6G (red) and shows the morphology of the hepatocyte. APAP overdose caused liver necrosis and the release of the DNA was predominately (ii) in Sytox Green after 12 hours. The DNase enzyme catalyzes the cleaved phosphodiester links in the DNA, therefore destroying the DNA. It was directly added to the liver site and DNA degradation was noted by the reduction in the green color after 5 minutes of one administration of the DNase. The CIVM also clearly assesses the in-vivo neutrophil migration to the necrotic site over the period (from 0 to 4 hours).

Confocal microscopy requires a longer scan time to produce images with a higher resolution. It increases the total amount of laser light exposure to the fluorophore-tagged cells or biomolecules. The maximal fluorophore excitation signal is amplified by increased laser power. When a living cell with fluorophore is exposed to a point-scanning laser, the heat triggers photo-oxidative cell death. Increased fluorophores enable longer exposure times and prevent cell damage, but their imaging data are ineffective because of the greater signal-to-noise ratio. Photobleaching is another serious issue that can arise as a result of prolonged laser exposure to cells. The fluorophore loses all fluorescence properties due to the cleavage of the bonds with the cells after exposure. Therefore, careful consideration must be given to the imaging parameters of the cells while using confocal microscopy to minimize the amount of damage caused.

11.2.2 TOTAL INTERNAL REFLECTION FLUORESCENCE MICROSCOPY (TIRFM)

The signal-to-noise ratio can be considerably enhanced by methodically removing background fluorescence from outside the focal plane using selective activation of the fluorophores [54]. Total Internal Reflection Fluorescence Microscopy (TIRFM) is one such method for selectively activating fluorophores in an aqueous environment adjacent to a solid surface [55]. TIRFM makes use of the distinctive qualities of an evanescent wave or field generated in a small region of a specimen near the interface of two media with different refractive indices. Figure 11.6, demonstrates the whole configuration of the TIRFM system, including the placement of the excitation source, filters, mirrors, detection locations, and a monitoring system. When

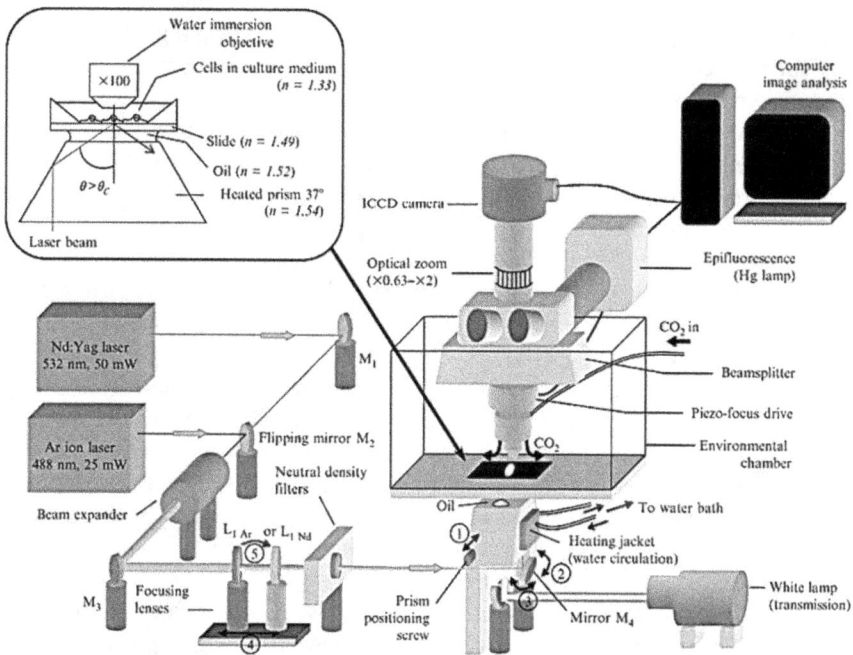

FIG. 1. Schematic diagram of the TIRF setup described in the text.

FIGURE 11.6 A graphic representation of the TIRFM arrangement, showing its various components. (The image was reproduced with permission from the reference [57].)

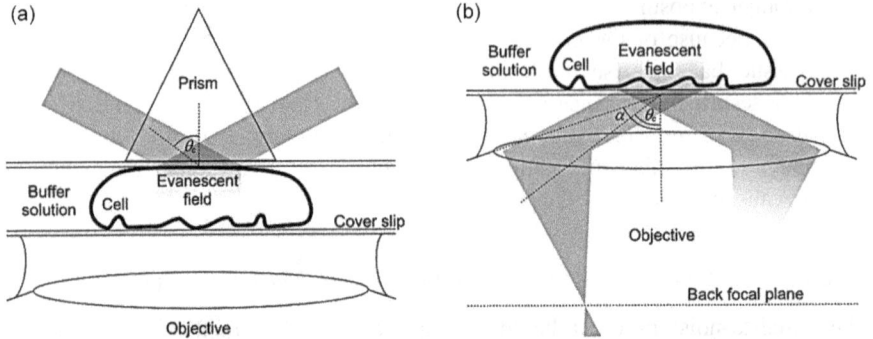

FIGURE 11.7 Types of TIRFM, including prism and objective operation. (a) The evanescent field is formed on the sample's other side from the objective lens that gathers the fluorescence in prism-type TIRFM. (b) Fluorescence from the sample is gathered on the same side of the beam as it is excited in objective-type TIRFM. Focusing the excitation beam off-axis in the back focal plane results in a collimated beam. (The images are adapted from the reference [60] with the license by CC BY 3.0.)

a sinusoidal wave is reflected off an interface at an angle greater than the critical angle, evanescent waves are generated. Interaction between the specimen and the glass coverslip or tissue culture container is the most used interface in TIRFM. The TIRFM laser eliminates sample contact with the glass coverslip region. As a result, only adherent cells or biomolecules may be viewed with this approach, while non-adherent samples are not seen [56].

A TIRFM instrument is constructed either using the prism approach or the objective lens method (Figure 11.7) [58]. In the prism method, a focused laser beam is directed through a prism that is affixed to the coverslip of a microscope. The laser beam can be focused on the area of interest by adjusting the incidence angle such that it is equal to the critical angle (Figure 11.7a). The prism technique has been applied in biological applications for over two decades. However, it has never been a popular research tool due to a variety of limitations, most significantly due to geometric constraints on specimen manipulation. Prisms occur in a wide variety of shapes and sizes, but all have some constraints on access to the specimen. Fluid injection into the specimen compartment and monitoring of physiological parameters are still challenging tasks [59]. The objective lens approach, often known as through-the-lens illumination, excludes many of the drawbacks of prisms. In objective TIRFM no bending or reflection of light happens, it absorbs all the light emitted by the tissues or biomolecule samples. A further advantage of objective-based TIRFM is simpler and less expensive than other types of TIRF (Figure 11.7b) [60].

The numerical aperture of an objective describes the lens's capacity to capture light. High numerical aperture objectives enable incidence angles greater than the critical angle. The objective's numerical aperture specifies the range of angles at which light could be emitted. The cellular refractive index is approximately 1.38, hence an objective numerical aperture should have more than the refractive index value [61]. It is preferable to use objectives with a wider numerical aperture because they allow for greater fine adjustment of angles that are beyond the critical angle range. The laser focal point distance from the lens axis increases when the critical angle is exceeded. As the laser's focal point moves away from the lens' axis, the evanescent field's penetration depth decreases consistently [62]. Based on this precise configuration, TIRFM can be used to analyze the pattern of biomolecules by detecting the position, size, composition, and mobility of the contact zone with a fluorescent substance [63].

TIRFM allows real-time imaging of plasma membrane events in adhering cells with nanometer axial resolution, low phototoxicity, and virtually no fluorescent background [64]. TIRFM has developed unique uses with the development of live cell imaging employing fluorescent proteins or probes. Jean-Baptiste Manneville studied the quantitative information on the organization of the cytoskeleton in both fixed and live migrating cells [57]. The cytoskeleton is a framework that aids cells in maintaining their shape and internal structure, as well as providing mechanical support for cell processes including division and mobility. Actin filaments, microtubules (MT), and intermediate filaments are the three main types of filaments that make up the cytoskeleton [65]. Cell-substrate adhesion studies using TIRFM have largely focused on the involvement of the actin cytoskeleton. The actin cytoskeleton is a crucial regulator of a variety of physiological processes that are essential for successful wound healing [66]. The TIRFM technique was used to observe the actin and microtubule cytoskeleton in astrocytes during wound healing–induced migration. It is conceivable to employ a prism-based TIRFM configuration that permits dual-color imaging for both fixed and living samples [67]. Sources of TIRFM are Argon (Ar) and Nd: YAG laser beams with 488 and 532 nm excitation wavelengths, respectively. During migration, astrocyte protrusions can grow over 100 m in length, necessitating the use of a 40x objective to view the entire cell. The glass-bottom dish is used to cultivate and maintain confluent primary rat astrocytes. Stains like rhodamine-phalloidin, a bicyclic peptide that is extremely selective for actin filaments (also known as F-actin). Rhodamine phalloidin bright red-orange dye having excitation at 540 nm and emission at 565 nm [68]. Green Fluorescent Protein (GFP) is incorporated to track the MT which has two excitation peaks 395 and 475 nm (blue), with an emission peak at 509 nm (green). Microtubules and actin filaments in the cytoskeleton of the migratory astrocyte are depicted in Figure 11.8. The TIRFM image (color-coded in green) of fixed cells stained for actin and microtubule (Epi, color-coded in red) can be superimposed with the corresponding epifluorescence (Epi) image (Figure 11.8b). In the microtubule projection with actin filament, the moving cells are in close contact with the substrate and are seen using TIRFM.

The TIRFM technique can also be used to monitor the dynamic build-up of microtubules development. One of the important regulatory sites affecting microtubule cytoskeleton architecture and function is the dynamic plus-ends of microtubules. Microtubule plus-end tracking proteins (+TIPs) are essential for the connections between microtubule ends and other cellular structures. +TIPs can bind microtubule ends to the cell surface by interacting with plasma membrane-associated proteins, in the case of some +TIPs, or directly with actin fibers [69]. Ram Dixit and Jennifer L. Ross investigated the +TIP activity mechanism at the single-molecule level [70]. Cellular activities such as microtubule dynamics, cell migration, and intracellular

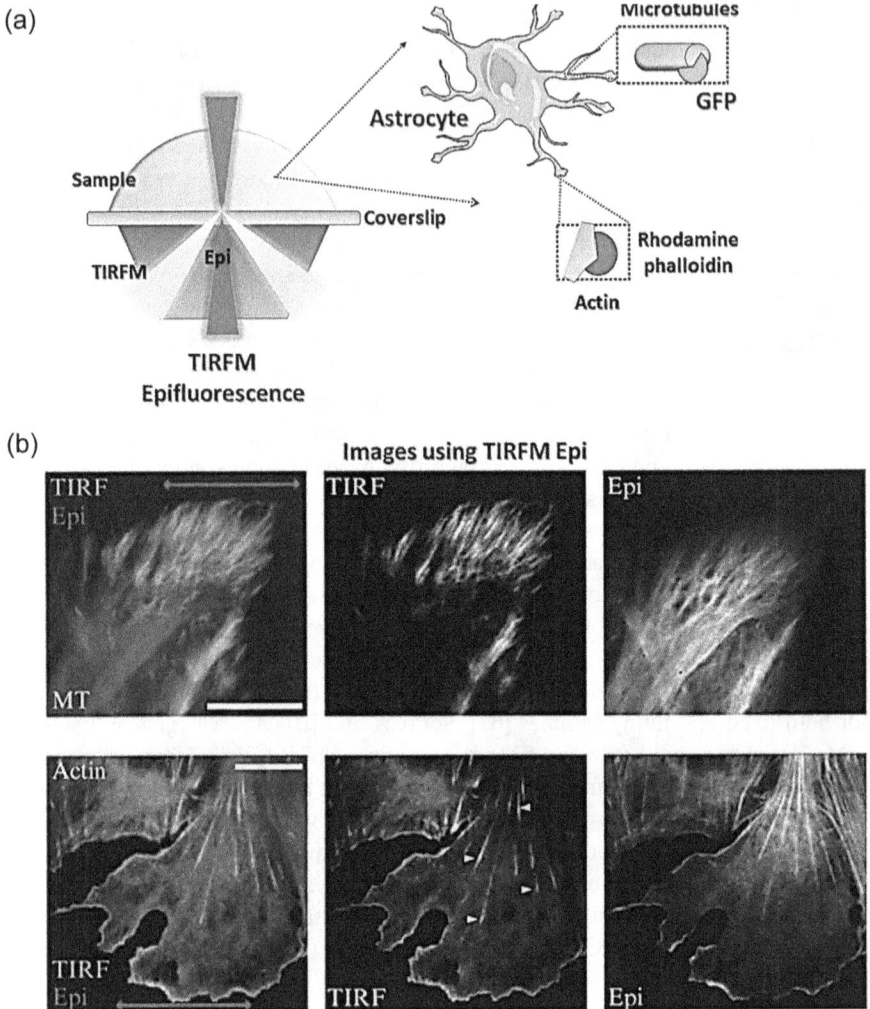

FIGURE 11.8 (a) Diagrammatic illustration of the TIRFM and epifluorescence image of the actin and microtubule of the astrocyte's cells tagged with fluorophores. (b) The microtubules and actin of the astrocyte cells are produced using TIRFM and epi imaging techniques. (The astrocyte image in A was obtained from smartservier.com, while the image in B was reprinted with permission from reference [57].)

transport are all affected by the cytoplasmic linker protein 170 (CLIP170). CLIP170 is a CAP-Gly domain-containing protein found at the plus end of growing microtubules [71]. End-binding protein 1 (EB1) is the primary regulator of the +TIP networks that expand the microtubule ends. The TIRFM is used to demonstrate variations of Mammalian EB1 and CLIP170 protein kinetics. The inclusion of rhodamine-labeled tubulin subunits along with the presence of Green Fluorescent Protein (GFP) and

Alex fluor-488 monitors microtubule assembly. CLIP170 fragment with a green fluorescent protein (GFP) tag is compatible with microtubule-binding and +TIP activity. However, GFP is a 28 kDa protein that looks like a cylinder with dimensions of 4.2 and 2.4 nm [72]. Due to its vast size, tagging the EB1 with GPF can significantly impact its ability to interact with other +TIPs [73]. So, use of small fluorophores like Alexa fluor-488 was employed to investigate dynamic microtubules. Alexa Fluor® 488 (AF488) is a commercially available green fluorophore with absorption and emission peaks in the 496–519 nm range.

The behavior of single +TIP molecules is evaluated in the presence of normal +TIP concentration and an AF488 labeled pattern at microtubule plus ends. A comet-like localization pattern of +TIPs at increasing microtubule ends indicates plus-end tracking activity. EB1 is an independent +TIP that is only tracked during the polymerization of microtubule plus ends and not during depolymerization using TIRFM (Figure 11.9a). CLIP170, on the other hand, does not contain its plus-end track. Instead, it attaches to microtubules and moves in a single dimension (Figure 11.9b). In the presence of normal EB1, CLIP170-GFP demonstrates strong +TIP activity (Figure 11.9c). Thus, TIRFM provided novel insights into the cytoskeleton's actin filament and microtubule dynamics.

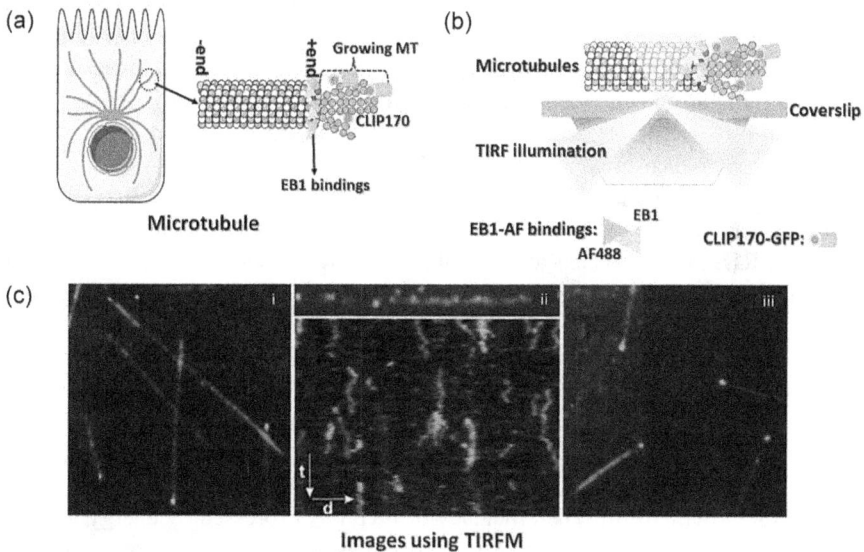

FIGURE 11.9 (a) Schematic representation of the microtubule's preparation for +TIP detection of the proteins EB1 and CLIP170 with GFP and AF488 fluorotags. (b) Diagrammatic representation of fluorescent-tagged microtubule imaging using TIRFM. (c) Images obtained from the TIRFM; (i) are the ends of developing microtubules highlighted by AF488-EB1's brilliant spots. (ii) is the CLIP170-GFP binds along the length of microtubules and exhibits diffusive mobility along the microtubule lattice (as illustrated in the kymograph below). (iii) is the specific localization of the growing end of microtubules in the presence of CLIP170-GFP alone. (The microtubules image in A was obtained from smartservier.com, while the image in B was reprinted with permission from reference [70].)

The scattering induced by the laser source's excitation frequently distorted the image quality of TIRF images. The scatter in TIRFM is caused when the laser illuminates the sample at an angle close to the critical angle. Also, the dense organelles within the cells have a high index of refraction, resulting in light dispersion from the sample [56]. As stated previously, the limitation of TIRF microscopy is that only slices of adherent cultivated cells can be employed in investigations. This results in TIRFM imaging applicable in understanding the activity that takes place at the cell's cytoskeleton levels. Moreover, integrating TIRFM with other techniques such as fluorescence recovery after photobleaching (FRAP), fluorescence correlation spectroscopy (FCS), fluorescence resonance energy transfer (FRET), or atomic force microscopy (AFM) can provide a vast array of mapping information on the molecular dynamics of living cells [74].

11.2.3 Fluorescence Resonance Energy Transfer (FRET)

Cell biology research has entered a new era because of its ability to understand complicated biological interactions at the single-molecule level. Fluorophore-tagged interactions aid in the visualization of the changes that occurred during the biochemical process [75,76]. Different imaging techniques are used to study the outcomes of cell alterations with fluorophores, either in fixed cells or live cells. The topic of need is understanding the dynamic interaction confirmation and co-localization of the suspected biomolecular process using an *in-situ* technique [77]. Imaging approaches for examining the aforementioned occurrences are usually hampered by the low resolution and biological noise caused by the complex biomolecular structure of cells. However, closer measurements of the molecules' distances from one another are required to better understand the physical interactions between protein, DNA, and RNA partners in a typical biomolecular process. The fluorescence resonance energy transfer (FRET) mechanism is an incredibly effective method for studying molecular interactions within cells [78]. FRET is employed to investigate protein-protein interactions, ligand binding to a receptor and molecular dimerization.

The FRET mechanism requires both the donor and acceptor fluorophore [79]. Energy is transmitted between the two fluorophores such that the emission wavelength of the donor fluorophore is the excitation wavelength of the acceptor fluorophore. FRET is a non-radiative energy transfer that occurs as a result of non-radiative dipole-dipole coupling. FRET occurs only when the distance between two fluorophores is less than 10 nm. The efficiency of FRET is expressed by energy transfer, which is inversely proportional to the distance between the donor and acceptor [80]. As the distance rises, the energy transfer reduces, and the FRET mechanism cannot take place. FRET requires the construction of fluorophores and targets within cells. The resonance energy transfer happens due to the fluorophore pair's significantly overlapping emission spectra. Fluorophores utilized in FRET come from a particular class of fluorescent proteins (FPs). Only a certain number of donors and acceptors can work together effectively as a pair. The underlying molecular pathways can be investigated using a number of different fluorescent proteins that exhibit FRET [81]. The most efficient FRET fluorophore combination is cyan fluorescent proteins (CFP) and yellow fluorescent proteins (YFP). CFP and YFP are effective donor-acceptors

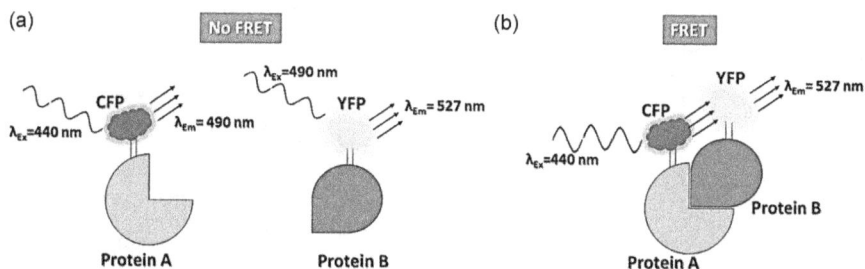

FIGURE 11.10 The FRET mechanism is depicted schematically with an illustration of the fundamental protein interactions.

for FRET because their spectrum properties allow them to last longer [82]. Protein-protein interaction mapping with FPs illustrates the concept of FRET and provides insight into the key mechanism. The fluorophores CFP and YFP are attached to macromolecular proteins A and B, respectively. In the native condition, proteins A and B are kept far apart from one another to prevent interaction. When proteins A and B are excited at their respective peak wavelengths, the emission wavelength between them remains unchanged (Figure 11.10). When the protein undergoes conformational alterations that reduce the distance between two macro proteins to < 10 nanometers. The interaction of the proteins of two fluorophores is brought close, they can engage in the FRET molecular interaction. The excited-state energy of donor fluorescent molecules linked with protein A is transferred to acceptor fluorescent molecules associated with protein B.

As a non-radiative quantum mechanical process, resonance energy transfer does not involve a collision or the generation of heat. The quantum yield is impacted by the orientation of the donor and acceptor transition dipole moments and the distance between the fluorophores [83]. The Jablonski transition is used in FRET to demonstrate the linked transitions between CFP donor emission and YFP acceptor absorption (Figure 11.11). The excited state energy is transferred directly to the acceptor from the donor through the electronic transitions that are mediated by the photon emission of the donor. The quantity of overlap between the donor emission and acceptor absorption spectra influences energy transfer efficiency. Figure 11.11bs depicts the features of the CFP and YFP excitation and emission wavelength spectra. A cross-over of the emission channel is indicated by the presence of a green area in the representation of the region that contains the overlap between the emission of CFP and YFP. The fluorophores' cross-over signals revealed the structural changes that occurred as a result of protein-protein interactions [84]. Though FPs are commonly used for FRET evaluation, their huge size (more than 25 kD) disturbs the structure of target proteins and impairs their ability to carry out vital functions. FPs should be built specifically for biomolecular FRET sensing to reduce unintended consequences. Recent genetic code expansion employing unnatural amino acids (UAAs) enables site-specific installation of novel and essential chemical functionalities, such as fluorescence or ligand binding or cross-linking, or photo caging, at

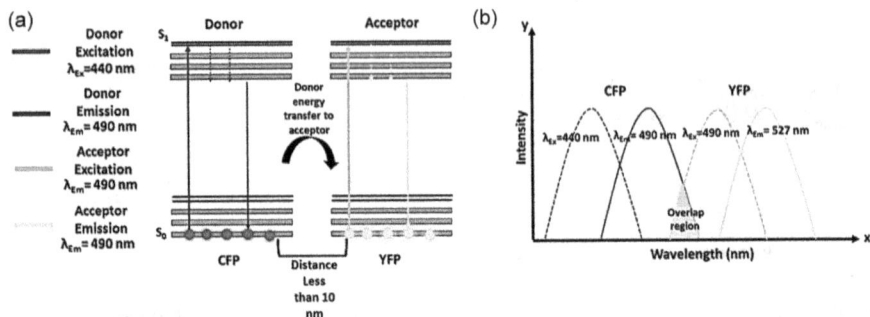

FIGURE 11.11 (a) The process of resonance energy transfer that occurs during protein-protein interaction is depicted by the Jablonski diagram. (b) The graph shows the portions of the spectrum that overlap between donor proteins and acceptor proteins.

regions of interest [85]. This technique can be used to fuse small FPs to either the target protein's N or C terminus to understand its interactions' changes. Using this technique, various unknown dynamic mechanisms can be tested with FRET at a small scale without affecting the structure of target proteins.

The stimulation of apoptosis in cancer cells is believed to be crucial to the efficacy of cancer therapy. However, the role of certain proteins in apoptosis has been unclear. The Bcl-2 Associated X-protein (Bax) is a protein that is a member of the Bcl-2 gene family and is known to promote cell death [86]. This protein regulates intrinsic apoptosis by influencing mitochondrial-initiated pathways. Bax can create a pore by inserting its alpha-helical structures into the outer membrane of the mitochondria [87]. During activation, monomeric Bax undergoes a conformational shift that exposes a Bcl-2 Homology 3 (BH3) domain and a transmembrane domain outside. Activated Bax oligomerizes in the outer membrane of the mitochondria, releases cytochrome c, and induces apoptosis. The fate of a cell is determined by the dynamic equilibrium between anti-apoptotic membranes and pro-apoptotic family members [88]. For instance, the apoptosis-inhibiting activity of heat shock protein (Hsp70) is maintained in normal conditions through direct binding of the ATPase domain to Bax. In conditions of cancer, significantly increased Hsp70 protects tumor cells from apoptosis by serving anti-apoptotic pathways, including the inhibition of Bax [89]. This occurs when Bax detached from Hsp70 and translocates to the mitochondria to induce mitochondria-associated apoptosis. The mechanism of Bax detachment from Hsp70 is not been well elucidated. The Centre for Bio-functional Molecules, led by Injae Shin, employed the FRET system with a genetic code expansion technique to analyze the effects of various apoptosis-inducing substances on the interaction between Bax and Hsp70 in single cells [90]. Genetic code expansion technology was used to incorporate the small fluorescent amino acid 6-acetyl (naphthalene-2-yl amino)-2- amino propanoic acid (ANAP) into three specific locations in the Bax protein (G11, D86, and F105) (Figure 11.12c). The genetically inserted ANAP serves as a fluorescent donor in the range of 410–6530 nm, with excitation at 405 nm (Figure 11.12a). The fluorescent acceptor was chosen to be YFP attached to the C-terminus of Hsp70 because the emission spectrum of ANAP overlaps with the

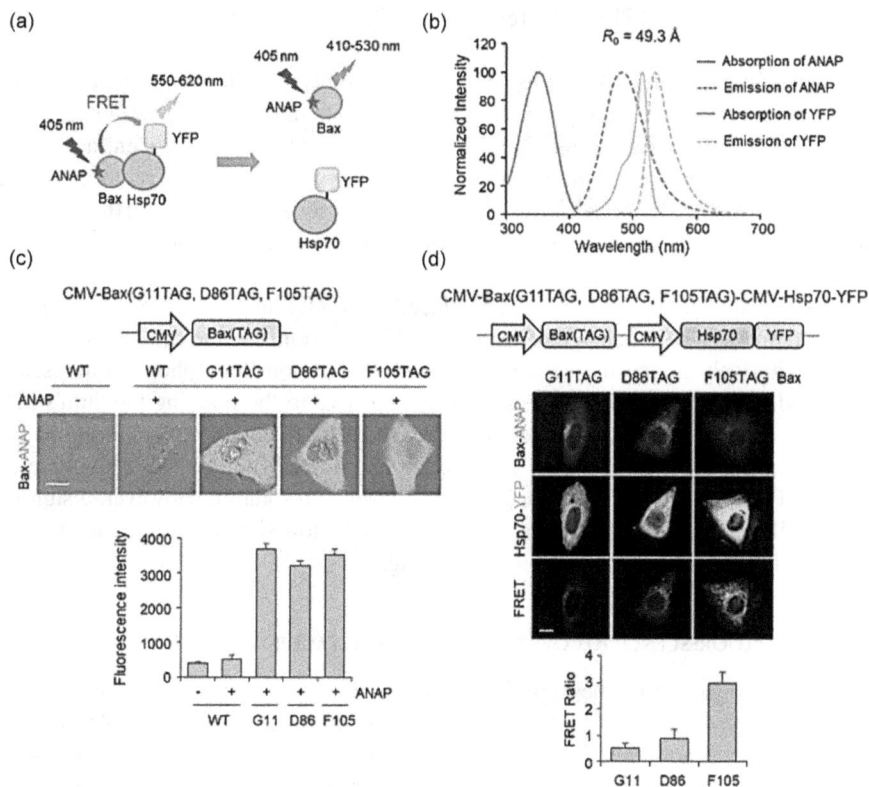

FIGURE 11.12 Efficiency of the FRET process involving Bax-ANAP and Hsp70-YFP. (a) The diagrammatic model shows the FRET which occurs as Bax interacts with Hsp70 between Bax-ANAP and Hsp70-YFP. (b) The graph showed the absorption and emission spectra of the ANAP and the YFP. (c) The image and graph depict the fluorescence intensity efficiency of ANAP in the mutant Bax genes G11TAG, D86TAG, and F105TAG after 24 hours. (d) The image characterization of FRET was performed with Bax-ANAP and Hsp70-YFP while using G11TAG, D86TAG, and F105TAG. The FRET ratio for the mutant Bax gene is shown in the graph. (The image is reprinted from the reference [90] licensed under the Copyright © 2019, American Chemical Society.)

absorption spectrum of YFP (Figure 11.12b). Bax-ANAP and Hsp70-YFP are separated by 49.3 Å, resulting in an energy transfer efficiency of 50% upon interaction. Interactions between Bax-ANAP and Hsp70-YFP demonstrated that the F105-labeled site of ANAP in Bax had the highest FRET signal ratio (Figure 11.12d). The effectiveness of the interaction can be measured by co-incubating HeLa cell lines with pro-apoptotic Bax activators and blockers. The activators include beta-amylase 7 (BAM7) and Beta-tubulin-specific alpha-helical activator 1 (BTSA1) function to stimulate alpha-helical sites on Bax. The small molecule Bax agonist 1 (SMBA1) blocks the phosphorylation of serine (S) 163 by binding to the S184 site at the protein's C terminus.

The decrease in FRET singles revealed that Bax activators that bind to Bax trigger sites affected the interaction between Bax and Hsp70. The Bax blocker that inhibits S184 phosphorylation by attaching to a location near Bax's C terminus did not affect this interaction. These findings imply that FRET systems involving proteins containing ANAP as fluorescent donors and YFP fusion proteins as fluorescent acceptors will be valuable tools for a study aiming to elucidate various forms of protein-protein interactions at the single-cell level. This detailed description is noteworthy because the FRET technique can be utilized as a cell-based screening tool to identify even small molecules.

Even though many restrictions are placed on FRET techniques because of the physical process, parameters like distance and orientation, and the equipment needed to quantitatively image the activity [91]. FRET relies on fluorophores, fluorescent tags, and labels, all of which are extremely sensitive to the imaging medium's pH, ionic concentrations, oxidation, temperature, and refractive index. Combining imaging modalities like the confocal microscope and TIRFM may make it possible to execute FRET in three dimensions at a high level of resolution. However, a significant challenge to obtaining reliable FRET data is the low signal-to-noise ratio (SNR) commonly encountered during FRET imaging.

11.2.4 Fluorescence Recovery after Photobleaching (FRAP)

Photobleaching is yet another significant issue that can arise when the cells are exposed to lasers for an extended period. Fluorescent proteins can be permanently bleached by bright light into non-fluorescent molecules without causing damage to the intracellular components. By exploiting this flaw as part of a selective bleaching process, the mobility of biological proteins in the plasma membrane can be observed under physiological conditions. Bright light is shone on the fluorescently labeled protein at the site of interest to bleach the fluorophore. The migration of unbleached fluorescent-tagged proteins causes the recovery of the fluorescently bleached area. This technique known as Fluorescence recovery after photobleaching (FRAP) is a tool for estimating the kinetics of single-molecule diffusion [92,93]. The transport of protein molecules through the plasma membrane was the focus of the first research using the FRAP technique [94] (Figure 11.13). In subsequent studies, diffusion rates and interactions with additional physiological components were incorporated. FRAP is considered an effective method for evaluating protein mobility, cytoskeleton dynamics, cell adhesion, mitosis, chromatin structure, Transcription & Translation of mRNAs, and intracellular transport using fluorescently tagged proteins for ROI [95–97].

The fundamental process of biology is the transcription and translation of genetic information within cells, which results in the production of proteins [98]. The transcription and translation processes are two-step procedures involving the cytoplasmic organelles Nucleus, Endoplasmic reticulum, Golgi bodies, and Ribosomes (Figure 11.14a). The initial process involves the transcription function, which duplicates the sequence of one DNA strand into the form of mRNA molecules in the nucleus. The endoplasmic reticulum serves as a channel for RNA transport into the cytoplasm. During the translation step, an mRNA molecule serves as the template for the synthesis of amino acids into the desired protein [99]. It is necessary

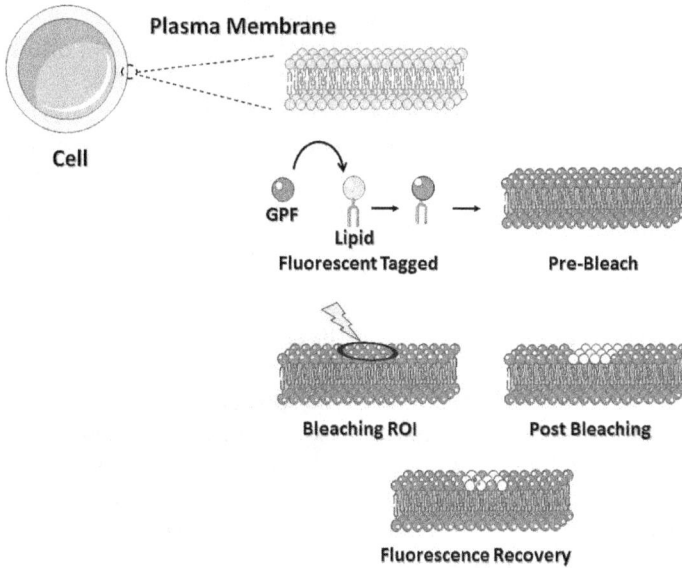

FIGURE 11.13 The mechanism of Fluorescence Recovery After Photobleaching (FRAP) is depicted schematically utilizing the plasma membrane of cells containing GFP-tagged lipids. (The art is adapted from smartservier.com.)

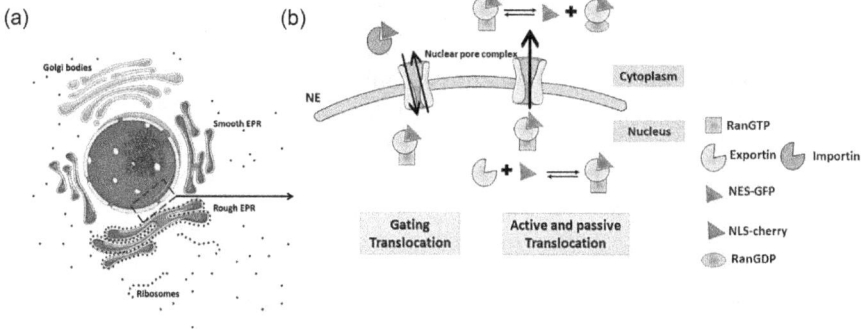

FIGURE 11.14 Schematic representation of the mechanism of the nucleocytoplasmic translocation. (a) Depicting the subcellular organelles near the nucleus region of a cell. (b) The cargo transport is mediated by the active, passive, and gating-based translocation through the nuclear pore complex. (The art is adapted from smartservier.com.)

to investigate the process that facilitates the transfer of molecules from the nucleus to the cytoplasm to ease the exchange of biomolecules [100]. A well-known point is that the nuclear envelope (NE) acts as a barrier that regulates the transport of RNA or proteins [101–103]. The NE's nuclear pore complexes (NPCs) act as channels for the movement of chemicals from the nucleus to the cytoplasm. Existing research on the biomolecule transport mechanism across the NPC is extensive. Small molecules with a diameter of less than 10 nm, such as ions and metabolites, diffuse directly through NPCs [104]. Large molecules such as proteins and RNAs are transported selectively by receptor-mediated transport. The molecules exported from the nucleus carry cargo peptides that can bind to a specific nuclear localization signal (NLS) or nuclear export signal (NES) to undergo conformational change for transport [105]. Despite the existence of different transport pathways through NE, the ability to quantitatively study the export and import of biomolecules is limited. Francesco Cardarelli and his team employed the FRAP technique to quantify carrier-mediated Nucleocytoplasmic Transport [106].

The movement of biomolecules within the cell is facilitated by carrier molecules. The karyopherin family is one form of carrier molecule that binds directly to the cargo. Two types of karyopherins contain the NPC binding domain (Figure 11.14b): import carriers (also called importins) and export carriers (also known as exportins). The Ran is responsible for controlling the dissociation and association of cargo to karyopherin. Ran, a 25 kDa protein, plays a role in cellular transport into and out of the nucleus [107]. Ran is expressed in both GDP-bound and GTP-bound nucleotide-bound forms. RanGTP and RanGDP interact depending on a gradient in the nucleus and cytoplasm to facilitate transport in both directions. Importins bind to the cherry-labeled NLS (NLS-cherry) and exportins bind to the GFP-labeled NES in the cyto-plasm (NES-GFP) [108]. The low RanGTP gradient in the cytoplasm enhances the active transport of cargo into the nucleus which is mediated by RanGDP. This pro-cess allows for both active and passive transfer of complex importin exportin through diffusion. Fluorophore tagging aided the FRAP-based study of the dynamic proper-ties of the NES and NLS. The transport of GFP guided by NES allowed us to deter-mine the translocation of nucleus biomolecules (Figure 11.15).

Confocal microscopy was used to examine the subcellular localization of NES-GFP using a CHO hamster cell line (CHO-k1) (Figure 11.15a). The nucleus region was photobleached, and the recovery time was observed to be 110 seconds (Figure 11.15b). It is anticipated that NES-GFP monitoring will give a one-way pro-cess in which no recovery will occur. However, the time-based recovery after pho-tobleaching suggested that active export and import transport occur at NPC at about the same time. Thus, the translocation efficiency of the biomolecule at the subcellular level can be gained insight through the quantitative measurement of the differential gating properties at the level of NPC utilizing FRAP-based assays.

The FRAP technique is an effective method for elucidating fundamental biologi-cal concerns by examining molecular dynamics [109]. The photo-bleachable fluo-rescent molecules are the source of the difficulty involved with the approach. When preparing samples for FRAP, it is difficult to predict how many fluorescent molecules would fade when exposed to light. Thus, the collecting and processing of FRAP data vary from laboratory to laboratory [110]. Recent developments in determining the intensity difference between the monitoring of patterns in dark bleached and unbleached areas. Fluorescence recovery after pattern photobleaching (FRAPP) is a method for correcting the inaccuracies caused by photobleaching. When com-bined with other microscopic techniques, the FRAP method can assist researchers to understand the dynamic activities occurring in single cells [84].

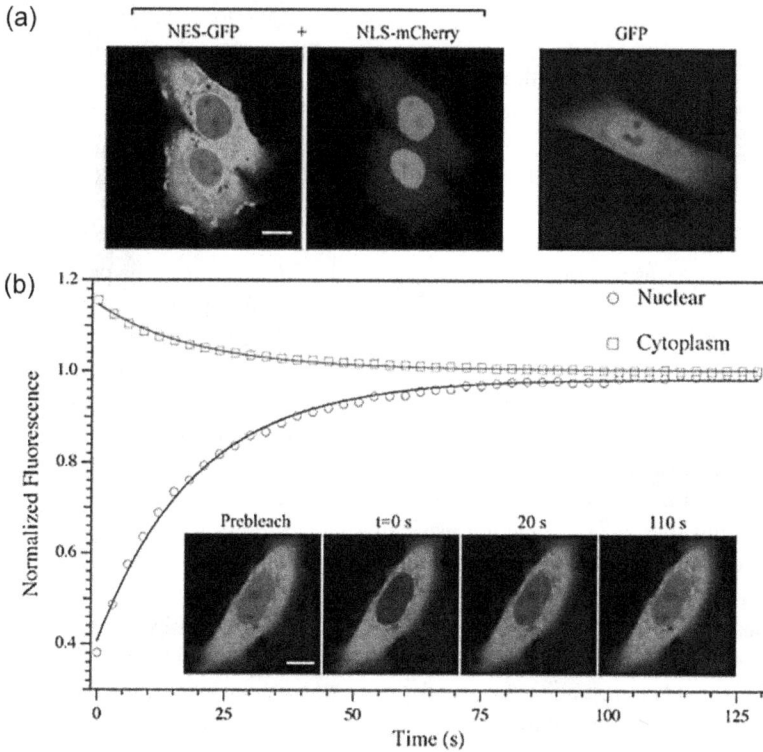

FIGURE 11.15 Nucleocytoplasmic Translocation examination using FRAP technique. (a) Cellular localization of the NES-GFP, NLS-Cherry was imaged using confocal microscopy. (b) The photobleached nucleus region with NES-GFP recovery time was observed between 0 and 110 seconds, and the graph depicts the nuclear region and cytoplasm recovery times. (The images are reprinted from the reference [106] with the license by CC BY 4.0.)

11.2.5 SUPER-RESOLUTION MICROSCOPY

Super-resolution microscopy (SRM) is a fast-emerging field that exceeds the diffraction limit to produce higher-quality images of biological specimens [111]. The lateral spatial resolution of SRM is up to 200 nm, and the axial resolution is at a minimum of 500 nm. Recent advancements in the SRM have resulted in image resolutions as low as 5 nm [112]. The benefits of SRM include live cell imaging, labeling specificity, and high throughput that provide a comprehensive understanding of cell interactions. Super-resolution microscopy can be used in conjunction with traditional microscopy techniques such as wide-field or confocal microscopy. When it comes to monitoring the movement of individual molecules, these methods are largely ineffective. However, modern technology has spawned independent SRM techniques that can detect single molecules. These include photoactivation localization microscopy (PALM), structured illumination microscopy (SIM), stimulated emission depletion (STED) microscopy, and stochastic optical reconstruction microscopy (STORM)

[113]. Each of these methods relies on a unique strategy for breaking the diffraction limit. They all come with their own set of benefits and drawbacks when applied to various scientific problems in biology.

PALM and STORM are single-molecule localized SRM techniques that use image stacking to create sub-diffraction limit images [114]. The principles underlying these two techniques are identical. The distinction between PALM and STORM is based on the fluorophores utilized in the experiment and the mechanism of bright-to-dark state conversion. PALM employs photo-switchable FPs, while STORM makes use of organic dyes as a fluorescence imaging probe (Figure 11.16). The primary goal of PALM or STORM is to image the activity of single fluorophores in response to the regulated activation of subsets of sample molecules [115]. It permits the observation of single molecules with almost tenfold the resolution of conventional microscopes. The diffraction limit makes it difficult to examine molecular-scale biological processes. PALM can visualize images with 10–50 nanometer resolution to comprehend the spatial intricacies of molecules packed closely together. PALM operates on the sample at a high density, but only on the visualization at a certain low density in the imaging frame. The detection wavelength acquired at UV light determined the effectiveness. The imaging will be erased and recreated once the activation has taken place. The point separation function (PSF) achieves nanoscale precision mostly with the assistance of a few molecules within a specific frame. PALM/STORM produced artifacts in overexpressed samples, similar to other traditional techniques [116].

STED is a technique that falls under the category of deterministic methods and is dependent on the modulation of the PSF size to allow emission or excitation in the center [118,119]. STED-stimulated emission is created by the excitation of two laser pulses to localize fluorescence at each focal location, as indicated in Figure 11.17. The first pulse is the excitation of the fluorophore, while the second pulse is the long-wavelength STED beam [120]. The STED beam acts as a quencher for the exciting beam's overlapping zone around the PSF's periphery [121]. The effect of these

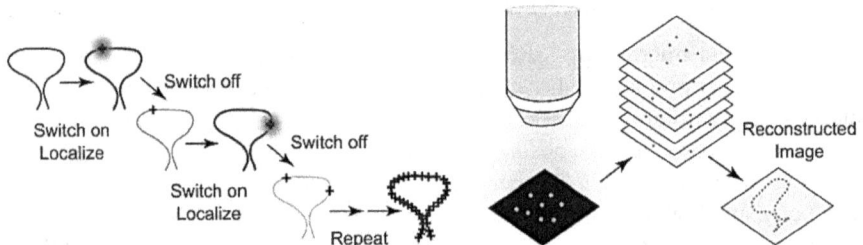

FIGURE 11.16 A graphical illustration of the PALM/STORM techniques. The single molecules are triggered in a randomized fashion by a stochastic process. In the sample subset, the photo-switchable fluorophore is activated by localized excitation and deactivated throughout the cycle. The emission of every single region of the molecule is precisely detected, and a high-resolution image is obtained after reconstruction. (The image is reprinted with permission from the reference [117].)

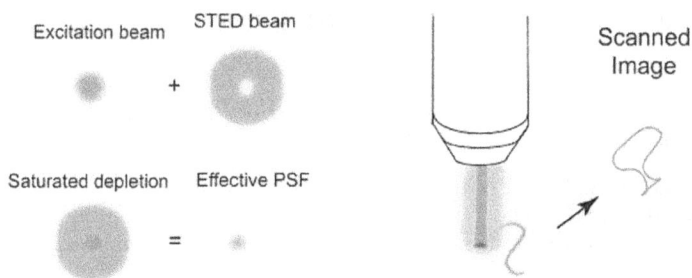

FIGURE 11.17 A diagram depicting the STED functioning mechanism. The initial excitation laser beam passes through the fluorophore-tagged sample, and pixel-by-pixel emission data is captured. The second laser beam is a STED laser beam, whitewashing the extra excitation lights. When the intensity of the STED laser reaches its maximum saturation point, depletion occurs and all the excited molecules return to their ground state before emission. Thus, the effective PSF of excitation with a high lateral resolution is obtained. (The image is reprinted with permission from the reference [117].)

two pulses cause fluorophore-ON (bright) and fluorophore-OFF (dark) switching, resulting in a high spatial gradient. The spatial gradient brings about an increase in resolution along both the X and Y directions. The spatial resolution of the single-molecule samples is often stated to be between 20 and 50 nm in the XY dimension. Enhancing the STED beam's pinpointing to the PSF can improve the Z-dimension resolution [122].

In general, SRM has a tremendous amount of power and ability, yet its higher resolution also presents some obstacles. The SRM image encompasses finer biomolecular features, and the lower sensitivities of the fluorophores necessitated the invention of new fluorophores with a high quantum yield [123]. The SRM is a less adaptable system that cannot readily accommodate hardware and experimental protocol changes. The SRM method is more expensive than other high-resolution imaging methods. SRM gives details in the image acquisition part, but it is less crucial in determining the fundamental functioning of single molecules [124].

11.3 CONCLUSION

Single-molecule approaches are frequently used in biochemistry, biophysics, and biology for quantitative study. The shift to digital microscopy, advancements in camera technology, and the introduction of new labeling and imaging procedures are all contributing to an environment in which it is becoming significantly simpler to derive meaningful quantitative data from images [125]. Single-molecule approaches are less intrusive and more rapid than traditional assay methods. The chapter concludes by emphasizing that each technique, mechanism, and application for biomolecule detection has its peculiarities and limitations. The ongoing improvement of the instrumentation might make it possible to improve the conjugation of two methods, which would result in improved detection of molecular behavior, structure, and dynamics in live systems [126].

REFERENCES

1. Elsasser, W.M., Outline of a theory of cellular heterogeneity. *Proc Natl Acad Sci U S A*, 1984. **81**(16): 5126–9.
2. Rubin, H., The significance of biological heterogeneity. *Cancer Metastasis Rev*, 1990. **9**(1): 1–20.
3. Leake, M.C., The physics of life: One molecule at a time. *Philos Trans R Soc Lond B Biol Sci*, 2013. **368**(1611): 20120248.
4. Perkel, J.M., Single-cell biology: The power of one. *Science*, 2015. **350**(6261): 696–8.
5. Leake, M.C., Shining the spotlight on functional molecular complexes: The new science of single-molecule cell biology. *Commun Integr Biol*, 2010. **3**(5): 415–8.
6. Leake, M.C., *Single-Molecule Cellular Biophysics*. Cambridge: Cambridge University Press, 2013.
7. Hinterdorfer, P. and A. Oijen, *Handbook of Single-Molecule Biophysics*. New York: Springer, 2009.
8. Deniz, A.A., S. Mukhopadhyay, and E.A. Lemke, *Single-molecule biophysics: At the interface of biology, physics and chemistry. J R Soc Interface*, 2008. **5**(18): 15–45.
9. Walla, P.J., *Modern Biophysical Chemistry*. 2014. doi: 10.1002/9783527683505.
10. Leake, M.C., *Biophysics: Tools and Techniques*. Boca Raton, FL: CRC Press, Taylor and Francis Group, 2016.
11. Lenn, T. and M.C. Leake, Experimental approaches for addressing fundamental biological questions in living, functioning cells with single molecule precision. *Open Biol*, 2012. **2**(6): 120090.
12. Shashkova, S. and M.C. Leake, Single-molecule fluorescence microscopy review: Shedding new light on old problems. *Biosci Rep*, 2017. **37**(4). doi: 10.1042/BSR20170031.
13. Walt, D.R., Optical methods for single molecule detection and analysis. *Anal Chem*, 2013. **85**(3): 1258–63.
14. Fernandez-Suarez, M. and A.Y. Ting, Fluorescent probes for super-resolution imaging in living cells. *Nat Rev Mol Cell Biol*, 2008. **9**(12): 929–43.
15. Leake, M.C., Analytical tools for single-molecule fluorescence imaging in cellulo. *Phys Chem Chem Phys*, 2014. **16**(25): 12635–47.
16. Godbey, W.T., Fluorescence, in *Biotechnology and Its Applications*. Cambridge, MA: Academic Press, 2022. 187–201.
17. Heilemann, M., et al., Super-resolution imaging with small organic fluorophores. *Angew Chem Int Ed Engl*, 2009. **48**(37): 6903–8.
18. Nirmal, M., et al., Fluorescence intermittency in single cadmium selenide nanocrystals. *Nature*, 1996. **383**(6603): 802–4.
19. Bopp, M.A., et al., Fluorescence and photobleaching dynamics of single light-harvesting complexes. *Proc Natl Acad Sci U S A*, 1997. **94**(20): 10630–5.
20. Shimomura, O., F.H. Johnson, and Y. Saiga, Extraction, purification and properties of aequorin, a bioluminescent protein from the luminous hydromedusan, Aequorea. *J Cell Comp Physiol*, 1962. **59**: 223–39.
21. Soboleski, M.R., J. Oaks, and W.P. Halford, Green fluorescent protein is a quantitative reporter of gene expression in individual eukaryotic cells. *FASEB J*, 2005. **19**(3): 440–2.
22. Stephens, D.J. and V.J. Allan, Light microscopy techniques for live cell imaging. *Science*, 2003. **300**(5616): 82–6.
23. Miller, H., et al., Single-molecule techniques in biophysics: A review of the progress in methods and applications. *Rep Prog Phys*, 2018. **81**(2): 024601.

24. Hall, C.E., Method for the observation of macromolecules with the electron microscope illustrated with micrographs of DNA. *J Biophys Biochem Cytol*, 1956. **2**(5): 625–8.

25. Rotman, B., Measurement of activity of single molecules of beta-D-galactosidase. *Proc Natl Acad Sci U S A*, 1961. **47**: 1981–91.

26. Hirschfeld, T., Optical microscopic observation of single small molecules. *Appl Opt*, 1976. **15**(12): 2965–6.

27. Barak, L.S. and W.W. Webb, Diffusion of low density lipoprotein-receptor complex on human fibroblasts. *J Cell Biol*, 1982. **95**(3): 846–52.

28. Moerner, W.E. and L. Kador, Optical detection and spectroscopy of single molecules in a solid. *Phys Rev Lett*, 1989. **62**(21): 2535–8.

29. Dickson, R.M., et al., On/off blinking and switching behaviour of single molecules of green fluorescent protein. *Nature*, 1997. **388**(6640): 355–8.

30. Sanderson, M.J., et al., Fluorescence microscopy. *Cold Spring Harb Protoc*, 2014. **2014**(10). doi: 10.1101/pdb.top071795.

31. Banz, A., et al., Sensitivity of single-molecule array assays for detection of clostridium difficile toxins in comparison to conventional laboratory testing algorithms. *J Clin Microbiol*, 2018. **56**(8): e00452.

32. Canette, A. and R. Briandet, MICROSCOPY I Confocal laser scanning microscopy, in Batt C.A., Tortorello M.L. (eds.) *Encyclopedia of Food Microbiology*. 2014. Cambridge, MA: Academic Press: 676–83.

33. Elliott, A.D., Confocal microscopy: Principles and modern practices. *Curr Protoc Cytom*, 2020. **92**(1): e68.

34. St Croix, C.M., S.H. Shand, and S.C. Watkins, Confocal microscopy: Comparisons, applications, and problems. *Biotechniques*, 2005. **39**(6 Suppl): S2–5.

35. White, J.G., W.B. Amos, and M. Fordham, An evaluation of confocal versus conventional imaging of biological structures by fluorescence light microscopy. *J Cell Biol*, 1987. **105**(1): 41–8.

36. Diaspro, A., et al., Confocal laser scanning fluorescence microscopy, in Roberts, G.C.K. (ed.) *Encyclopedia of Biophysics*. 2013. Heidelberg: Springer, 362–6.

37. Gratton, E. and M.J. vandeVen, Laser sources for confocal microscopy, in Pawley, J. (ed.) *Handbook of Biological Confocal Microscopy*. 2006. Boston, MA: Springer, 80–125.

38. Chiu, L.D., et al., Use of a white light supercontinuum laser for confocal interference-reflection microscopy. *J Microsc*, 2012. **246**(2): 153–9.

39. Diaspro, A., et al., Multi-photon excitation microscopy. *Biomed Eng Online*, 2006. **5**: 36.

40. Claxton, N.S., T.J. Fellers, and M.W. Davidson, Microscopy, Confocal, in Webster, J. G. (ed.) *Encyclopedia of Medical Devices and Instrumentation*. 2006. https://doi.org/10.1002/0471732877.emd291

41. Johnson, I., Fluorescent probes for living cells. *Histochem J*, 1998. **30**(3): 123–40.

42. Wood, E.J., Molecular probes: Handbook of fluorescent probes and research chemicals. *Biochem Educ*, 1994. **22**(2): 83.

43. Lemasters, J.J., et al., Confocal imaging of Ca^{2+}, pH, electrical potential, and membrane permeability in single living cells, in Conn, P. M. (ed.) *Green Fluorescent Protein*. 1999. New York: Academic Press, 341–58.

44. Wessendorf, M.W. and T.C. Brelje, Which fluorophore is brightest? A comparison of the staining obtained using fluorescein, tetramethylrhodamine, lissamine rhodamine, Texas red, and cyanine 3.18. *Histochemistry*, 1992. **98**(2): 81–5.

45. Waring, M.J., Complex formation between ethidium bromide and nucleic acids. *J Mol Biol*, 1965. **13**(1): 269–82.

46. Yang, Z. and K.K. Wang, Glial fibrillary acidic protein: From intermediate filament assembly and gliosis to neurobiomarker. *Trends Neurosci*, 2015. **38**(6): 364–74.

47. Segers-Nolten, G.M., et al., Scanning confocal fluorescence microscopy for single molecule analysis of nucleotide excision repair complexes. *Nucleic Acids Res*, 2002. **30**(21): 4720–7.

48. Spivak, G., Nucleotide excision repair in humans. *DNA Repair (Amst)*, 2015. **36**: 13–18.
49. Cinelli, R.A.G., et al., The enhanced green fluorescent protein as a tool for the analysis of protein dynamics and localization: Local fluorescence study at the single-molecule level. *Photochem Photobiol*, 2000. **71**(6): 771–6.
50. Ziarani, G.M., et al., Cyanine dyes, in *Metal-Free Synthetic Organic Dyes*. 2018. 127–52.
51. Pinaud, F. and M. Dahan, Targeting and imaging single biomolecules in living cells by complementation-activated light microscopy with split-fluorescent proteins. *Proc Natl Acad Sci USA*, 2011. **108**(24): E201–10.
52. Choo, Y.W., J. Jeong, and K. Jung, Recent advances in intravital microscopy for investigation of dynamic cellular behavior *in vivo*. *BMB Rep*, 2020. **53**(7): 357–66.
53. Marques, P.E., et al., Imaging liver biology in vivo using conventional confocal microscopy. *Nat Protoc*, 2015. **10**(2): 258–68.
54. Reck-Peterson, S.L., N.D. Derr, and N. Stuurman, Imaging single molecules using total internal reflection fluorescence microscopy (TIRFM). *Cold Spring Harb Protoc*, 2010. **2010**(3). doi: 10.1101/pdb.top73.
55. Axelrod, D., Total internal reflection fluorescence microscopy, in *Encyclopedia of Cell Biology*. 2016. 2: 62–9.
56. Mattheyses, A.L., S.M. Simon, and J.Z. Rappoport, Imaging with total internal reflection fluorescence microscopy for the cell biologist. *J Cell Sci*, 2010. **123**(Pt 21): 3621–8.
57. Manneville, J.B., Use of TIRF microscopy to visualize actin and microtubules in migrating cells, in *Regulators and Effectors of Small GTPases: Rho Family*. 2006. 406: 520–32.
58. Zhang, Y., Application of TIRFM in biomolecule research and clinical medicine. *Adv Biol Chem*, 2021. **11**(05): 206–19.
59. Loerke, D., et al., Super-resolution measurements with evanescent-wave fluorescence excitation using variable beam incidence. *J Biomed Opt*, 2000. **5**(1): 23–30.
60. Martin-Fernandez, M.L., C.J. Tynan, and S.E. Webb, A 'pocket guide' to total internal reflection fluorescence. *J Microsc*, 2013. **252**(1): 16–22.
61. Poulter, N.S., et al., The physical basis of total internal reflection fluorescence (TIRF) microscopy and its cellular applications. *Methods Mol Biol*, 2015. **1251**: 1–23.
62. Fish, K.N., Total internal reflection fluorescence (TIRF) microscopy. *Curr Protoc Cytom*, 2009. doi: 10.1002/0471142956.cy1218s50.
63. Joos, U., et al., Investigation of cell adhesion to structured surfaces using total internal reflection fluorescence and confocal laser scanning microscopy. *Eur J Cell Biol*, 2006. **85**(3–4): 225–8.
64. Trache, A. and G.A. Meininger, Total internal reflection fluorescence (TIRF) microscopy. *Curr Protoc Microbiol*, 2008. **Chapter 2**: Unit 2A 2 1–2A 2 22. doi: 10.1002/0471142956.cy1218s50.
65. Fletcher, D.A. and R.D. Mullins, Cell mechanics and the cytoskeleton. *Nature*, 2010. **463**(7280): 485–92.
66. Kopecki, Z. and A.J. Cowin, The role of actin remodelling proteins in wound healing and tissue regeneration, in *Wound Healing - New insights into Ancient Challenges*. 2016. doi: 10.5772/64673.
67. Manneville, J.B., et al., Interaction of the actin cytoskeleton with microtubules regulates secretory organelle movement near the plasma membrane in human endothelial cells. *J Cell Sci*, 2003. **116**(Pt 19): 3927–38.
68. Clark Brelje, T., M.W. Wessendorf, and R.L. Sorenson, Multicolor laser scanning confocal immunofluorescence microscopy: Practical application and limitations, in *Cell Biological Applications of Confocal Microscopy*. 2002. 70: 165–244.
69. Akhmanova, A. and M.O. Steinmetz, *Microtubule +TIPs at a glance. J Cell Sci*, 2010. **123**(Pt 20): 3415–9.

70. Dixit, R. and J.L. Ross, Studying plus-end tracking at single molecule resolution using TIRF microscopy, in Wilson L., Correia J.J. (eds.) *Microtubules, In Vitro.* 2010. 543–54. Doi: 10.1016/S0091-679X(10)95027-9.

71. Sun, X., et al., Microtubule-binding protein CLIP-170 is a mediator of paclitaxel sensitivity. *J Pathol,* 2012. **226**(4): 666–73.

72. Hink, M.A., et al., Structural dynamics of green fluorescent protein alone and fused with a single chain Fv protein. *J Biol Chem,* 2000. **275**(23): 17556–60.

73. Song, Y., et al., The microtubule end-binding affinity of EB1 is enhanced by a dimeric organization that is susceptible to phosphorylation. *J Cell Sci,* 2020. **133**(9). doi: 10.1242/jcs.241216.

74. Sankaran, J. and T. Wohland, Fluorescence strategies for mapping cell membrane dynamics and structures. *APL Bioeng,* 2020. **4**(2): 020901.

75. Toseland, C.P., Fluorescent labeling and modification of proteins. *J Chem Biol,* 2013. **6**(3): 85–95.

76. Winzeler, E.A., M. Schena, and R.W. Davis, Fluorescence-based expression monitoring using microarrays, in *Expression of Recombinant Genes in Eukaryotic Systems.* 1999. 306: 3–6.

77. Jares-Erijman, E.A. and T.M. Jovin, FRET imaging. *Nat Biotechnol,* 2003. **21**(11): 1387–95.

78. Sekar, R.B. and A. Periasamy, Fluorescence resonance energy transfer (FRET) microscopy imaging of live cell protein localizations. *J Cell Biol,* 2003. **160**(5): 629–33.

79. Hochreiter, B., A.P. Garcia, and J.A. Schmid, Fluorescent proteins as genetically encoded FRET biosensors in life sciences. *Sensors (Basel),* 2015. **15**(10): 26281–314.

80. Chen, H., et al., Measurement of FRET efficiency and ratio of donor to acceptor concentration in living cells. *Biophys J,* 2006. **91**(5): L39–41.

81. Muller, S.M., et al., Quantification of Forster resonance energy transfer by monitoring sensitized emission in living plant cells. *Front Plant Sci,* 2013. **4**: 413.

82. Bajar, B.T., et al., A Guide to Fluorescent Protein FRET Pairs. *Sensors (Basel),* 2016. **16**(9): 1488.

83. Shrestha, D., et al., Understanding FRET as a research tool for cellular studies. *Int J Mol Sci,* 2015. **16**(4): 6718–56.

84. Ishikawa-Ankerhold, H.C., R. Ankerhold, and G.P. Drummen, Advanced fluorescence microscopy techniques--FRAP, FLIP, FLAP, FRET and FLIM. *Molecules,* 2012. **17**(4): 4047–132.

85. Shandell, M.A., Z. Tan, and V.W. Cornish, Genetic code expansion: A brief history and perspective. *Biochemistry,* 2021. **60**(46): 3455–69.

86. Zhang, L., et al., Role of BAX in the apoptotic response to anticancer agents. *Science,* 2000. **290**(5493): 989–92.

87. Karbowski, M., et al., Role of Bax and Bak in mitochondrial morphogenesis. *Nature,* 2006. **443**(7112): 658–62.

88. Shamas-Din, A., et al., Mechanisms of action of Bcl-2 family proteins. *Cold Spring Harb Perspect Biol,* 2013. **5**(4): a008714.

89. De Maio, A., Extracellular Hsp70: Export and function. *Curr Protein Pept Sci,* 2014. **15**(3): 225–31.

90. Park, S.H., et al., Analysis of protein-protein interaction in a single live cell by using a FRET system based on genetic code expansion technology. *J Am Chem Soc,* 2019. **141**(10): 4273–81.

91. Leavesley, S.J. and T.C. Rich, Overcoming limitations of FRET measurements. *Cytometry A,* 2016. **89**(4): 325–7.

92. Axelrod, D., et al., Mobility measurement by analysis of fluorescence photobleaching recovery kinetics. *Biophys J,* 1976. **16**(9): 1055–69.

93. Lippincott-Schwartz, J., E.L. Snapp, and R.D. Phair, The development and enhancement of FRAP as a key tool for investigating protein dynamics. *Biophys J*, 2018. **115**(7): 1146–55.

94. Pincet, F., et al., FRAP to characterize molecular diffusion and interaction in various membrane environments. *PLoS One*, 2016. **11**(7): e0158457.

95. White, J., Photobleaching GFP reveals protein dynamics inside live cells. *Trends Cell Biol*, 1999. **9**(2): 61–5.

96. van Royen, M.E., et al., Fluorescence recovery after photobleaching (FRAP) to study nuclear protein dynamics in living cells. *Methods Mol Biol*, 2009. **464**: 363–85.

97. Sarkar, P. and A. Chattopadhyay, Exploring membrane lipid and protein diffusion by FRAP, in Prasad, R., Singh, A. (eds.) *Analysis of Membrane Lipids*. 2020. New York: Springer, 119–41.

98. Bartoli, V., et al., Tunable genetic devices through simultaneous control of transcription and translation. *Nat Commun*, 2020. **11**(1): 2095.

99. de Klerk, E. and P.A. t Hoen, Alternative mRNA transcription, processing, and translation: Insights from RNA sequencing. *Trends Genet*, 2015. **31**(3): 128–39.

100. Stewart, M., Function of the nuclear transport machinery in maintaining the distinctive compositions of the nucleus and cytoplasm. *Int J Mol Sci*, 2022. **23**(5): 2578.

101. Gorlich, D. and U. Kutay, Transport between the cell nucleus and the cytoplasm. *Annu Rev Cell Dev Biol*, 1999. **15**: 607–60.

102. Weis, K., Regulating access to the genome. *Cell*, 2003. **112**(4): 441–51.

103. Fahrenkrog, B. and U. Aebi, The nuclear pore complex: Nucleocytoplasmic transport and beyond. *Nat Rev Mol Cell Biol*, 2003. **4**(10): 757–66.

104. Pemberton, L.F. and B.M. Paschal, Mechanisms of receptor-mediated nuclear import and nuclear export. *Traffic*, 2005. **6**(3): 187–98.

105. Kimura, M., et al., Novel approaches for the identification of nuclear transport receptor substrates. *Methods Cell Biol*, 2014. **122**: 353–78.

106. Cardarelli, F., et al., Fluorescent recovery after photobleaching (FRAP) analysis of nuclear export rates identifies intrinsic features of nucleocytoplasmic transport. *J Biol Chem*, 2012. **287**(8): 5554–61.

107. Macara, I.G., Transport into and out of the nucleus. *Microbiol Mol Biol Rev*, 2001. **65**(4): 570–94, table of contents.

108. Yoneda, Y., Nucleocytoplasmic protein traffic and its significance to cell function. *Genes Cells*, 2000. **5**(10): 777–87.

109. Rayan, G., et al., Recent applications of fluorescence recovery after photobleaching (FRAP) to membrane bio-macromolecules. *Sensors (Basel)*, 2010. **10**(6): 5927–48.

110. Blumenthal, D., et al., Universal approach to FRAP analysis of arbitrary bleaching patterns. *Sci Rep*, 2015. **5**: 11655.

111. Prakash, K., et al., Super-resolution microscopy: A brief history and new avenues. *Philos Trans A Math Phys Eng Sci*, 2022. **380**(2220): 20210110.

112. Gu, L., et al., Author Correction: Molecular resolution imaging by repetitive optical selective exposure. *Nat Methods*, 2019. **16**(11): 1193.

113. Thorley, J.A., J. Pike, and J.Z. Rappoport, Super-resolution Microscopy, in *Fluorescence Microscopy*. 2014. 199–212. doi: 10.1016/B978-0-12-409513-7.00014-2.

114. Rust, M.J., M. Bates, and X. Zhuang, Sub-diffraction-limit imaging by stochastic optical reconstruction microscopy (STORM). *Nat Methods*, 2006. **3**(10): 793–5.

115. Shroff, H., et al., Live-cell photoactivated localization microscopy of nanoscale adhesion dynamics. *Nat Methods*, 2008. **5**(5): 417–23.

116. Bates, M., S.A. Jones, and X. Zhuang, Stochastic optical reconstruction microscopy (STORM): A method for superresolution fluorescence imaging. *Cold Spring Harb Protoc*, 2013. **2013**(6): 498–520.

117. Egger, B. and S.G. Sprecher, Super-resolution STED and STORM/PALM microscopy for brain imaging, in Kao, F.J., Keiser, G., Gogoi, A. (eds.) *Advanced Optical Methods for Brain Imaging.* 2019. Singapore: Springer, 245–58.

118. Hell, S.W. and J. Wichmann, Breaking the diffraction resolution limit by stimulated emission: Stimulated-emission-depletion fluorescence microscopy. *Opt Lett*, 1994. **19** (11): 780–2.

119. Wu, Y., et al., Ultrastable fluorescent polymer dots for stimulated emission depletion bioimaging. *Adv Opt Mater*, 2018. **6**(19). doi: 10.1002/adom.201800333.

120. Vicidomini, G., P. Bianchini, and A. Diaspro, STED super-resolved microscopy. *Nat Methods*, 2018. **15**(3): 173–82.

121. Shi, J., et al., Coaxial illumination module of the stimulated-emission-depletion nanoscope. *Opt Express*, 2022. **30**(8): 13481–90.

122. Wildanger, D., et al., A compact STED microscope providing 3D nanoscale resolution. *J Microsc*, 2009. **236**(1): 35–43.

123. Henriques, R., et al., PALM and STORM: Unlocking live-cell super-resolution. *Biopolymers*, 2011. **95**(5): 322–31.

124. Vogelsang, J., et al., Make them blink: Probes for super-resolution microscopy. *Chemphyschem*, 2010. **11**(12): 2475–90.

125. The quest for quantitative microscopy. *Nat Methods*, 2012. **9**(7): 627. https://doi.org/10.1038/nmeth.2102

126. Croop, B., et al., Recent advancement of light-based single-molecule approaches for studying biomolecules. *Wiley Interdiscip Rev Syst Biol Med*, 2019. 11(4): e1445.

Index

Note: **Bold** page numbers refer to tables; *italic* page numbers refer to figures.

For Product Safety Concerns and Information please contact our EU
representative GPSR@taylorandfrancis.com
Taylor & Francis Verlag GmbH, Kaufingerstraße 24, 80331 München, Germany

www.ingramcontent.com/pod-product-compliance
Lightning Source LLC
Chambersburg PA
CBHW060804220326
41598CB00022B/2536

9 7 8 1 0 3 2 5 2 9 6 7 7